UNITEXT for Physics

UNITEXT for Physics series, formerly UNITEXT Collana di Fisica e Astronomia, publishes textbooks and monographs in Physics and Astronomy, mainly in English language, characterized of a didactic style and comprehensiveness. The books published in UNITEXT for Physics series are addressed to graduate and advanced graduate students, but also to scientists and researchers as important resources for their education, knowledge and teaching.

More information about this series at http://www.springer.com/series/13351

Michele Cini

Elements of Classical
and Quantum Physics

 Springer

Michele Cini
Università di Roma Tor Vergata
Rome
Italy

ISSN 2198-7882   ISSN 2198-7890   (electronic)
UNITEXT for Physics
ISBN 978-3-319-89063-0   ISBN 978-3-319-71330-4   (eBook)
https://doi.org/10.1007/978-3-319-71330-4

Printed on acid-free paper

This Springer imprint is published by the registered company Springer International Publishing AG
part of Springer Nature
The registered company address is: Gewerbestrasse 11, 6330 Cham, Switzerland

To my wife Anna and to my son Massimo

Preface

This book originates out of many years of teaching different courses at the University of Roma Tor Vergata, although it includes so many applications for diverse branches of Physics that I hope that many professional physicists will find it of interest as well. It can serve as a textbook for second- and third-year students of Physics and related disciplines that require being introduced to Theoretical Physics at the level of the short degree. I develop all the mathematical methods, but the main focus is on the physical meaning of the formalisms. I tried to make the book fun and interesting by stimulating the reader's curiosity about many more physical effects than is usual in textbooks. These range from the measurement of stellar radii to anyons, quantum pumping, entanglement, frame dragging, teleportation, black holes, superconductivity and more. Physics is the only science in which theory and experiment have comparable importance. Like in any other natural science, experiment discerns what is true and what is false; however, in countless cases, important discoveries are due to theoreticians making detailed predictions that guided experimenters to verify new effects—antimatter was discovered theoretically by Dirac; many gravitational effects were predicted by Einstein; in quantum electrodynamics, the progress of both theory and experiment pushed the agreement to more than 10 significant digits. Thanks to theory, Physics is not just an immense inventory of data, but has a coherent logic which can only be understood in terms of ingenious and beautiful Mathematics.

More exercises can be found in a book by Michele Cini, Francesco Fucito and Mauro Sbragaglia, *Solved Problems in Quantum and Statistical Physics*, Springer Verlag Italia 2012. A more advanced treatment of Solid State Theory is presented in Michele Cini, *Topics and Methods in Condensed Matter Theory*, Springer 2007.

Rome, Italy   Michele Cini
October 2017

Contents

Chapter 1
Theoretical Physics and Mathematics

> *Salvini:*
> *It is true that the Copernican System perturbs Aristotle's*
> *Universe, but we are dealing with our Universe, the true, real*
> *one.*
> *From Galileo Galilei, Dialogue Concerning the Two Chief*
> *World Systems*

Following the method of Galileo, Theoretical Physics uses Mathematics as natural and essential language to describe reality. But even Galileo would probably be surprised by the degree of success of the mathematical description of the world. Simple elegant laws like the Planck radiation formula agree with reality accurately over all the electromagnetic spectrum; on the other hand, the need for abstract mathematical concepts stemming from complex analysis, such as topology, Group Theory, and infinite-dimensional spaces shows that the fabric of the Universe is highly nontrivial. *Is God a mathematician?* is the title of an interesting book by Mario Livio. The Nobel Laureate in Physics Eugene Wigner in 1960 spoke about the *unreasonable effectiveness* of Mathematics in the understanding of Reality.[1] Like Mathematics, Physics involves a depth of thought that is unknown to most people; only well-posed questions are permissible, and there is no room for puns, since scientists deal with nothing but of verifiable facts, which one cannot simply adjust to make them fit with one's statements. In fact, there is progress, and every scientist contributes to building something definitive, even if further progress may build on the new findings.

However, Theoretical Physics is not Mathematics. It has its own method of investigation and is subtle and concrete. While a Mathematician seeks the maximum of

---

[1]This opens the question as to whether the findings of mathematicians must be considered as inventions like those of an artist or rather discoveries like those of a scientist. Personally I believe they are discoveries, since there is an internal logic in the development of Mathematics, and Mathematics is the logic of Nature.

© Springer International Publishing AG, part of Springer Nature 2018
M. Cini, *Elements of Classical and Quantum Physics*,
UNITEXT for Physics, https://doi.org/10.1007/978-3-319-71330-4_1

generality, the Theoretical Physicist must strive to give an operational meaning to all the new statements. Experiment has the final word on controversies. All the statements of a theorist make sense if they can be verified by experiment. This is not so, for example, in the case of recurrent theories about the Multiverse, which are often proposed but not of physical interest. Even the most abstract-looking aspects of the theory have an operational meaning, at least in principle. One must make an exception for Astrophysics, since a direct measurement of distances is not feasible and we rely on the known Laws of Physics and the consistency of clues.

The philosopher Carl Popper has gained wide popularity by stressing that Science comes from experiment, so each of its conclusions is provisional, and at risk of being proved wrong the next morning by another experiment. This seems to remove Science as a source of a sound knowledge. But scientists have always known that, and have no orthodoxy to defend, and thus even try to test existing theories with ever increasing precision measurements. Some discoveries have been completely unexpected and new surprises are possible. Relativity and Quantum Mechanics have shown that the extrapolation of physical laws to new ranges of phenomena is at risk. Besides, the borderline between classical and Quantum Physics is still a matter of research. There are quantum phenomena, like entanglement, that have been tested over distances on the order of many kilometers in satellite experiments. Whatever new high energy physics may be discovered, the Maxwell equations and the laws of Thermodynamics are not in danger in regard to the wide range of phenomena for which they are known to work fine. Nevertheless, the 10 digits precision of the agreement between theory and experiment in Quantum Electrodynamics means that, besides the single experiment, there is detailed and coherent understanding of many interdependent facts. The Truth exists in Religions. By far, a great number of the conclusions in Physics rank among the most sound knowledge that can exist. If somebody claims the invention of the perpetual motion, or superluminal flight, the claim can be safely ignored. The power of Science to increase the real grasp of Man and his ability to solve problems cannot be questioned.

Let us start with some constants (Table 1.1).[2] In the current Theories, some of the above values are really fundamental ($\hbar$, c, G, e) and enter as empirical constants. Nobody knows why the constants have such values, but a Universe with different values (if possible) should look very different. Others are known combinations of the former, while still others can be obtained through theories that are well established or in progress. For example, QCD calculations of the mass of the proton have been reported, in terms of quark masses, but such topics are outside the scope of the present book.

G was the first constant to be discovered (by Newton) and measured (by Cavendish, in 1798); Galileo tried to measure the speed of light with inadequate means; after the order-of magnitude measurement by Rømer, we now know the much more accurate

---

[2]I am using both the International Units and the Gauss Units, since both are commonly used.

Table 1.1 Some constants of Physics

Name	Symbol	Value	Units
Avogadro number	N_A	$6,022169 \ 10^{23}$	mol^{-1}
Gravity constant	G	$6,6732 \ 10^{-11}$	$N \ m^2/Kg^2$
Boltzmann constant	K_B	$1,380622 \ 10^{-23}$	$J/^0K$
Stefan-Boltzmann constant	σ	$5,66961 \ 10^{-8}$	$W/m^2 \times^0 K^4$
Speed of light	c	$2,99792458 \ 10^8$	m/s
Classical radius of the electron	$r_e = \frac{e^2}{2\,mc^2}$	$2,819489 \ 10^{-15}$	m
Electron mass	m_e	$9,109558 \ 10^{-31}$	Kg
Proton mass	m_p	$1,672614 \ 10^{-27}$	Kg
Neutron mass	m_n	$1,674920 \ 10^{-27}$	Kg
Planck's constant	h	$6,626196 \ 10^{-34}$	J s
Fine structure constant	$\alpha = \frac{e^2}{\hbar c}$	$\frac{1}{137,03602}$	Pure number, Gauss units
Fine structure constant	$\alpha = \frac{e^2}{2\epsilon_0 hc}$	$\frac{1}{137,03602}$	Pure number, SI units
Flux quantum	$\frac{hc}{e}$	4×10^{-7}	$Gausscm^2$
Electronvolt	eV	$1,6 \ 10^{-19}$	J
Proton magnetic moment	μ_p	$1,4106203 \ 10^{-26}$	J/T

value $c = 299792458$ m/s; the Maxwell equations, after a long crisis, led to Relativity. The introduction of the Boltzmann constant started Statistical Mechanics and led to a microscopic understanding of the Thermodynamic principles; the introduction of h by Planck started Quantum Mechanics.

The introduction of these constants has marked the great conceptual conquests that are sometimes called scientific revolutions. However, the term is too extreme.

It is true that Galileo tossed the Aristotle's Physics into the basket, but Einstein's Relativity did not throw out Galileo, and Quantum Mechanics did not abolish Classical Physics.

The Mechanics of Galileo and Newton and its developments continue to be used in the daily life; it describes well and correctly the motions of planets and stars; the Solar System is explored through the classical equations. Simply put, it cannot be extrapolated to high energies and to phenomena that are far from the experimental conditions for which it was formulated. When c or h enter, we are outside the range of validity.

Notably, the generalized theories are not obtained by fixing details of the old ones, but require a new formalism and a set of new concepts; the older theories remain as limiting cases of the general theories. Understanding of the physical laws proceeds by successive layers. But we shall find that a deep part of the theory proceeds from the special to the general theory without any discontinuity.

Paradoxically, Science today has detractors and very active adversaries. But it remains the best result of man's ingenuity, our most important resource. The intuitions of the great scientists are unknown to most people but the effort wich is needed to understand them is very rewarding.

Part I
Classical Physics-Complements

Chapter 2
Analytical Mechanics

In the mathematical formulation of the classical theory, we meet key concepts that are needed in its extensions (Relativity, Quantum Mechanics). Theoretical Physics is a large body of knowledge, yet it is deeply unitary. No doubt, it begins here.

2.1 Galileo's Revolution and Newton's $F = ma$

Galileo[1] contradicted all the ancient wisdom and doctrine when he established the principle of inertia, clearly stated the principle of Relativity and measured the acceleration of gravity and the circular motion around the year 1600. His concept of Mechanics contrasted with that of Aristotle, who wrote that a force was needed to keep a body in motion; and Aristotle's Physics had been given a holy status by the Counter-Reformation. Then, Newton[2] established the basic law of motion

---

[1] Galileo Galilei (Pisa 1564- Arcetri 1642) was the father of Modern Science. Professor of Mathematics in the University of Pisa since 1589, he moved to Padova and back to Pisa in 1611. He conceived and stated the scientific method, and was equally gifted for experiment, theory and publication of the results. He also tried to measure the speed of light and built the first astronomical telescope, becoming the founder of modern Astronomy. His *Sidereus Nuncius* (1610) is also a literary masterpiece for its clear and very readable style. When he reported the discovery of the *Pianeti Medicei* (Jupiter's moons) and of the sunspots he was suspected of heresy. The foundations of Mechanics are in his *Discorsi e dimostrazioni matematiche intorno a due nuove scienze, attinenti alla meccanica ed ai meccanismi locali*, Leiden (1636). The book was published abroad because of the persecutions by the Church, whose teaching was contrary to any motion of the Earth. He avoided the stake thanks to his prudence and to good relations with the catholic hierarchy. He was a friend of Pope Urbano VIII Barberini. Nevertheless he was taken to trial, forced to retract and imprisoned for the rest of his life in Arcetri. Thereafter, life was very hard for scientists in Italy for quite a long time.

[2] Isaac Newton (Woolsthorpe 1642-London 1727) was a professor in Cambridge since 1669; then he had already invented Calculus, although his book *De Methodis Serierum et Fluxionum* was published in 1671. The basic work about Mechanics is entitled *Philosophiae naturalis principia mathematica*, Iussu societatis regiae, London (1687). He subsequently discovered the law of gravity, and made discoveries in Optics. He also invented the calculus of variations. Unlike Galileo, he was

© Springer International Publishing AG, part of Springer Nature 2018
M. Cini, *Elements of Classical and Quantum Physics*,
UNITEXT for Physics, https://doi.org/10.1007/978-3-319-71330-4_2

$$m \vec{a} = \vec{F},$$

where $\vec{a}$ is the acceleration of a material point mass m under the action of a force $\vec{F}$. He also invented Calculus and gave this *equation of motion* the well known differential form. Newton's equation in one dimension reads:

$$m\frac{d^2 x(t)}{dt^2} = F_x \tag{2.1}$$

that is, $m\ddot{x} = F_x$. In terms of the momentum $p_x = m\dot{x}$,

$$\dot{p}_x = F_x. \tag{2.2}$$

If F is known, this determines the *law of motion* $x(t)$ as a function of the initial conditions $x(0)$, $\dot{x}(0)$. The motion is then determined at all times. The laws of classical mechanics are these; the reader at this point might suspect that the rest of the chapter is just a series of examples of solution; instead there is a beautiful theory that we must build. The initial motivation for developing a theory was that the equation involves the second derivative and is not amenable to quadratures.

The *conservative* forces, which derive from a potential $V(x)$ according to

$$F_x = -\frac{dV(x)}{dx}, \tag{2.3}$$

are a promising special case with which to start an investigation, since they are easier. Along a trajectory $x(t)$, the potential $V(x(t))$ varies at a rate $\dot{V} = \frac{dV}{dx}\dot{x}$; then, multiplying (2.1) by $\dot{x}$, one finds $\dot{x}m\ddot{x} - \dot{x}F_x = \frac{dE}{dt}$, where E is the energy; this is the law of energy conservation. It implies that

$$E = T + V \tag{2.4}$$

is fixed. While $V(x(t))$ varies, the kinetic energy

$$T = \frac{m}{2}\dot{x}^2$$

also varies and the sum is constant. Equation (2.4) yields the *integral of the motion*

$$E = \frac{m}{2}\dot{x}^2 + V(x);$$

it links the first derivative $\dot{x}$ and the constant E, and is much simpler to solve than (2.1).

---

not persecuted, but highly honored in his own country, receiving prestigious commissions from the Government.

For a point mass in 3 dimensions, Newton's equation is a vector equation, and one writes: $\vec{F}(\vec{r}) = -\vec{\nabla}V(\vec{r})$: for a system of N point masses, the equations are $3N$; interactions between particles produce coupled equations. Kinetic energies add:

$$T = \sum_i^N \sum_\alpha^3 \frac{m_i}{2}\dot{x}_{i,\alpha}^2$$

(i runs over points and α over components of $\vec{r}_i$).

Analytical mechanics allows us to apply the laws of Newton effectively to any problem, even when the forces are not given at the start; above all, it reveals a mathematical structure underlying the equations of motion that is much more general than classical mechanics itself, as we shall see. The Theory of Relativity and Quantum Mechanics are generalizations of classical mechanics rooted in analytical mechanics.

In many interesting problems, the motion is limited by constraints. For example, a point particle may be constrained to move on a surface, or along a line. A constraint like this that can be expressed as an equation of the form $F(x_1, \cdots, x_N) = 0$ is called holonomic and reduces the number of unknowns and equations. For a point mass, one should solve 3 equations of motion, but if the particle is known to move on a circle of radius R, an angle is sufficient to locate the point. There is, however, the additional difficulty is that the forces resulting from the constraints are known only after solving the problem.

For $N = 2$ independent points, one has 6 equations of motion to solve. Now suppose we have a system of two masses m_1, m_2 constrained to remain at a distance r_0 at the ends of a rigid stick of negligible mass, hinged at the origin of coordinates of the center of gravity. Such a system is called a *rigid rotor* in 3 dimensions. The center of gravity of the two masses is the point

$$\vec{R} = \frac{m_1\vec{r}_1 + m_2\vec{r}_2}{m_1 + m_2}. \tag{2.5}$$

In this case, $\vec{R} = \vec{0}$ and $\vec{r}_1 = -\frac{m_1}{m_2}\vec{r}_2$. Denoting r_1, r_2 the fixed distances from the origin, $r_1 = \frac{m_1}{m_2}r_2$. Moreover, $r_1 + r_2 = r_0$, and so $r_1 = \frac{m_2 r_0}{m_1 + m_2}$, $r_2 = \frac{m_1 r_0}{m_1 + m_2}$.

When the rotor moves, the stick acts on the masses with unknown forces. Are we lost? No. The general method for writing the equations of motion is due to Giuseppe Luigi Lagrange (Torino 1736 - Paris 1813), the great Italian-French mathematician, and (independently) to the great Swiss Euler.[3] Next, we shall see the Euler–Lagrange theory, assuming smooth constraints (that is, they should not make work and should not dissipate energy during the motion.

---

[3]Leonhard Euler (Basel, Switzerland 1707- S. Petersburg, Russia 1783) was probably the greatest mathematician of all times. He contributed to all fields of analysis, number theory and geometry. He also wrote the book *Mechanica* about his results on the equations of motion.

The rotor has two degrees of freedom, and we can choose the angles θ and ϕ of one mass in polar coordinates. We do not need 6 equations, but only 2. Before proceeding, I introduce the general method terms.

2.2   Lagrangian Formalism

Constraints and Lagrangian Coordinates

N material points in 3 dimensions are described in Cartesian coordinates by

$$\{x_1, y_1, z_1, x_2 \ldots z_N\} \equiv \{x_{i\alpha}\}, \quad 0 < i \leq N, \quad 1 \leq \alpha \leq 3.$$

We consider constraints, such as those requiring the points to move on certain surfaces, or to maintain some fixed distances. The force on the particle i can be decomposed:

$$\vec{F}_i = \vec{F}_i^{appl} + \vec{F}_i^{vinc},$$

where the first term (applied force) is known (we can think of springs, gravitational forces, electric fields, assigned by the problem), but the constraint reaction is not. Suppose that the applied forces are conservative, i.e., they arise from a potential energy $V\{x_1, y_1, z_1, x_2 \ldots z_N\}$ according to

$$F_{i,\alpha}^{appl} = -\frac{\partial V}{\partial x_{i,\alpha}}. \tag{2.6}$$

The $-$ sign tells us that the force pulls towards a decreased potential energy. We assume smooth constraints. Therefore, the reaction of the constraints is orthogonal to the infinitesimal shift of particle i $\delta \vec{R}_i$ and does not work:

$$\vec{F}_i^{vinc} \cdot \delta \vec{R}_i = 0. \tag{2.7}$$

The force due to the constraint is perpendicular to the shift of particle i. The work δL done by the applied forces is given by

$$-\delta V = \delta L = \sum_i \vec{F}_i \cdot \delta \vec{R}_i \equiv \sum_i \vec{F}_i^{appl} \cdot \delta \vec{R}_i, \tag{2.8}$$

where i runs over the point masses; so

$$-\delta V = \sum_{i,\alpha} F_{i,\alpha}^{appl} \delta x_{i,\alpha}. \tag{2.9}$$

Note that we are assuming a potential that depends on positions and not on velocities. This restriction will be removed when dealing with the Lorentz force. Since the Cartesian shift $\{\delta x_1, \delta y_1, \delta z_1, \delta x_2 \ldots\} \equiv \{\delta x_{i\alpha}\}$ is not allowed by the constraints, the Lagrangian coordinates q_β, $1 \leq \beta \leq s$ are more convenient to represent $\delta \vec{R}_i$ provided that every δq_β represents a possible motion. One substantial advantage is that the number s of degrees of freedom is reduced: $s < 3N$. The Cartesian components of an infinitesimal motion allowed by the constraints is:

$$\delta x_{i,\alpha} = \sum_\beta \frac{\partial x_{i,\alpha}}{\partial q_\beta} \delta q_\beta. \tag{2.10}$$

From (2.9), one gets:

$$-\frac{\partial V}{\partial q_\beta} = \sum_{i,\alpha} F_{i,\alpha}^{appl} \frac{\partial x_{i,\alpha}}{\partial q_\beta}. \tag{2.11}$$

The infinitesimal work done by the applied force F^{appl} is:

$$\delta V = \sum_\beta \frac{\partial V}{\partial q_\beta} \delta q_\beta. \tag{2.12}$$

In analogy with (2.6),

$$Q_\beta = -\frac{\partial V}{\partial q_\beta} \tag{2.13}$$

is called the generalized force; this can be written in terms of the applied force using (2.11):

$$Q_\beta = \sum_{i,\alpha} F_{i,\alpha}^{appl} \frac{\partial x_{i,\alpha}}{\partial q_\beta}. \tag{2.14}$$

The Euler–Lagrange Equations

Starting with the equations of motion and the constraints (if any) written in Cartesian coordinates, we want to rewrite everything in terms of Lagrangian coordinates q_α, $\alpha = 1, \cdots s$, in order to have as many coordinates as the degrees of freedom, and in order to take into account the constraints automatically. The scalar product of the Cartesian equation

$$\vec{F}_i = \frac{d\vec{p}_i}{dt} \tag{2.15}$$

with an infinitesimal displacement $\delta \vec{R}$ compatible with the constraints, yields:

$$-\delta V = \sum_i \vec{F}_i \cdot \delta \vec{R}_i = \sum_i \frac{d\vec{p}_i}{dt} \cdot \delta \vec{R}_i \implies -\delta V = \sum_{i,\alpha} \dot{p}_{i,\alpha} \delta x_{i,\alpha}. \tag{2.16}$$

The first term is (2.12); The first equality introduces the assumption that the constraints do not make work. We change variables in the last term, using (2.10).

$$\sum_{i,\alpha} \dot{p}_{i,\alpha}\delta x_{i\alpha} = \sum_{i,\alpha} \dot{p}_{i,\alpha} \sum_{\beta} \frac{\partial x_{i\alpha}}{\partial q_\beta}\delta q_\beta = \sum_{i,\alpha} m_i\ddot{x}_{i,\alpha} \sum_{\beta} \frac{\partial x_{i\alpha}}{\partial q_\beta}\delta q_\beta;$$

exchanging the summations in order to highlight δq_β, one finds that:

$$-\delta V = \sum_{\beta} \delta q_\beta \sum_{i,\alpha} m_i\ddot{x}_{i,\alpha} \frac{\partial x_{i\alpha}}{\partial q_\beta},$$

hence, we obtain the generalized force:

$$-\frac{\partial V}{\partial q_\beta} = \sum_{i,\alpha} m_i\ddot{x}_{i,\alpha} \frac{\partial x_{i\alpha}}{\partial q_\beta}.$$

The potential term was easy to rewrite in Lagrangian coordinates; to transform the kinetic term in the Eq. 2.17, we note that

$$\ddot{x}_{i,\alpha} \frac{\partial x_{i\alpha}}{\partial q_\beta} = \frac{d}{dt}\left(\dot{x}_{i,\alpha} \frac{\partial x_{i\alpha}}{\partial q_\beta}\right) - \dot{x}_{i,\alpha}\frac{d}{dt} \frac{\partial x_{i,\alpha}}{\partial q_\beta},$$

that is, exchanging derivatives in the last term,

$$\ddot{x}_{i,\alpha} \frac{\partial x_{i\alpha}}{\partial q_\beta} = \frac{d}{dt}\left(\dot{x}_{i,\alpha} \frac{\partial x_{i\alpha}}{\partial q_\beta}\right) - \dot{x}_{i,\alpha} \frac{\partial \dot{x}_{i,\alpha}}{\partial q_\beta}. \tag{2.17}$$

So,

$$\sum_{i,\alpha} m_i\left[\frac{d}{dt}\left(\dot{x}_{i\alpha} \frac{\partial x_{ia}}{\partial q_\beta}\right) - \dot{x}_{i\alpha}\frac{\partial}{\partial q_\beta}\dot{x}_{i\alpha}\right] = -\frac{\partial V}{\partial q_\beta}. \tag{2.18}$$

The second term in the l.h.s. is

$$-\sum_{i,\alpha} m_i\dot{x}_{i\alpha}\frac{\partial}{\partial q_\beta}\dot{x}_{i\alpha} = -\frac{\partial T}{\partial q_\beta}, \tag{2.19}$$

where T is the kinetic energy. Now the equation of motion reads as:

$$\frac{d}{dt}\sum_{i,\alpha} m_i\left(\dot{x}_{i\alpha} \frac{\partial x_{ia}}{\partial q_\beta}\right) = -\frac{\partial}{\partial q_\beta}(T - V). \tag{2.20}$$

The Time Derivative

We wish to eliminate the Cartesian coordinates from the l.h.s., and this is highly nontrivial. First of all, dividing both sides of (2.10) by dt,

$$\dot{x}_{i\alpha} = \sum_{\beta} \frac{\partial x_{i\alpha}}{\partial q_{\beta}} \dot{q}_{\beta}. \tag{2.21}$$

At this point, the components of the velocity are to be thought of as functions of all the q and the time, but we decide that they also depend on the $\dot{q}_{\beta}$, which we take as independent variables. Differentiating $\dot{x}_{i\alpha}$ with respect to $\dot{q}_{\beta}$, one finds that:

$$\frac{\partial \dot{x}_{i\alpha}}{\partial \dot{q}_{\beta}} = \frac{\partial x_{i\alpha}}{\partial q_{\beta}}. \tag{2.22}$$

Now the l.h.s. of (2.20) becomes

$$\frac{d}{dt} \sum_{i,\alpha} m_i \left(\dot{x}_{i\alpha} \frac{\partial \dot{x}_{ia}}{\partial \dot{q}_{\beta}} \right) = \frac{d}{dt} \frac{\partial T}{\partial \dot{q}_{\beta}}. \tag{2.23}$$

Substitution in in (2.20) yields:

$$\frac{d}{dt} \frac{\partial T}{\partial \dot{q}_{\beta}} = \frac{\partial T}{\partial q_{\beta}} - \frac{\partial V}{\partial q_{\beta}}. \tag{2.24}$$

Then, we introduce the Lagrangian function, or simply the Lagrangian

$$L(q, \dot{q}, t) = T(q, \dot{q}, t) - V(q, t), \tag{2.25}$$

where q stands for the set of all coordinates, $\dot{q}$ for the set of velocities, and so on. The Lagrangian equations of motion, valid in every reference and coordinate system, are of the same form for all systems, namely,

$$\boxed{\frac{d}{dt} \frac{\partial L}{\partial \dot{q}_{\beta}} = \frac{\partial L}{\partial q_{\beta}}.} \tag{2.26}$$

To appreciate the reasons why $L = L(q, \dot{q}, t)$, we must pause a moment to understand the decisive step taken in Eq. (2.22). The fact that the velocity components are to be treated as independent variables in that particular way is not a physically evident fact, but it is the nontrivial point, the one that required the work of great mathematicians. Indeed, $\dot{q}_{\beta}$ is determined if we know q_{β} versus time t, so one could have taken $\dot{q}$ as a function of q and t. This would have led to a mathematically different functional dependence and to an impasse. The Lagrangian formalism is achieved only by considering the velocities as independent variables for T and L.

The *generalized momenta* are defined by

$$p_i = \frac{\partial L}{\partial \dot{q}_i} \qquad (2.27)$$

and are functions of the independent variables of L, namely,

$$p_i = p_i(q, \dot{q}, t), \qquad (2.28)$$

(recall that here, $q, \dot{q}$ stand for the set of all the coordinates and velocities). If, for a particular q, $\frac{\partial L}{\partial q} = 0$, then q is *cyclic* and Eq. (2.26) says that the corresponding momentum (which is called the conjugate momentum) is conserved. If an angle is cyclic, a component of the angular momentum is conserved; if x is cyclic, p_x is conserved, and so on.

If L has no explicit dependence on time, the energy is conserved. Indeed, the total derivative is

$$\frac{dL}{dt} = \sum_\alpha \frac{\partial L}{\partial q_\alpha} \dot{q}_\alpha + \sum_\alpha \frac{\partial L}{\partial \dot{q}_\alpha} \ddot{q}_\alpha, \qquad (2.29)$$

and one finds that

$$\frac{dL}{dt} = \sum_\alpha \dot{q}_\alpha \frac{d}{dt} \frac{\partial L}{\partial \dot{q}_\beta} + \sum_\alpha \frac{\partial L}{\partial \dot{q}_\alpha} \ddot{q}_\alpha = \frac{d}{dt} \sum_\alpha \frac{\partial L}{\partial \dot{q}_\alpha} \dot{q}_\alpha,$$

which implies the conservation of the energy

$$\sum_\alpha \frac{\partial L}{\partial \dot{q}_\alpha} \dot{q}_\alpha - L = \sum_\alpha p_\alpha \dot{q}_\alpha - L = E. \qquad (2.30)$$

We have a great freedom of choice of Lagrangian coordinates q, and the equations of motion are always correct. In other words, the theory is invariant under *point transformations* to new coordinates $Q = Q(q, t)$. These transformations include the change of reference to an arbitrarily moving frame. If one knows a Lagrangian $L(q, \dot{q}, t)$ and wants to switch to a new Lagrangian $L(Q, \dot{Q}, t)$, all that is needed is to express the old variables in terms of the new, i.e., to make a change of variables. The reason for this statement will be clear when we introduce the action, below.

2.2.1   Rotating Platform

Let us write the equations of motion of a material point in a platform rotating in a clockwise direction. Since the Earth is a rotating frame, this example is relevant to everyday life. In the inertial frame, the Lagrangian reds

$$L(x_0, y_0, \dot{x}_0, \dot{y}_0) = \frac{1}{2} m(\dot{x}_0^2 + \dot{y}_0^2), \qquad (2.31)$$

and the point transformation is of the form

$$x = x_0 \cos(\omega t) - y_0 \sin(\omega t), \quad y = y_0 \cos(\omega t) + x_0 \sin(\omega t). \qquad (2.32)$$

We find that the square of the velocity is $\dot{x}_0^2 + \dot{y}_0^2 = (\dot{x} + \omega y)^2 + (\dot{y} - \omega x)^2$. Thus, the Lagrangian in the rotating frame becomes:

$$L(x, y, \dot{x}, \dot{y}) = \frac{1}{2} m[\dot{x}^2 + \dot{y}^2 + \omega^2(x^2 + y^2) - 2\omega(x\dot{y} - y\dot{x})]. \qquad (2.33)$$

The equations of motion are as follows:

$$m\ddot{x} = m\omega^2 x - 2m\omega\dot{y}, \qquad m\ddot{y} = m\omega^2 y + 2m\omega\dot{x}. \qquad (2.34)$$

In the rotating system, we experience the obvious centrifugal force

$$\vec{F} = (F_x, F_y) = m\omega^2(x, y) \qquad (2.35)$$

and the less obvious velocity-dependent Coriolis force

$$\omega(-\dot{y}, \dot{x}). \qquad (2.36)$$

Problem 1 Write the Lagrangian and the equations of motion for the one-dimensional oscillator of mass m and force constant k, that is, $F = -kx$.

Solution 1

$$L = \frac{1}{2} m\dot{x}^2 - \frac{1}{2} kx^2.$$

Problem 2 A plane pendulum has mass m and length l; the acceleration of gravity is parallel to the vertical axis z is g. Write the Lagrangian and the equation of motion.

Solution 2 The speed is $l\dot{\phi}$, and the kinetic energy is $T = \frac{1}{2} ml^2 \dot{\phi}^2$. Since $V = mz = g - mgl \cos(\phi)$, one finds $L = \frac{1}{2} ml^2 \dot{\phi}^2 + mgl \cos(\phi)$. Therefore, $\frac{\partial L}{\partial \dot{\phi}} = ml^2 \dot{\phi}$, $\frac{\partial L}{\partial \phi} = -mgl \sin(\phi)$ and the equation of motion is $l\ddot{\phi} = -g \sin(\phi)$.

Problem 3 A plane pendulum of mass m_2 and length l is linked to a mass m_1, which can move on a horizontal axis x. The acceleration of gravity is parallel to the vertical axis z and its magnitude is g. Write the Lagrangian (Fig. 2.1).

Solution 3 Since $x_2 = x + l \sin(\phi)$, $z_2 = -l \cos(\phi)$, $\dot{x}_2 = \dot{x} + l \cos(\phi)\dot{\phi}$, $\dot{z}_2 = l \sin(\phi)\dot{\phi}$, and substituting into $T = \frac{1}{2}[m_1\dot{x}^2 + m_2(\dot{x}_2^2 + \dot{z}_2^2)]$ one finds $T = \frac{1}{2}(m_1 + m_2)\dot{x}^2 + \frac{m_2}{2}(l^2\dot{\phi}^2 + 2l\dot{x}\dot{\phi}\cos(\phi))$. So,

Fig. 2.1 A pendulum
hanged to a mobile mass

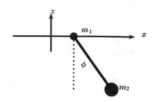

$$L = \frac{1}{2}(m_1 + m_2)\dot{x}^2 + \frac{m_2}{2}(l^2\dot{\phi}^2 + 2l\dot{x}\dot{\phi}\cos(\phi)) + m_2gl\,\cos(\phi).$$

Hence, the equations of motion follow immediately.

Problem 4 (*Free-falling oscillator*) In a free-falling lift, (the acceleration of gravity
is g), a harmonic oscillator (mass $= m$, elastic constant $= k$, force $F = -kz$) is
aligned to the vertical z axis of a local Cartesian system (x, y, z), which is fixed in
the lift. The oscillator starts from $z = 0$ at time $t = 0$. At $t = 0$, the local system
(x, y, z) coincides with a stationary one (XYZ) fixed in the ground and the relative
velocity is 0. Write the Lagrangian, equation of motion and its general solution in
the (XYZ) frame.

Solution 4 The Lagrangian in the system of the lift is $L = \frac{1}{2}m\dot{z}^2 - \frac{1}{2}kz^2$. The origin
of the falling system is at La $Z = -\frac{1}{2}gt^2$ and the oscillator is at $Z(t) = z(t) - \frac{1}{2}gt^2$.
Changing the coordinates,

$$L = \frac{1}{2}m\dot{z}^2 - \frac{1}{2}kz^2 = \frac{1}{2}m\left(\dot{Z} + gt\right)^2 - \frac{1}{2}k\left(Z + \frac{1}{2}gt^2\right)^2,$$

the equation of motion reads as:

$$m(\ddot{Z} + g) = -k\left(Z + \frac{1}{2}gt^2\right).$$

The general solution is $Z(t)A\sin(\omega t + \varphi) - \frac{1}{2}gt^2$.

Problem 5 Rigid rotor. Above, we have defined the rotor as a system of two masses
m_1, m_2 constrained to remain at a distance r_0 at the ends of a rigid stick of negligible
mass. Find the equations of motion for the rotor.

Solution 5 The coordinate transformation is:

$$x_1 = r_1\,\sin\theta\,\cos\phi,$$
$$y_1 = r_1\,\sin\theta\,\sin\phi,$$
$$z_1 = r_1\,\cos\theta,$$

with $x_2 = -\frac{r_2}{r_1}x_1 = -\frac{m_1}{m_2}x_1$, and so on. The Lagrangian coincides with the kinetic energy, that is, in obvious notation,

$$T = \frac{1}{2}(m_1 \vec{v}_1^2 + m_2 \vec{v}_2^2).$$

From the velocity components

$$\dot{x}_1 = r_1(\cos\theta\cos\phi\dot{\theta} - \sin\theta\sin\phi\,\dot{\phi}),$$
$$\dot{y}_1 = r_1 (\cos\theta\sin\phi\dot{\theta} + \sin\theta\,\cos\phi\dot{\phi}),$$
$$\dot{z}_1 = -r_1\,\sin\theta\,\dot{\theta},$$

one finds $\vec{v}_1^2 = \dot{x}_1^2 + \dot{y}_1^2 + \dot{z}_1^2 r_1^2(\dot{\theta}^2 + \sin^2\theta\dot{\phi}^2)$, $\vec{v}_2^2 r_2^2(\dot{\theta}^2 + \sin^2\theta\dot{\phi}^2)$. Therefore,

$$T = \frac{1}{2}I(\dot{\theta}^2 + \sin^2\theta\dot{\phi}^2),$$

where

$$I = m_1 r_1^2 + m_2 r_2^2,$$

is the *moment of inertia* of the rotor. Setting

$$p_\theta = \frac{\partial L}{\partial\dot{\theta}} = I\dot{\theta},$$
$$p_\phi = \frac{\partial L}{\partial\dot{\phi}} = I\sin^2\theta\dot{\phi},$$

we may write the equation of motion

$$\dot{p}_\theta = I\sin\theta\cos\theta\dot{\phi}^2,$$
$$\dot{p}_\phi = 0.$$

Problem 6 A point mass m moves along the curve $z(x) = H\sin(\frac{x}{L})$ on the vertical zx plane, where H is some length. There is no friction and the acceleration of gravity is g. One can take x as the only Lagrangian coordinate of this one-dimensional motion. Write the Lagrangian $L(x, \dot{x})$ and the kinetic momentum p_x.

Solution 6 $\dot{z} = \cos(\frac{x}{H})\dot{x}$, therefore setting $c(x) \equiv \cos\left(\frac{x}{H}\right)$,

$$T = \frac{1}{2}m\dot{x}^2\left[1 + (x)c^2\right], \, L = T - mgH\sin\left(\frac{x}{H}\right).$$

Hence, $p_x = \frac{\partial L}{\partial\dot{x}} = m\dot{x}\left[1 + c^2\right].$

2.2.2   Kepler Problem

The two-body problem, in which the two bodies interact through a central force F that varies as the inverse square of the distance r between them is known as the Kepler problem. In Sect. 2.5.1, we shall see that a two-body problem of this kind can be reduced to the problem of a single effective mass in an external potential. Here, we review the latter problem, first solved by I. Newton.

In cylindric coordinates, the square length element is $dl^2 = dr^2 + r^2 d\phi^2 + dz^2$, and therefore the kinetic energy of a point mass m is $K = \frac{1}{2}m(\dot{r}^2 + r^2\dot{\phi}^2 + \dot{z}^2)$. Letting the orbit of a planet of mass m lay on the z plane, the Lagrangian is, in obvious notation,

$$L = \frac{1}{2}m\left[\dot{r}^2 + r^2\dot{\phi}^2\right] + G\frac{Mm}{r}. \tag{2.37}$$

One readily obtains from the Lagrange formalism the conservation of the angular momentum directed along z $J = \frac{\partial L}{\partial \dot{\phi}} = mr^2\dot{\phi}$ and of the energy $E = \frac{1}{2}m(\dot{r}^2 + r^2\dot{\phi}^2) - \frac{GMm}{r}$. These conditions determine the orbits, but we need to do some work to extract them. We divide both sides by

$$\dot{\phi}^2 = \left(\frac{J}{mr^2}\right)^2 \tag{2.38}$$

and make the substitution

$$\frac{\frac{dr}{dt}}{\frac{d\phi}{dt}} \rightarrow \frac{dr}{d\phi};$$

we arrive at

$$\frac{mr^4 E}{J^2} = \frac{m^2 r^4}{2J^2}\left(\frac{dr}{d\phi}\frac{d\phi}{dt}\right)^2 + \frac{m^2 r^6}{2J^2}\left(\frac{d\phi}{dt}\right)^2 - \frac{Gm^2 Mr^3}{J^2}. \tag{2.39}$$

Next, let $\xi = \frac{1}{r}$, which implies $\frac{dr}{d\phi} = -\frac{1}{\xi^2}\frac{d\xi}{d\phi}$, and use Eq. (2.38); this gives us

$$\frac{mE}{J^2} = \frac{1}{2}\left(\frac{d\xi}{d\phi}\right)^2 + \frac{\xi^2}{2} - \frac{GMm^2\xi}{J^2}. \tag{2.40}$$

Next, we apply $\frac{\partial}{\partial\phi}$ and simplify, obtaining

$$\frac{d^2\xi}{d\phi^2} + \xi - \frac{GMm^2}{J^2} = 0. \tag{2.41}$$

This leads us to

$$\xi = \frac{GMm^2}{J^2}[1 + e\cos(\phi - \phi_0)]. \tag{2.42}$$

Then, the orbit is given by

$$r = \frac{a(1 - e^2)}{1 + e\cos(\phi - \phi_0)}, \tag{2.43}$$

with $e = \sqrt{1 + \frac{2EJ^2}{G^2M^2m^3}}$, $a(1 - e^2) = \frac{J^2}{GMm^2}$. If the constant of integration e is less than 1, the orbit is an ellipse with eccentricity e and major axis $2a$. If $e > 1$ it is an hyperbola, and for $e = 1$ it is a parabola.

2.3 The Path Between t_1 and t_2 and the Action Integral

The Lagrangian contains the necessary information for writing the equation of motion, but it is not a measurable quantity, and every change of coordinates changes it as a function of time. Is there any quantity that can be used to characterize a given motion and is invariant for punctual transformations? To answer this question, we reformulate the problem of Mechanics as follows. Suppose we know the forces that act in a mechanical system, and we observe that in the time interval (t_1, t_2), the coordinate q goes[4] from $q(t_1)$ to $q(t_2)$. In a sense we have two *snapshots*. Can we use this information to retrace what happened in between? The answer must be sought with the help of the Lagrangian. A priori, i.e. before solving the problem of motion, one knows only that the physical law $q(t)$ belongs to a large set $\mathcal{E}$ of functions $q(t)$ which have nothing in common except that they all satisfy the same boundary conditions; these trajectories laws are called *virtual paths*. Solving the equations of motion means choosing between the virtual paths $q(t) \in \mathcal{E}$ the few possible ones. For every virtual path $q(t)$, it turns out that $L(q(t), \dot{q}(t), t)$ is a function of time. Integrating this function of time we can calculate the *action* S:

$$S = \int_{t_1}^{t_2} dt\, L(q(t), \dot{q}(t), t). \tag{2.44}$$

Dimensionally, S is energy $\times$ time and is a *functional* of the virtual path, i.e. a function of all the values of $q(t)$ at each t in the path under consideration. In this way, S depends on an infinite number of variables. For instance, for a point mass moving at constant speed $v = \frac{x_b - x_a}{t_2 - t_1}$ from x_a to x_b,

---

[4]As usual, we speak about one variable q for short, but the argument is intended for a set of s variables.

$$S = \frac{m}{2} \int_{t_1}^{t_2} dt = \frac{(x_b - x_a)^2}{(t_2 - t1)^2} = \frac{m}{2} \frac{(x_b - x_a)^2}{t_2 - t_1}.$$

There is an infinity of alternative virtual paths $x(t)$, $x(t_1) = x_a$, $x(t_2) = x_b$, and one can calculate S for each of them.

The reason why S is interesting is that it does not depend on the choice of Lagrangian coordinates used to describe the virtual path, but only on the path itself. In fact, the value that S takes at time t does not change if we make a point transformation, which is simply a change of variables in an integral. The fact that for a given path, S is not arbitrary suggests that it must have a physical meaning, and indeed, we shall see that it offers the opportunity to make considerable further progress.

Problem 7 For an harmonic oscillator of mass m and constant k, $L = \frac{m}{2}(\dot{x}^2 - \omega^2 x^2)$, where $\omega^2 = \frac{k}{m}$. Calculate S for a physical path from $x = 0, t = 0$ to $x = X, t = T$.

Solution 7 Set $x = A\sin(\omega t)$, $X = A\sin(\omega T)$, $L = \frac{m}{2} A^2 \omega^2 \cos(2\omega t)$; one finds:

$$S = \frac{m}{2} A^2 \omega \sin(\omega T) \cos(\omega T) = \frac{m}{2} X^2 \omega \cot g(\omega T).$$

2.3.1 Principle of Least Action

There is an infinity of virtual paths between the *snapshots* $q(t_1)$ and $q(t_2)$, and one should distinguish those that are physically possible (if any) from the infinity of absurd ones. The action integral assigns a number (with the dimensions of energy × time, or angular momentum) to each virtual path, and so we can compare the action integrals of different paths with the initial and final configurations $q(t_1)$ and $q(t_2)$. Two virtual paths that differ little in position and velocity also give close results when one computes S. In the infinite-dimensional space of paths, one has the means to compare two of them and also define the concept of the extreme, such as a maximum or a local minimum. For instance, a path is a minimum if any "small" change produces a increase of S. In order to define rigorously an arbitrary variation, we define an arbitrary function $\eta_i(t)$ such that $\eta_i(t_1) = \eta_i(t_2) = 0$ and vary the path by $q_i(t) \rightarrow q_i(t) + \eta_i(t)$. The varied Lagrangian is

$$L(q_i(t), \dot{q}_i(t), t) \rightarrow L(q_i(t) + \eta_i(t), \dot{q}_i(t) + \dot{\eta}_i(t), t).$$

We are interested in small arbitrary variations, therefore we instead put

$$q_i(t) \rightarrow q_i(t) + \alpha_i \eta_i(t),$$

where α_i is a parameter that allows us to adjust the magnitude of the change. We can always (provided that $\eta_i(t)$ is not too bad) choose α_i small enough to obtain a small change of S. For a generic $q(t)$, with $\alpha_i \ll 1$ we find, when we vary paths,

$$S(\alpha_i) - S(0) = \delta S^{(1)} \alpha_i + \delta S^{(2)} \alpha_i^2 + \ldots;$$

and the variational principle is:

$$\delta S^{(1)} = \frac{\partial S}{\partial \alpha_i} d\alpha_i = 0. \tag{2.45}$$

This condition makes the action stationary; the first-order correction vanishes.
Since

$$S(\alpha_i) = \int_{t_1}^{t_2} dt L(q_i(t) + \alpha_i \eta_i(t), \dot{q}_i(t) + \alpha_i \dot{\eta}_i(t), t) dt,$$

the condition is

$$\frac{\partial S}{\partial \alpha_i}|_{\alpha_i \to 0} = 0 = \int_{t_1}^{t_2} dt \left[\frac{\partial L}{\partial q_i} \eta_i(t) + \frac{\partial L}{\partial \dot{q}_i} \dot{\eta}_i(t) \right]. \tag{2.46}$$

We must transform $\dot{\eta}$, since we do not know its behaviour at t_1 and t_2, so we integrate by parts:

$$\frac{\partial S}{\partial \alpha_i} = 0 = \left[\frac{\partial L}{\partial \dot{q}_i} \eta_i(t) \right]_{t_1}^{t_2} + \int_{t_1}^{t_2} dt \left[\frac{\partial L}{\partial q_i} \eta_i(t) - \eta_i(t) \frac{d}{dt} \frac{\partial L}{\partial \dot{q}_i} \right]. \tag{2.47}$$

Since $\eta_i(t)$ vanishes at t_1 and t_2,

$$\int_{t_1}^{t_2} dt \eta_i(t) \left[\frac{\partial L}{\partial q_i} - \frac{d}{dt} \frac{\partial L}{\partial \dot{q}_i} \right] = 0. \tag{2.48}$$

This is possible with arbitrary η_i only if the Euler–Lagrange equations hold. Therefore, the principle of least action is a particularly concise way to reformulate the laws of mechanics.

The fascination for variational principles has always been strong. Pierre Louis de Maupertuis, who wrote about this principle in 1744, felt that "Nature is thrifty in all its actions." Earlier, Pierre de Fermat (1601–1665) found a famous variational principle in optics, which is presented in Sect. 4.4. In Theoretical Physics, variational principles are everywhere. They are important in Electromagnetism, General Relativity and Quantum Mechanics. Besides a great aesthetic *appeal,* they are also important in practice, as we shall see.

It is not granted that the least action principle leads to a unique extremal path. It is not always possible to retrace what happened in between two *snapshots* giving the configuration of a system at different times, i.e., $q(t_1)$ and $q(t_2)$. If the two snapshots show a pendulum in the vertical position, it could have always remained still, or it could have been in oscillation provided that $t_2 - t_1$ is a multiple of the period of oscillation. In general, this sort of information selects a class of possibilities, but the motions that are not compatible with the Euler–Lagrange equations are impossible.[5] One learns from this example that the physical motions correspond to stationary paths but not necessarily to the minimum.

Suppose we add the total time derivative of an arbitrary function to the Lagrangian:

$$L(q, \dot{q}, t) \rightarrow L(q, \dot{q}, t) + \frac{d}{dt} F(q, t). \tag{2.49}$$

This addition changes S by $F(q(t_2), t_2) - F(q(t_1), t_1)$. The equations of motion are obtained by varying the paths, however, the initial and final 'snapshots' are not changed. So, one is free to add the total time derivative of an arbitrary function to the Lagrangian, since this does not alter δS. The lagrangian is largely arbitrary, and one can exploit this fact.

While $\dot{q}$ are independent variables for L, their variation is a consequence of the variation of q. If $q_i(t, \alpha_i) = q_i(t) + \alpha_i \eta_i(t)$, then $\dot{q}_i(t, \alpha_i) = \dot{q}_i(t) + \alpha_i \dot{\eta}_i(t)$. So, there is some redundancy in the process.

The variational principle contains another remarkable result, if instead of the virtual paths from $q_1(t_1)$ to $q_2(t_2,)$ we consider the physical paths, without fixing $q_2(t_2)$. Then, δS does not vanish any more, but instead of Eq. 2.47 one finds:

$$\frac{\partial S}{\partial \alpha_i} = \left[\frac{\partial L}{\partial \dot{q}_i} \eta_i(t) \right]_{t_1}^{t_2} + \int_{t_1}^{t_2} dt \left[\frac{\partial L}{\partial q_i} \eta_i(t) - \eta_i(t) \frac{d}{dt} \frac{\partial L}{\partial \dot{q}_i} \right], \tag{2.50}$$

where the integral now vanishes; since $\eta_i(t_2) = \frac{\partial q_i}{\partial \alpha_1}|_0$, we may write $\delta S = \sum_i \frac{\partial L}{\partial \dot{q}_i} \delta q_i(t_2)$, that is,

$$\frac{\partial S}{\partial q_i} = p_i. \tag{2.51}$$

This result will be useful below.

2.4 Legendre Transformation

Let $f(x, y)$ be a good function defined in the plane. Then, its differential is

$$df = u dx + v dy, \tag{2.52}$$

where

$$u = \frac{\partial f}{\partial x}, v = \frac{\partial f}{\partial y}. \tag{2.53}$$

---

[5]We shall see that in Quantum Mechanics, this is no longer true.

It occurred to Legendre[6] that if one introduces a new function

$$\gamma(x, y) = f(x, y) - u(x, y)x, \tag{2.54}$$

then

$$d\gamma(x, y) = df(x, y) - u(x, y)dx - xdu(x, y) = v(x, y)dy - xdu(x, y). \tag{2.55}$$

The differential dx has disappeared, by the rules of differential calculus. But Legendre was able to see more than that. Indeed, $vdy - xdu$ is the differential of a function $g(u, y)$, provided that instead of treating u as a function of x, y, we consider x as a function of u, y. Explicitly, the differential is

$$dg(u, y) = v(x(u, y), y)dy - x(u, y)du(x(u, y), y).$$

This inverse function $x(u, y)$ can be found uniquely if there is a one-to-one correspondence between x and u; this is granted if $u = \frac{\partial f(x,y)}{\partial x}$ increases (or decreases) with increasing x, or, equivalently, if f is convex. The x-independent $g(u, y)$ is well defined; when $u = u(x, y)$, it takes all the values of $\gamma(x, y)$, but the functional dependence of the two functions g and γ on the respective independent variables is different.

In Eq. (2.54,) one gets γ in terms of f, but to find f from γ, one should solve a partial differential equation because of the presence of $u = \frac{\partial f}{\partial x}$. After the change of independent variable, we have

$$g(u, y) = \phi(u, y) - ux(u, y) \tag{2.56}$$

where

$$\phi(u(x, y), y) = f(x, y). \tag{2.57}$$

We see that

$$\frac{\partial g}{\partial u} = -x, \quad \frac{\partial g}{\partial y} = v. \tag{2.58}$$

The similarity with Eq. (2.53) confirms the duality of f and g. Now the knowledge of $g(u, y)$ is fully equivalent to the knowledge of $f(x, y)$, and it is a matter of convenience as to which function to use.

The reason why this change of independent variables is useful is as follows. Sometimes u is a physical quantity that one can measure more easily than x, and in this sense, the transformation is analogous to the Fourier transformation (one can

---

[6] Adrien Marie Legendre (Paris 1752- Paris 1833) academic of Sciences and one of the great mathematicians of his age.

decide to write all quantities in terms of time if the equipment is based on clocks or in terms of frequency, if the equipment is based on antennas.) The inverse transformation $f = g + xu$ is just (2.56). As in the case of Fourier, if you apply the transform twice, you are back at the original form.

Legendre's method is simple and general and can be applied directly to functions of many variables $f(x, y, z, w, \ldots)$ since the transformation can be applied independently to each variable.

2.5   The Hamiltonian

Building on the Lagrangian formulation, William Rowan Hamilton[7] found a new, more effective framework. The Hamiltonian formulation allows us to solve difficult problems and deepens the general understanding of the workings of the theory. The main idea is a change of the independent variables. No longer $q, \dot{q}$, but q_i and $p_i = \frac{\partial L}{\partial \dot{q}_i}$. At first sight, this does not appear to be a real novelty, since for a point mass in Cartesian coordinates velocity and momentum are proportional, $\vec{p} = m \frac{d\vec{x}}{dt}$. However, when using other Lagrangian coordinates, there is no such trivial relation between momentum and velocity, and further complications arise when introducing a vector potential (see Sect. 2.5.4). The change from the independent variables $\dot{q}_i$ to p_i is a Legendre transformation

$$H(p, q, t) = \sum_i p_i \dot{q}_i - L(q, \dot{q}, t), \tag{2.59}$$

where $H(p, q, t)$ is the Hamiltonian, which does not depend on the velocities:

$$\frac{\partial H}{\partial \dot{q}_k} = p_k - \frac{\partial L}{\partial \dot{q}_k} = 0. \tag{2.60}$$

If L does not depend on time explicitly, we know from (2.30) that $H = E$ is the energy and is conserved along the physical evolution of the system; then, we can say that *the Hamiltonian is the energy as a function of q and p*. The equations of motion were obtained by Hamilton from the variational principle $\delta S = 0$, with S now a functional of the path written as $p(t), q(t)$.

$$S = \int_{t_1}^{t_2} dt \, L(q, \dot{q}, t) = \int_{t_1}^{t_2} dt \left[\sum_i p_i \dot{q}_i - H(p, q, t) \right]. \tag{2.61}$$

---

[7]This early Irish genius (Dublin 1805 - Dublin 1865), at the age of 22, went to Trinity College to explain his reformulation of Mechanics, and immediately got a chair. He continued for the rest of his life to produce high-level results; he also worked on the quaternions, which are essentially the spin matrices that we will discuss later on.

Here, $\dot{q}(t)$ is thought of as a function of $q(t)$ and $p(t)$ along the virtual path. Note that $q(t)$ e $p(t)$ can now be varied independently, while in the Lagrangian formalism, $\dot{q}$ is fixed by $q(t)$). We consider variations

$$q_i(t, \alpha_i) = q_i(t, 0) + \alpha_i \eta_i(t),$$
$$p_i(t, \alpha_i) = p_i(t, 0) + \alpha_i \xi_i(t).$$

where $\eta_i(t)$, $\xi_i(t)$ are arbitrary independent functions. As in the Lagrangian formalism, q_i is fixed at t_1 and t_2,

$$\eta_i(t_1) = \eta_i(t_2) = 0. \tag{2.62}$$

Note, however, that no such constraint limits ξ_i. In Lagrangian formalism, there was no prescription on $\dot{q}(t_1)$ and $\dot{q}(t_2)$; in the same way, $p(t_1)$ and $p(t_2)$ are left arbitrary. So, (2.61) becomes:

$$S = \int_{t_1}^{t_2} dt \left[\sum_i [p_i + \alpha_i \xi_i][\dot{q}_i + \alpha_i \dot{\eta}_i] - H(p_i + \alpha_i \xi_i, q_i + \alpha_i \eta_i, t) \right].$$

Neglecting terms in α_i^2,

$$\frac{dS}{d\alpha_i} = \int_{t_1}^{t_2} dt \left(p_i \dot{\eta}_i + \dot{q}_i \xi_i - \frac{\partial H}{\partial q_i} \eta_i - \frac{\partial H}{\partial p_i} \xi_i \right) = \int_{t_1}^{t_2} dt \left(p_i \dot{\eta}_i - \frac{\partial H}{\partial q_i} \eta_i + \xi_i \left[\dot{q}_i - \frac{\partial H}{\partial p_i} \right] \right).$$

An integration by parts eliminates $\dot{\eta}$:

$$\int_{t_1}^{t_2} dt \, p_i \dot{\eta}_i \, p_i \eta_i |_{t_1}^{t_2} - \int_{t_1}^{t_2} dt \, \dot{p}_i \eta_i = - \int_{t_1}^{t_2} dt \, \dot{p}_i \eta_i,$$

and we are left with

$$\int_{t_1}^{t_2} dt \left[\eta_i \left(-\dot{p}_i - \frac{\partial H}{\partial q_i} \right) + \xi_i \left(\dot{q}_i - \frac{\partial H}{\partial p_i} \right) \right] = 0.$$

Hence, one obtains the Hamiltonian equations (or *canonical equations*)

$$\dot{p}_i = -\frac{\partial H}{\partial q_i}, \tag{2.63}$$

$$\dot{q}_i = \frac{\partial H}{\partial p_i}. \tag{2.64}$$

The p, q that appear in (2.63), (2.64) are canonically conjugate variables. The canonical equations also follow from Lagrange's equations, but to see that, starting from the definition (2.59) of the Hamiltonian, one must be careful about the independent variables.

For L they are q, $\dot{q}$, but for $H(p, q, t) = \sum_i p_i \dot{q}_i - L(q, \dot{q}, t)$, $\dot{q}$ must be rewritten in terms of p and q. Thus,

$$\frac{\partial H}{\partial q_k} = \sum_i p_i \frac{\partial \dot{q}_i}{\partial q_k} - \frac{\partial L}{\partial q_k} - \sum_i \frac{\partial L}{\partial \dot{q}_i} \frac{\partial \dot{q}_i}{\partial q_k}. \tag{2.65}$$

The first term yields 0 with the third and one gets, using Lagrange's equations,

$$\frac{\partial H}{\partial q_k} - \frac{\partial L}{\partial q_k} = -\frac{d}{dt} \frac{\partial L}{\partial \dot{q}_k} = -\dot{p}_k. \tag{2.66}$$

This verifies (2.63). On the other hand, differentiating (2.59),

$$\frac{\partial H}{\partial p_k} = \dot{q}_k + \sum_i p_i \frac{\partial \dot{q}_i}{\partial p_k} - \sum_i \frac{\partial L}{\partial \dot{q}_i} \frac{\partial \dot{q}_i}{\partial p_k};$$

simplifying the last two terms one is left with (2.64). Along a physical trajectory, $H = H(p(t), q(t), t)$ depends on the time in such a way that

$$\frac{dH}{dt} = \frac{\partial H}{\partial t}. \tag{2.67}$$

Indeed,

$$\frac{dH}{dt} = \frac{\partial H}{\partial t} + \sum_i \left(\frac{\partial H}{\partial q_i} \dot{q}_i + \frac{\partial H}{\partial p_i} \dot{p}_i \right),$$

but the sum vanishes by the canonical equations. So, when H is not explicitly a function of time, then $H = E$ is a constant.

2.5.1 Reduced Mass

For $N = 2$ point masses m_1, m_2 interacting via an instantaneous potential energy $V(r_{12})$ depending on their distance, one can start from the observation that $m_1 \ddot{\vec{r}}_1 = -\nabla_1 V(r_{12})$ and $m_2 \ddot{\vec{r}}_2 = -\nabla_2 V(r_{12})$ are opposite; in other words, the instantaneous forces felt by the two masses are opposite. Consequently, their center of mass

$$\vec{R} = \frac{m_1 \vec{r}_1 + m_2 \vec{r}_2}{m_1 + m_2} \tag{2.68}$$

obeys $\ddot{\vec{R}} = 0$, that is, makes a free-particle motion. Thus, the Hamiltonian

$$\frac{p_1^2}{2m_1} + \frac{p_2^2}{2m_2} + V(r_{12}) \tag{2.69}$$

can be simplified, by separating the center-of-mass degrees of freedom from the more interesting relative motion. Setting $\rho = r_{12}$ and $\mathbf{p}_r = \mathbf{p}_1 - \mathbf{p}_2$, one can change variables with

$$\vec{p}_1 = \vec{p}_\rho + \frac{m_1}{m_1 + m_2} \vec{p}_R, \qquad \vec{p}_2 = -\vec{p}_\rho + \frac{m_2}{m_1 + m_2} \vec{p}_R.$$

Computing the squares one realizes that

$$\frac{p_1^2}{m_1} + \frac{p_2^2}{m_2} = \frac{p_R^2}{m_1 + m_2} + \left[\frac{1}{m_1} + \frac{1}{m_2} \right] p_\rho^2. \tag{2.70}$$

One can solve the problem in terms of a single effective particle having the reduced mass μ such that

$$\frac{1}{\mu} = \frac{1}{m_1} + \frac{1}{m_2}. \tag{2.71}$$

Since $\vec{p}_R$ is conserved, we can separate the center-of-mass motion and reduce the two-body problem to an effective one-body problem with Hamiltonian

$$H = \frac{p_\rho^2}{2\mu} + V(\rho).$$

We already used this separation in Sect. 2.2.2. Once the relative motion is determined, one can use $\rho(t)$ and $\mathbf{R}(t)$ to find $\mathbf{r}_1(t)$ and $\mathbf{r}_2(t)$. This argument is easily extended to reduce any N body problem to $N - 1$: for $N = 3$ one can separate the motion of the center of mass

$$\vec{R} = \frac{m_1 \vec{r}_1 + m_2 \vec{r}_2 + m_3 \vec{r}_3}{m_1 + m_2 + m_3} \tag{2.72}$$

and reduce the problem to 2 effective particles.

2.5.2 Canonical Transformations

Lagrangian formalism allows us to use any choice of the s coordinates q that specify the configuration of the system, and this freedom helps in the set-up and solution of equations. The change from one set of coordinates q to a new set Q is a *point transformation*. The change in the velocities does not add anything. The Hamiltonian formalism replaces the s second-order Lagrange equations with $2s$ first-order Eqs. 2.63, 2.64 and enlarges the class of transformations from the point transformations to the *canonical transformations*. Such transformations take from coordinates q_i and canonically conjugate momenta p_i, $i = 1 \ldots s$, to new coordinates and momenta Q_i and P_i. In short we may say that the transformation is $q, p \rightarrow Q, P$. Both choices correspond to different pictures or representations of the same physical reality. The new independent variables Q, P are assumed to be in one-to-one correspondence with q, p via some invertible functional dependence such that $P = P(p, q, t)$ and $Q = Q(p, q, t)$, and must be canonically conjugate to a new Hamiltonian $\tilde{H}(P, Q, t)$. Each virtual path $p(t), q(t)$ is also represented by $P(t), Q(t)$, and the physical paths correspond to solutions of the Hamilton equations of motions in both pictures. It is clear that the point transformations are a special case, but Hamilton's formalism is much more rewarding both in the solution of mechanical problems and in the clarification of the inner mathematical structure of the theory.

It is desirable to develop a general technique that, starting from $H(p, q, t)$, generates canonical transformations to $P(p, q, t)$, $Q(p, q, t)$, and the corresponding new Hamiltonians $\tilde{H}(P, Q, t)$. If we assume that the Lagrangian is unaffected by the transformation, that is,

$$\sum_i p_i \dot{q}_i - H(p, q, t) \equiv \sum_i P_i \dot{Q}_i - \tilde{H}(P, Q, t),$$

we are granted that the action is the same for each virtual path and the condition $\delta S = 0$ gives the same paths in the two representations. This is correct, but too restrictive, since the Lagrangian is largely arbitrary, and we can allow it to be different in the two pictures. What we really want is that a path which satisfies $\delta S = 0$ in the (p, q) representation is mapped to a minimal path in the (P, Q) representation. In other words, we just need that for all the variations with fixed end points of any virtual path,

$$\delta S = \delta \int_{t_1}^{t_2} dt \left[\sum_i p_i \dot{q}_i - H \right] = \delta \int_{t_1}^{t_2} dt \left[\sum_i P_i \dot{Q}_i - \tilde{H} \right]. \tag{2.73}$$

The condition (2.73) is fulfilled if the integrands differ by the total derivative $\frac{dF}{dt}$ of a function $F = F(p, q, P, Q, t)$. The total derivative inserted into the integral gives some constant, and then δ yields 0. F is called the *generating function of the canonical transformation*. Evidently, $F = F(p, q, P, Q, t)$ is an action.

Indeed,

$$\delta \int_{t_1}^{t_2} dt \frac{dF}{dt} = \delta [F(t_2) - F(t_1)] = 0.$$

Therefore, a condition that ensures a canonical transformation is:

$$\sum_i p_i \dot{q}_i - H = \sum_i P_i \dot{Q}_i - \tilde{H} + \frac{dF}{dt}. \tag{2.74}$$

Here, p, q are functions of t along a path, while $Q(t)$, $P(t)$ are a different representation of the same path. A clever choice of F can allow for drastic simplifications of
hard problems. However we cannot be too euphoric: there is no method for finding
F, and it is difficult to invent a function of many variables. So in a sense the method
is too general to be practical. Normally one assumes that $F = F(p, q, P, Q, t)$ does
not really depend on all its variables. $F = F(q, Q, t)$ leads to

$$\sum_i p_i \dot{q}_i - H = \sum_i P_i \dot{Q}_i - \tilde{H} + \frac{\partial F}{\partial t} + \sum_i [\frac{\partial F}{\partial q_i} \dot{q}_i + \frac{\partial F}{\partial Q_i} \dot{Q}_i]. \tag{2.75}$$

The coefficients of $\dot{q}_i$ on both sides must be equal,

$$p_i = \frac{\partial F(q, Q, t)}{\partial q_i}. \tag{2.76}$$

The same argument applies to $\dot{Q}_i$:

$$P_i = -\frac{\partial F(q, Q, t)}{\partial Q_i}. \tag{2.77}$$

The transformation is canonical provided that:

$$\tilde{H} = H + \frac{\partial F(q, Q, t)}{\partial t}. \tag{2.78}$$

Thus, unlike the Lagrangian, $\tilde{H}$ is not obtained from H with a simple change of
variables; $\tilde{H} = H$ only if F is time-independent. Any $F(q, Q, t)$ that one can
invent will lead to some $\tilde{H}$ (although a smart choice is needed to find a useful
transformation); however, we are not finished, since $\tilde{H}$ deserves to be considered
the new Hamiltonian only after eliminating p, q in favor of P, Q. Thus, we must be
able to do the transformation $(p, q) \rightarrow (P, Q)$. We must be able to solve Eqs. 2.76
and 2.77 for P e Q in terms di p, q. Once we know $P(p, q, t)$ and $Q(p, q, t)$
and the inverse functions $p(P, Q, t)$ and $q(P, Q, t)$, then $\tilde{H}$ is achieved and we
have a transformed problem to solve. If we can solve $\tilde{H}$, then the transformation
$(P, Q) \rightarrow (p, q)$ solves the original problem.

Example 1 The generating function

$$F = \sum_i \lambda_i q_i Q_i$$

yields

$$p_i = \lambda_i Q_i, \quad P_i = -\lambda_i q_i.$$

There is no reason to talk about momenta and coordinates any longer. One would do better to speak about canonically conjugate variables. (Dimensionally pq is an action, and this remains true, while the dimensions of the coordinates and momenta may change.) For example, a harmonic oscillator is sent by the transformation into a harmonic oscillator.

Problem 8 Let

$$H(p, q, t) = \alpha p t + \mu \left(q - \frac{1}{2} \alpha t^2 \right)^2, \tag{2.79}$$

with constant α, μ. Solve the mechanical problem with the canonical transformation specified by the generating function

$$F(q, Q, t) = Q \left(q - \frac{1}{2} \alpha t^2 \right). \tag{2.80}$$

Solution 8 One finds

$$p = Q, \quad P = -\left(q - \frac{1}{2} \alpha t^2 \right), \quad \frac{\partial F}{\partial t} = -\alpha t Q,$$

and so

$$\tilde{H}(P, Q, t) = \mu P^2.$$

Every motion described by (2.79) is the image of a free motion.

Example 2 An important kind of transformations is generated by

$$F = S(q, P, t) - \sum_i P_i Q_i.$$

Now it is S which is commonly called the generating function. Then, (2.74) becomes

$$\sum_i p_i \dot{q}_i - H = \sum_i P_i \dot{Q}_i - \tilde{H} + \frac{\partial S}{\partial t}$$
$$+ \sum_i \left[\frac{\partial S}{\partial q_i} \dot{q}_i + \frac{\partial S}{\partial P_i} \dot{P}_i - P_i \dot{Q}_i - Q_i \dot{P}_i \right]. \tag{2.81}$$

Equating the coefficients of $\dot{q}_i$ one finds:

$$p_i = \frac{\partial S}{\partial q_i}$$

this is just Eq. (2.51); S is the action as a function of the old coordinates and the new momenta at the upper integration limit. Equating the coefficients of one gets: $\dot{P}_i$

$$Q_i = \frac{\partial S}{\partial P_i},$$
$$\tilde{H} = H + \frac{\partial S}{\partial t}.$$

Example 3 $S = \sum_k q_k P_k,$ $F = S(q, P, t) - \sum_i P_i Q_i$ generates the identity transformation $P = p,\ Q = q$.

Problem 9 Given

$$H(p, q) = vp + \left(\frac{A}{p}\right)^2 + B^2 p^4 (q - vt)^2,$$

where A, B, v are constants, transform the problem according to the generating function

$$F(q, Q, t) = \frac{q - vt}{Q}.$$

Solution 9 Since $p = \frac{\partial F}{\partial q} = \frac{1}{Q},\ P = -\frac{\partial F}{\partial Q} = \frac{q-vt}{Q^2},\ \frac{\partial F}{\partial t} = -\frac{v}{Q}$, one finds

$$\tilde{H}(P, Q) = H + \frac{\partial F}{\partial t} = A^2 Q^2 + B^2 P^2,$$

which describes a harmonic oscillator with pulsation $\omega = A$ and mass $m = \frac{1}{2B^2}$.

2.5.3 Hamilton–Jacobi Equation

We have seen in the last Section that a generating function $F = S(q, P, t) - \sum_i P_i Q_i$ with $p_i = \frac{\partial S}{\partial q_i}$ leads to a transformed Hamiltonian $\tilde{H} = H + \frac{\partial S}{\partial t}$, and this is clever if we are able to solve the problem with $\tilde{H}$. Indeed, S can be chosen such that

$$\tilde{H}(P, Q, t) = H(p, q, t) + \frac{\partial S}{\partial t} = 0. \tag{2.82}$$

Then, $\frac{\partial P}{\partial t} = -\frac{\partial}{\partial Q}\tilde{H}(P, Q, t) = 0$ implies that P are constant and $\frac{\partial Q}{\partial t} = -\frac{\partial}{\partial P}\tilde{H}(P, Q, t) = 0$ implies that Q are constant, too. Putting $p_i = \frac{\partial S}{\partial q_i}$ into (2.82) one arrives at the celebrated Hamilton–Jacobi equation

$$H\left(\frac{\partial S}{\partial q}, q, t\right) + \frac{\partial S}{\partial t} = 0. \tag{2.83}$$

Actually, the unknown S is an old acquaintance. From the differential $dS = \sum_i \frac{\partial S}{\partial q_i} dq_i + \frac{\partial S}{\partial t} dt$ and from $p_i = \frac{\partial S}{\partial q_i}$, one obtains

$$\frac{dS}{dt} = \sum_i p_i \frac{dq_i}{dt} + \frac{\partial S}{\partial t}.$$

Substituting the Hamilton–Jacobi equation one concludes that

$$\frac{dS}{dt} = \sum_i p_i \frac{dq_i}{dt} - H = L,$$

that is, S is the action. If H does not depend on time, (2.82) ensures that $\frac{\partial S}{\partial t} = -E$ is a constant, and then

$$H(\frac{\partial S}{\partial q}, q, t) = E. \tag{2.84}$$

The mechanical problem is reformulated in terms of a partial differential equation, like the propagation of a nonlinear wave. This is an important independent method of solution and something of a premonition of Quantum Mechanics.

2.5.4   Point Charge in an Electromagnetic Field

A point charge q of mass m in an electromagnetic field $\vec{E}$, $\vec{B}$ feels the Lorentz force. In the MKSA system, this reads as:

$$\vec{F} = q \left[\vec{E} + \vec{v} \wedge \vec{B} \right]. \tag{2.85}$$

In the Gauss cgs system,[8] one writes:

$$\vec{F} = q \left[\vec{E} + \frac{\vec{v}}{c} \wedge \vec{B} \right].$$

In this section, we do the calculations in the International System. Instead of working with 2 fields (6 components), it is better to express everything in terms of potentials $\vec{A}$ and ϕ (4 components overall); the magnetic induction field is:

$$\vec{B} = \vec{\nabla} \wedge \vec{A} = (\partial_2 A_3 - \partial_3 A_2, \partial_3 A_1 - \partial_1 A_3, \partial_1 A_2 - \partial_2 A_1),$$

---

[8]This is popular in scientific literature, see e.g. John D. Jackson, "Classical Electrodynamics", John Wiley and Sons (1962).

and the electric field comes from:

$$\vec{E} = -\vec{\nabla}\phi - \frac{\partial \vec{A}}{\partial t};$$ (2.86)

so, the equation of motion becomes

$$m\ddot{\vec{r}} = -q\vec{\nabla}\phi - q\frac{\partial \vec{A}}{\partial t} + q\dot{\vec{r}} \wedge \vec{B}.$$ (2.87)

We deduced the Euler–Lagrange equations under the hypothesis that force was the gradient of a potential. The Lorentz force depends on the velocity, but don't worry: we will immediately see that a Lagrangian also exists in this case; Eq. (2.87) is obtained from the Lagrangian[9]

$$L(\vec{r}, \vec{v}, t) = \frac{1}{2}mv^2 - q\phi + q\vec{v} \cdot \vec{A}(\vec{r}, t).$$ (2.88)

Using Cartesian coordinates, one finds

$$\frac{\partial L}{\partial x_i} = -q\frac{\partial \phi}{\partial x_i} + q\sum_j \dot{x}_j \frac{\partial A_j}{\partial x_i},$$ (2.89)

while

$$p_i = \frac{\partial L}{\partial \dot{x}_i} = m\dot{x}_i + qA_i.$$ (2.90)

Since $\vec{A}(\vec{r}, t)$ must be understood as $\vec{A}(\vec{r}(t), t)$, where $\vec{r}(t)$ travels along the trajectory,

$$\dot{\vec{A}} = \left(\frac{\partial}{\partial t} + \dot{x}_1 \frac{\partial}{\partial x_1} + \dot{x}_2 \frac{\partial}{\partial x_2} + \dot{x}_3 \frac{\partial}{\partial x_3} \right) \vec{A}.$$

Therefore the Lagrange equations of motion are:

$$m\ddot{x}_i = q\left[-\frac{\partial \phi}{\partial x_i} - \frac{\partial A_i}{\partial t} + \sum_j \dot{x}_j \left(\frac{\partial A_j}{\partial x_i} - \frac{\partial A_i}{\partial x_j} \right) \right].$$ (2.91)

---

[9]Any other lagrangian that differs by a total derivative $\frac{dF(\vec{r},t)}{dt}$ is equivalent.

One can readily check that, expanding (2.87) in components, the result is the same. Since the force depends on the velocity, now the Lagrangian is $L = T - U$, with $T = \frac{1}{2}mv^2$ and $U = U(q, \dot{q}, t)$. There is a velocity-dependent generalized force

$$Q_i = -\frac{\partial U}{\partial q_i} + \frac{d}{dt}\frac{\partial U}{\partial \dot{q}_i}, \tag{2.92}$$

which is just the Lorentz one, and x_i is canonically conjugate to the *canonical momentum*

$$p_i = \frac{\partial L}{\partial \dot{x}_i} = m\dot{x}_i + qA_i,$$

which, however, is not observable (like A_i). It is the *mechanical momentum* $p^{mec} \equiv m\dot{x}_i$ that is, of course, observable. The energy,

$$\begin{aligned} E &= \sum_i p_i\dot{q}_i - L \\ &= \sum_i \left[(m\dot{x}_i + qA_i)\dot{x}_i - \left(\tfrac{1}{2}m\dot{x}_i^2 + q\dot{x}_i \cdot A_i\right)\right] + q\phi = \tfrac{1}{2}m\dot{r}^2 + q\phi \end{aligned} \tag{2.93}$$

does not depend on the vector potential; however, since it contains ϕ, one can measure neither, but one can measure the difference of ϕ between two space-time points and energy differences. However the Hamiltonian H must be a function of the canonical momentum $\vec{p}$. Therefore, the Hamiltonian is

$$H = \frac{1}{2m}(\vec{p} - q\vec{A})^2 + q\phi. \tag{2.94}$$

Thus, H contains **p** and **A**, which are not measurable, while the energy is defined up to an arbitrary constant. To sum up, the rule to include the electromagnetic field in a problem, is:

$$\boxed{E \rightarrow E + q\phi, \ \vec{p} \rightarrow \vec{p} - q\vec{A}.}$$

This rule (*minimal coupling*) remains true in relativistic, quantum and quantum relativistic theories.

Remark 1 In the Gauss system, the rule becomes:

$$E \rightarrow E + q\phi, \ \vec{p} \rightarrow \vec{p} - \frac{q}{c}\vec{A}.$$

2.5.5 Poisson Brackets

Let $F = F(p(t), q(t), t)$ be a measurable physical quantity, a function of the canonical variables; the values it takes at time t while $q(t)$ and $p(t)$ vary along a trajectory

must, of course, be invariant under all canonical transformations. The total time
derivative of F is:

$$\dot{F} = \frac{\partial F}{\partial t} + \sum_i \left(\frac{\partial F}{\partial q_i} \dot{q}_i + \frac{\partial F}{\partial p_i} \dot{p}_i \right).$$

Along the trajectory, $\dot{q} = \frac{\partial H}{\partial p}$, $\dot{p} = -\frac{\partial H}{\partial q}$, and one finds that:

$$\dot{F} = \frac{\partial F}{\partial t} + \sum_i \left(\frac{\partial F}{\partial q_i} \frac{\partial H}{\partial p_i} - \frac{\partial F}{\partial p_i} \frac{\partial H}{\partial q_i} \right) = \frac{\partial F}{\partial t} + \{F, H\}_{p,q}. \qquad (2.95)$$

In the last term, I have introduced the Poisson bracket; for any two functions $A(p, q)$
and $B(p, q)$ of the canonical variables, they are defined by:

$$\{A, B\}_{p,q} \equiv \sum_k \left(\frac{\partial A}{\partial q_k} \frac{\partial B}{\partial p_k} - \frac{\partial A}{\partial p_k} \frac{\partial B}{\partial q_k} \right) = -\{B, A\}_{p,q}. \qquad (2.96)$$

The symbol $\{A, B\}_{p,q}$ emphasizes that the derivatives are taken with respect to the
set (p, q). If F does not depend on time explicitly and $\{F, H\}_{p,q} = 0$, then F is a
constant of the motion. Anyhow, since $\dot{F} - \frac{\partial F}{\partial t}$ is a measurable, physical quantity,
it must be invariant, i.e., $\{F, H\}_{p,q} = \{F, H\}_{P,Q}$. The Poisson brackets with the
Hamiltonian are invariant.

If $H(p, q)$, which governs the time evolution, is replaced by any arbitrary function
of the same variables, the invariance of the Poisson bracket under canonical transformation remains. The transformation $\{A, B\}_{p,q} \to \{A, B\}_{P,Q}$ involves writing (p, q)
in terms of (P, Q) and then differentiating with respect to the new variables. In this
respect, H is not special. One could always consider B as a fictitious Hamiltonian
and justify the invariance as above. Therefore, we may change notation and simply
write $\{A, B\}$.

The following rules are readily seen:

$$\{A, \text{costant}\} = 0; \qquad (2.97)$$

$$\{A, B + C\} = \{A, B\} + \{A, C\}; \qquad (2.98)$$

$$\{A, BC\} = \{A, B\}\, C + \{A, C\}\, B. \qquad (2.99)$$

Moreover, since p and q are independent variables, and

$$\frac{\partial p_i}{\partial p_k} = \delta_{i,k}, \qquad \frac{\partial q_i}{\partial q_k} = \delta_{i,k}, \qquad \frac{\partial p_i}{\partial q_k} = 0,$$

one finds the *fundamental brackets*

$$\{q_i, p_j\} = \delta_{i,j}, \quad \{q_i, q_j\} = 0, \quad \{p_i, p_j\} = 0. \tag{2.100}$$

Now consider a canonical transformation to new variables $(p, q) \rightarrow (P, Q)$ The Hamiltonian $H(p, q)$ is transformed to a new Hamiltonian $\tilde{H}$. Then, P, Q have the fundamental brackets $\{Q_i, P_j\}_{P,Q} = \delta_{i,j}$, where $\{\}_{P,Q}$ implies differentiation with respect to the new variables. The fact that Poisson's brackets are invariant under canonical transformations implies that $\{Q_i, P_j\}_{p,q} = \delta_{i,j}$, and this is of primary practical importance; one can check the fundamental brackets using the old variables, and certify the canonical transformation without knowing the generating function, which remains unknown in most cases. If the problem is time-independent, one obtains $\tilde{H}$ as well by a direct change of variables.

The Poisson brackets also help to find new constants of the motion using a theorem credited to Poisson. One can show[10] that if f and g are integrals of the motion (i.e. constant quantities), then $\{f, g\}$ is also an integral of the motion. It could turn out to be a constant or some combination of f and g but sometimes it is a useful *new* conserved quantity. The proof is based on the Jacobi identity

$$\{f, \{g, h\}\} + \{g, \{h, f\}\} + \{h, \{g, f\}\} = 0. \tag{2.101}$$

Example 4 Consider the harmonic oscillator with Hamiltonian

$$H(p, q) = \frac{p^2}{2m} + \frac{1}{2}m\omega^2 q^2; \tag{2.102}$$

the equations of motion are solved by

$$q(t) = \sqrt{\frac{2A}{m\omega}} \sin(\omega t), \quad p = m\dot{q} = \sqrt{2Am\omega} \cos(\omega t).$$

Here, A is the amplitude of oscillation. The energy is $E = H(p, q) = A\omega$.

Through a canonical transformation, one obtains a new Hamiltonian

$$\tilde{H}(A, \phi) = A\omega, \tag{2.103}$$

where $\phi = \omega t$; so, the new variables are ϕ, A, with

$$A = \frac{p^2}{2m\omega} + \frac{1}{2}m\omega q^2.$$

Is this transformation canonical? One can verify this property by the Poisson brackets, using the fact that

---

[10]L.D. Landau e E.M. Lifšits, Mechanics (Pergamon Press) Chap. 7.

$$\tan(\phi) = \frac{mq\omega}{p}. \tag{2.104}$$

Indeed,

$$\frac{\partial A}{\partial q} = m\omega q, \qquad \frac{\partial A}{\partial p} = \frac{p}{m\omega}, \qquad \frac{\partial A}{\partial p} = \frac{p}{m\omega},$$

$$\frac{\partial \phi}{\partial q} = \frac{m\omega/p}{1 + (m\omega q/p)^2}, \; \frac{\partial \phi}{\partial p} = \frac{-m\omega q/p^2}{1 + (m\omega q/p)^2}$$

and

$$\{\phi, A\} = \frac{\partial \phi}{\partial q}\frac{\partial A}{\partial p} - \frac{\partial \phi}{\partial p}\frac{\partial A}{\partial q} = 1. \tag{2.105}$$

The coordinate is ϕ. The canonical equations of motion read as $\dot{A} = 0$, $\dot{\phi} = \omega$.

2.6 Delaunay Elements

We have just seen that the oscillator Hamiltonian (2.102) can be canonically transformed to the amazingly simple form $A\omega$ where A is an action, and $\omega = \dot{\phi}$, where ϕ is an angle variable. This property can be extended to any periodic motion in 1 dimension. Indeed, the equation $H(p, q) = E$ can be solved to find the contour C in phase space defined by $p = p(q, E)$. The action variable is then

$$A = \oint_C pdq, \tag{2.106}$$

where $pdq > 0$ along C, while the contour goes from a minimum $q = qmin$ to a maximum $q = qmax$ and back. Within C, $E > H$, so

$$A = \iint dpdq\theta(E - H(p, q)). \tag{2.107}$$

Thus,

$$\frac{\partial A}{\partial E} = \iint dpdq\delta(E - H(p, q)) = \int_{qmin}^{qmax} \left\{ \left[\frac{dq}{|\frac{\partial H}{\partial p}|}\right]_{p>0} + \left[\frac{dq}{|\frac{\partial H}{\partial p}|}\right]_{p<0} \right\}. \tag{2.108}$$

The Hamilton equation says that $\frac{\partial H}{\partial p} = \dot{q}$, and from each dq, one gets two positive contributions $\frac{dq}{|\dot{q}|}$. One can write

$$\frac{dA}{dE} = \oint \frac{dq}{\dot{q}} = \oint dt = T, \qquad (2.109)$$

where T is the period. One can conclude that the frequency is

$$\nu = \frac{\partial E}{\partial A}, \qquad (2.110)$$

which is a new Hamilton equation, with $E = E(A)$ as the new Hamiltonian and A the transformed momentum. Let ϕ denote the new coordinate, canonically conjugated to A; Hamilton's equations read as:

$$\frac{\partial E}{\partial A} = \dot{\phi} = \nu \qquad (2.111)$$

$$\frac{\partial E}{\partial \phi} = 0. \qquad (2.112)$$

This technique also lends itself to the solution of periodic motions with more than one degree of freedom, and, in this case, is easy to use provided that one can separate the variables. Then, for each (p, q) pair, there is a separate contribution $A_q = \oint p \, dq$ to the action. For the Kepler problem, which is separable, the Lagrangian is

$$L = \frac{1}{2}m \left[\dot{r}^2 + r^2 \dot{\phi}^2 \right] + G \frac{Mm}{r}, \qquad (2.113)$$

and since ϕ is cyclic, $p_\phi = mr^2 \dot{\phi}$ is just the angular momentum and is a constant of the motion. Hence, $A_\phi = \oint p_\phi d\phi = 2\pi p_\phi$. The Hamiltonian is:

$$H = \frac{p_r^2}{2m} + \frac{p_\phi^2}{2mr^2} - G \frac{Mm}{r} \qquad (2.114)$$

where, for an orbit with $H = E$,

$$p_r = m\dot{r} = \pm \sqrt{2mE + 2G \frac{Mm}{r} - \frac{p_\phi^2}{r^2}}. \qquad (2.115)$$

So,

$$A_r = \oint p_r dr \qquad (2.116)$$

is the area inside the curve in the (r, p_r) plane of equation $p_r^2 = -\frac{p_\phi^2}{r^2} + 2m \left(E + \frac{GmM}{r} \right)$. Solving this simple problem, it turns out that

$$A_r = -2\pi p_\phi + \pi G M m \sqrt{\frac{2m}{-E}}. \tag{2.117}$$

Solving for E, one finds the new Hamiltonian

$$E = -\frac{2\pi^2 G^2 M^2 m^3}{(A_r + A_\phi)^2}. \tag{2.118}$$

The two frequencies $\nu_r = \frac{\partial E}{\partial A_r}$ and $\nu_\phi = \frac{\partial E}{\partial A_\phi}$ are equal, in agreement with the well-known fact that the orbits are closed. This is related to the special form of the potential, which allows for the conservation of the Runge–Lenz vector

$$\mathbf{R} = \frac{\mathbf{p} \wedge \mathbf{L}}{m} - G m M \frac{\mathbf{r}}{r} \tag{2.119}$$

which is pinned at the aphelion-perihelion direction (except for circular orbits, when $\mathbf{R}$ vanishes). In order to verify that, one must compute the time derivative of the two terms of $\mathbf{R}$ using $\frac{d}{dt}\frac{\mathbf{r}}{r} = \frac{r^2 \mathbf{v} - \mathbf{r}(\mathbf{r}.\mathbf{v})}{r^3}$, where $\mathbf{v} = \frac{d\mathbf{r}}{dt}$, and the equation of motion. The French astronomer and mathematician Charles–Eugene Delaunay (1816–1872) demonstrated the usefulness of this method in the study of planetary motion. In addition, it is always possible to solve *integrable systems*. By the term integrable one refers to systems having N degrees of freedom and N constants of the motion I_k such that all the Poisson brackets vanish, i.e. $\{I_j, I_k\} = 0$. One would say that the conserved quantities are in involution.

2.7 Noether Theorem

We have noted that if $\frac{\partial L}{\partial q} = 0$, then the coordinate q is *cyclic,* and the Lagrange Equation says that the corresponding momentum (which is called the conjugate momentum) is conserved. This results admits an important generalization about system that have continuous symmetries. This means that there is a family of invertible transformations, such as translations or rotations, that leave the equations of motion unchanged. It is always possible to define a product of transformations as the result of applying the first after the second, and technically the family of symmetry transformation is always a Group, but we shall not use any Group property below. For notational convenience, I shall discuss this for one coordinate q, since the extension to several ones is trivial. Consider the symmetry transformation that changes the virtual paths as follows:

$$T_\alpha : q(t) \to Q[q(t), \alpha];$$

for $\alpha \to 0$ to the varied path $Q[q(t), \alpha]$ is close to $q(t)$. In the action integral,

$$L(q, \dot{q}, t) \to L(Q, \dot{Q}, t) = L(q, \dot{q}, t) + \alpha \left(\frac{\partial L}{\partial q} \frac{\partial Q}{\partial \alpha} + \frac{\partial L}{\partial \dot{q}} \frac{\partial \dot{Q}}{\partial \alpha} \right).$$

Since $\dot{Q} = \frac{\partial Q}{\partial q} \dot{q}$, we must insert

$$\frac{\partial \dot{Q}}{\partial \alpha} = \frac{\partial^2 Q}{\partial q \partial \alpha} \dot{q},$$

and since we are varying a physical path, we can replace $\frac{\partial L}{\partial q}$ by $\frac{d}{dt} \frac{\partial L}{\partial \dot{q}}$.
To sum up, an infinitesimal transformation changes L by

$$\Delta L = \alpha \left[\left(\frac{d}{dt} \frac{\partial L}{\partial \dot{q}} \right) \frac{\partial Q}{\partial \alpha} + \frac{\partial L}{\partial \dot{q}} \frac{d}{dt} \frac{\partial Q}{\partial \alpha} \right] = \alpha \frac{d}{dt} \left(\frac{\partial Q}{\partial \alpha} \frac{\partial L}{\partial \dot{q}} \right). \qquad (2.120)$$

So, the Lagrangian changes by a total derivative and the equations of motion are not modified by a generic infinitesimal transformation; however, the path is transformed by T_α, and so the action along the path $q(t)$ changes by

$$\Delta S = \alpha \int_{t_1}^{t_2} dt \frac{d}{dt} \left(\frac{\partial L}{\partial \dot{q}} \frac{\partial Q}{\partial \alpha} \right) = \Lambda(t_2) - \Lambda(t_1), \qquad (2.121)$$

where

$$\Lambda(t) = p(t) \frac{\partial Q(t)}{\partial \alpha} \qquad (2.122)$$

and $p = \frac{\partial L}{\partial \dot{q}}$ is the momentum conjugated with $q = Q(\alpha = 0)$. Since T_α is a symmetry, the transformed path must have identical S as the original path, therefore $\Delta S = 0$ and we may conclude that the $\Lambda(t)$ must be conserved. Any continuous symmetry leads to a conservation law. This reasoning extends to systems with n degrees of freedom, and thus

$$\Lambda(t) = \sum_n p_n(t) \frac{\partial Q_n(t)}{\partial \alpha}. \qquad (2.123)$$

It is straightforward to show that a translational symmetry leads to the conservation of angular momentum. For the motion in a central field, the transformation is a rotation $\phi \to \phi + \alpha, r \to r$ and the rotational symmetry leads to the conservation of angular momentum. The most important application is in field theory, in which the theorem gives the conserved currents.

2.8 Chaos

The great french mathematician Pierre-Simon Laplace was certainly inspired by Classical Mechanics in his influential concept of determinism. He wrote:

We may regard the present state of the universe as the effect of its past and the cause of its future.

Later thinkers have contrasted the use of probability in Quantum Mechanics with the determinism of Classical Mechanics. However, the presence of chaos and uncertainty in the classical events of real life is familiar to everybody as a consequence of the impossibility of exactly knowing the initial conditions. Any dynamical system with n degrees of freedom is described by a set of Hamilton first-order differential equations:

$$\frac{dx_i}{dt} = \phi_i(x_1, x_2, \cdots, x_n). \tag{2.124}$$

The equations of motion for slightly different initial conditions must result from $x_i \rightarrow x_i + \Delta x_i$. Using a first-order expansion of the Hamilton equations, we obtain

$$\frac{d\Delta x_i}{dt} = \sum_{j=1}^{n} \frac{\partial \phi_i}{\partial x_j} \Delta x_j. \tag{2.125}$$

Now we wish to know the evolution in time of the deviations Δx_i, and the result depends on the problem. For an integrable system that can be described in terms of independent constants of the motion, all the Δx_i must remain bound. More generally, one might assume that ϕ_i is an analytic function of its arguments and one can approximate $\frac{\partial \phi_i}{\partial x_j} = M_{ij}$ by a constant matrix, at least for small Δx_i. This leads to

$$\Delta x_i = \sum_{j} (e^{Mt})_{ij} \Delta x_j(0). \tag{2.126}$$

Now, if M has eigenvalues L_i, then e^{Mt} has eigenvalues $e^{L_i t}$; the real parts of L_i are called Lyapunov exponents and lead to an exponential (initial) growth of the deviations from the unperturbed problem. This is the butterfly effect, which makes it impossible to make long-term weather forecasts. By contrast, since in Quantum Theory there are no trajectories, there is no quantum chaos, but there are studies concerning the quantum behavior of classically chaotic systems.

Chapter 3
Dirac's Delta

The Dirac delta is the prototype distribution and
is an essential tool of Theoretical Physics.

3.1 Definition of the δ

Let us start with the Heaviside[1] θ discontinuous function, also known as the step function, defined by

$$\theta(x) = \begin{cases} 1 & \text{se } x > 0, \\ \frac{1}{2} & \text{if } x = 0, \\ 0 & \text{se } x < 0, \end{cases}$$

With it we can define a rectangular-shaped peak function, of width 2α,

$$\delta_\alpha(x) = \frac{\theta(\alpha^2 - x^2)}{2\alpha}, \tag{3.1}$$

such that

$$\int_{-\infty}^{\infty} \delta_\alpha(x)dx = 1. \tag{3.2}$$

Now pick an analytic[2] function ϕ and the integral $\int_{-\infty}^{\infty} \delta_\alpha(x)\phi(x)dx$ which gives an average of f for $x \in (-\alpha, \alpha)$ and if α is so small that $\phi(x)$ hardly varies in the interval le result is close to $\phi(0)$. In other words,

---

[1] Oliver Heaviside (1850–1925) was probably the first to use the δ before Dirac, and the work of George Green also implies the concept. Often the names are not historically fair.

[2] A function f is analytic in 0 if there is an interval including 0 in which the Taylor series converges to f.

© Springer International Publishing AG, part of Springer Nature 2018
M. Cini, *Elements of Classical and Quantum Physics*,
UNITEXT for Physics, https://doi.org/10.1007/978-3-319-71330-4_3

$$\lim_{\alpha \to 0} \int_{-\infty}^{\infty} \delta_\alpha(x)\phi(x)dx = \phi(0).$$

In this way we have justified the exchange of limit and integration

$$\int_{-\infty}^{\infty} \delta(x)\phi(x)dx = \phi(0), \tag{3.3}$$

where

$$\delta(x) \overset{?}{=} \lim_{\alpha \to 0} \delta_\alpha(x). \tag{3.4}$$

As a rule, this exchange of limits is earnestly forbidden; $\lim_{\xi \to 0} \lim_{\lambda \to 0} \frac{\lambda}{\lambda^2+\xi^2} = 0$
while $\lim_{\lambda \to 0} \lim_{\xi \to 0} \frac{\lambda}{\lambda^2+\xi^2}$ blows up. The sign of this fault is that Dirac's delta $\delta(x)$,
a very special function, which is zero everywhere except that it is ∞ in 0. Actually,
is too singular to meet the minimum of assumptions that are needed to define and
handle functions in Analysis. For instance, we should always be able to calculate
the square of a function, but $\delta(x)^2$ is meaningless. However, we can define $\delta(x)$ as a
distribution.[3] This means that this object exists only to be handled with special rules
(that we are going to see) and used under integral and this is the only way it can give
numbers at the end of the calculation. Despite such limitations, distributions are so
useful in Physics that we cannot simply do without them.

There is another way to define Dirac's δ[4] that is,

$$\delta(x) \equiv \frac{d}{dx}\theta(x), \quad \forall x, \tag{3.5}$$

Again, this does not exist when we work with ordinary functions. However, consider a function $\phi(x)$ that we assumed to be good ($\phi(x) \in C^\infty$) and the ordinary
integral

$$\int_{-\infty}^{\infty} dx\,\phi(x)\theta(x_0 - x) = \int_{-\infty}^{x_0} dx\,\phi(x).$$

Differentiating with respect to the upper limit,

$$\frac{d}{dx_0} \int_{-\infty}^{\infty} dx\,\phi(x)\theta(x_0 - x) = \phi(x_0).$$

---

[3]This procedure recalls the introduction of the negative numbers by the Chinese and the imaginary
numbers by Gerolamo Cardano and Rafael Bombelli in the Italian Renaissance. It is sometimes
necessary to invent new objects that break the old rules, and doing that properly can be very
rewarding. Thus, in order to be able to differentiate under the integral sign, one introduces the
distributions.

[4]P.A.M. Dirac (Bristol 1902 - Tallahassee, Florida, USA 1984), English physicist with a Swiss
father, wrote the relativistic equation for the electron. Dirac gave many fundamental contributions
to Physics, always with a taste for mathematical beauty. Among other achievements, he formulated
the theory of the magnetic monopole, predicted the existence of antimatter, proposed the bra-ket
notation, invented the technique of Second Quantization. He received the Nobel in 1933.

If we take the liberty to exchange differentiation and integration, we discover that the exchange is correct provided that

$$\int_{-\infty}^{\infty} dx \phi(x) \delta(x - x_0) \equiv \phi(x_0). \tag{3.6}$$

This is an alternative way to arrive at (3.3), starting from (3.5), and implies that

$$\int_{-\infty}^{x} du \delta(u - x_0) = \theta(x - x_0).$$

Evidently, $\delta(x - x_0) = 0$ for $x \neq x_0$, while x_0 *blows up*. Note that $x\delta(x) = 0$. We start to see that it is nice to find a proper way to exchange limits, and this is the reason for introducing the generalized functions, or distributions; the δ is just the prototype. We can think of the δ as the *limit* of sequences of ordinary functions δ_α for $\alpha \to 0$. The meaning is:

$$\lim_{\alpha \to 0} \int_{-\infty}^{\infty} dx \phi(x) \delta_\alpha(x - x_0) = \phi(x_0).$$

There are many simple functions, with the property (3.2) that are used as alternative δ_α. One is:

$$\frac{\sin^2\left(\frac{x}{\alpha}\right)}{\pi\left(\frac{x^2}{\alpha}\right).} \tag{3.7}$$

Here is a list of other $\delta_\alpha(x)$ that are often used:

$$\frac{1}{\pi}\frac{\alpha}{x^2 + \alpha^2}, \quad \frac{-1}{\pi}\text{Im}\frac{1}{x + i\alpha}, \quad \frac{1}{\alpha\sqrt{\pi}}\exp\left[-\frac{x^2}{\alpha^2}\right], \quad \frac{\sin\left(\frac{x}{\alpha}\right)}{\pi x}.$$

From the last example, one can derive an important integral representation of δ. Since

$$\int_{-\frac{1}{\alpha}}^{\frac{1}{\alpha}} e^{iqx} dq = 2\frac{\sin\left(\frac{x}{\alpha}\right)}{x} = 2\pi\delta_\alpha(x),$$

it follows that

$$\delta(x) = \frac{1}{2\pi}\int_{-\infty}^{\infty} e^{iqx} dq. \tag{3.8}$$

This implies that, given an arbitrary function $f = f(x)$ and its Fourier component

$$F(k) = \int_{-\infty}^{\infty} \frac{dx}{2\pi} e^{ikx} f(x),$$

one gets a one-shot proof of Fourier's theorem

$$\int_{-\infty}^{\infty} dk F(k) e^{-ikx} = \int_{-\infty}^{\infty} dk e^{-ikx} \int_{-\infty}^{\infty} \frac{dy}{2\pi} e^{iky} f(y)$$

$$= \int_{-\infty}^{\infty} \frac{dy}{2\pi} f(y) 2\pi \delta(x - y) = f(x).$$

The δ is the Fourier transform of $f(x) = 1$. Since $\delta(x)$ vanishes everywhere except at $x = 0$, it is evident that for each interval (a, b),

$$\int_a^b dx \delta(x - x_0) = \begin{cases} 1, & a < x_0 < b, \\ 0, & x_0 \notin (a, b). \end{cases}$$

Since $\delta(x)$ is real, the representation (3.8) implies that

$$\delta(-x) = \delta(x)^* = \delta(x);$$

so, δ is even, and for real a,

$$\delta(ax) = \delta(|a|x);$$

hence, $\int_{-\infty}^{\infty} dx \phi(x) \delta(ax) = \frac{1}{|a|} \int_{-\infty}^{\infty} d(|a|x) \phi(x) \delta(|a|x) = \frac{\phi(0)}{|a|}$, and so,

$$\delta(ax) = \frac{\delta(x)}{|a|}. \tag{3.9}$$

Now let $g(x)$ be a good function with an isolated zero in $x = x_0$; we may write in an interval including x_0 $g(x) \approx \frac{dg}{dx}\big|_{x_0} (x - x_0)$. Assuming that $g'(x_0) \neq 0$,

$$\delta(g(x)) = \delta\left[\frac{dg}{dx}\bigg|_{x_0} (x - x_0) \right] = \frac{\delta(x - x_0)}{\left| \frac{dg}{dx} \right|}. \tag{3.10}$$

If $g(x)$ has a countable set of zeros at $x = x_\alpha$, $\alpha = 1, 2, \cdots$, we may use the fact that around each zero, $g(x) \approx \frac{dg}{dx}\big|_{x_\alpha} (x - x_\alpha)$. Then,

$$\delta(g(x)) = \frac{\sum_\alpha \delta(x - x_\alpha)}{\left| \frac{dg}{dx} \right|}. \tag{3.11}$$

Some relations that are familiar among ordinary functions must be generalized to the space of distributions. Consider the equation

$$x f(x) = 1. \tag{3.12}$$

If f is a distribution, since $x\delta(x) = 0$, the general solution is

$$f(x) = P\frac{1}{x} + C\delta(x), \quad C = \text{constant}, \tag{3.13}$$

where P selects the *principal part* of the integral, defined by

$$P \int_{-\infty}^{\infty} \frac{1}{x}\phi(x)dx \equiv \lim_{\epsilon \to 0} \left(\int_{-\infty}^{-\epsilon} + \int_{\epsilon}^{\infty} \right) \frac{1}{x}\phi(x)dx. \tag{3.14}$$

P allows us to tackle some kinds of divergence in 0.

We also need the derivatives of the δ. To find out how

$$\delta'(x) = -\frac{d}{dx}\delta(x) \tag{3.15}$$

works, we integrate by parts

$$\int_{-\infty}^{\infty} dx\delta'(x)\phi(x) = -\int_{-\infty}^{\infty} dx\delta(x)\phi'(x) = -\phi'(0). \tag{3.16}$$

This can be iterated to give the action of the nth derivative,

$$\int_{-\infty}^{\infty} dx\delta^{(n)}(x)\phi(x) = (-1)^n \left. \phi^{(n)} \right|_{x=0}. \tag{3.17}$$

Moreover, the definition of δ has a natural extension to 3d space with

$$\int_{\Omega} d^3 r \delta^{(3)}(\vec{r} - \vec{r}_0) = \begin{cases} 1, \vec{r}_0 \in \Omega, \\ 0, \vec{r}_0 \notin \Omega. \end{cases} \tag{3.18}$$

In Cartesian coordinates,

$$\delta^{(3)}(\vec{r} - \vec{r}_0) \equiv \delta(\vec{r} - \vec{r}_0) = \delta(x - x_0)\delta(y - y_0)\delta(x - z_0). \tag{3.19}$$

We can go to curvilinear coordinates by introducing the Jacobian determinant, therefore, while $d^3x \to \left| \frac{\partial(x,y,z)}{\partial(\xi,\eta,\zeta)} \right| d\xi d\eta d\zeta$, the δ must be divided by the same Jacobian:

$$\phi(\vec{r}_0) = \phi(x_0, y_0, z_0) = \int \phi(\vec{r})\delta(\vec{r} - \vec{r}_0)dxdydz$$

becomes

$$\phi(\vec{r}_0) = \phi(\xi_0, \eta_0, \zeta_0)$$

$$= \int \phi(\vec{r}) \left\{ \frac{\delta(\xi - \xi_0)\delta(\eta - \eta_0)\delta(\zeta - \zeta_0)}{\left| \frac{\partial(x,y,z)}{\partial(\xi,\eta,\zeta)} \right|} \right\} \left| \frac{\partial(x,y,z)}{\partial(\xi,\eta,\zeta)} \right| d\xi d\eta d\zeta. \quad (3.20)$$

For example, in spherical coordinates,

$$\delta(\vec{r} - \vec{r}_0) = \frac{\delta(r - r_0)\delta\left[\cos(\theta) - \cos(\theta_0)\right]\delta(\phi - \phi_0)}{r^2}. \quad (3.21)$$

Using the δ, one can define a measure for a *hypersurface* in an N dimensional space. Take for example for $N = 2$, (xy plane) and consider the family of closed curves defined by

$$f(x, y) = C.$$

The *invariant measure* $w(C)$ of a member of the family can be defined by

$$w(C) = \int \int dx dy \delta(C - f(x, y)) = \frac{d}{dC} \Omega(C), \quad (3.22)$$

where

$$\Omega(C) = \int \int dx dy \theta(C - f(x, y)) \quad (3.23)$$

is the area inside $f = C$; in general, $w(C)$ does not coincide with the length, but is still independent of the system of coordinates.

3.1.1   Volume of the Hypersphere in N Dimensions

The equation of the Hypersphere of radius R in the N-dimensional space $\mathbb{R}^N$ is

$$\sqrt{\sum_i^N x_i^2} = R.$$

Its volume

$$\Omega_N(R) = \int_{\sum_i^N x_i^2 < R^2} dx_1 dx_2 \ldots dx_N = \int_{\mathbb{R}^N} d^N x \theta(R - r), \quad (3.24)$$

for dimensional reasons, is $\Omega_N(R) = c_N R^N$, where c_N is the volume of the Hypersphere with $R = 1$. A neat trick allows us to calculate c_N. Consider the following Gaussian integral:

$$I = \int_{-\infty}^{\infty} dx_1 \int_{-\infty}^{\infty} dx_2 \ldots \int_{-\infty}^{\infty} dx_N e^{-\sum_i^N x_i^2} = \left(\int_{-\infty}^{\infty} dx e^{-x^2}\right)^N = \pi^{\frac{N}{2}}.$$

Putting $\sum_i^N x_i^2 = r^2$, we may write

$$I = \int d^N x e^{-r^2}. \tag{3.25}$$

But $e^{-r^2} = \int_0^{\infty} dR e^{-R^2} \delta(R - r)$, therefore

$$I = \int d^N x \int_0^{\infty} dR e^{-R^2} \delta(R - r) = \int_0^{\infty} dR e^{-R^2} \frac{d}{dR} \int d^N x \theta(R - r).$$

So

$$I = \int_0^{\infty} dR e^{-R^2} \frac{d}{dR} \Omega_N(R). \tag{3.26}$$

Now the calculation is immediate. Inserting $\Omega_N(R) = c_N R^N$, one obtains:

$$\begin{aligned} I &= N c_N \int_0^{\infty} dR R^{N-1} e^{-R^2} = N c_N \int_0^{\infty} R dR R^{N-2} e^{-R^2} \\ &= \frac{N c_N}{2} \int_0^{\infty} dt t^{\frac{N-2}{2}} e^{-t} = \frac{N c_N}{2} \left(\frac{N}{2} - 1\right)! \end{aligned} \tag{3.27}$$

Here, we met the factorial function

$$x! = \int_0^{\infty} t^x e^{-t} dt, \tag{3.28}$$

which is $n! = \prod_k^n k$ for integer n, but $\left(\frac{1}{2}\right)! = \frac{\sqrt{\pi}}{2}$, $\left(\frac{3}{2}\right)! = \frac{3\sqrt{\pi}}{4}$, and so on. Equating this to $\pi^{N/2}$, we find

$$\Omega_N(R) = \frac{\pi^{N/2}}{\left(\frac{N}{2}\right)!} R^N. \tag{3.29}$$

Therefore, $\Omega_1 = 2R$, $\Omega_2 = \pi R^2$, $\Omega_3 = \frac{4}{3}\pi R^3$, $\Omega_4 = \frac{1}{2}\pi^2 R^4$, and so on. The measure of the surface of the Hypersphere as defined above is $\omega_N(R) = \frac{d}{dR}\Omega_N(R)$.

3.1.2 Plancherel Theorem

Let $\alpha(t)$, $\beta(t)$ be functions belonging to L^2 (that is, such that $\int_{-\infty}^{\infty} |\alpha(t)|^2 dt$ and $\int_{-\infty}^{\infty} |\beta(t)|^2 dt$ exist). Let us denote their Fourier transforms by $\alpha(\omega)$, $\beta(\omega)$. Then,

$$\int_{-\infty}^{\infty} \alpha(t)\beta^*(t)dt = \int_{-\infty}^{\infty} \alpha(\omega)\beta^*(\omega)d\left(\frac{\omega}{2\pi}\right).$$ (3.30)

By inserting the definition of the Fourier transforms on the r.h.s, exchanging the order of integration and using the δ functions, one can readily check this important result. One can note that the l.h.s. resembles a scalar product $\sum_i \alpha_i \beta_i^*$, where the discrete index i is replaced by a continuous one, and a similar interpretation is possible for the r.h.s. So, in a sense, this theorem is related to the invariance of scalar products under unitary transformations.

Chapter 4
Some Consequences of Maxwell's Equations

Maxwell published his Treatise in 1873 and his equations have not been revised, continuing to be unharmed through the revolutionary changes produced by Relativity and Quantum Mechanics. Nevertheless, the quantum theory of light and the use of lasers have revealed many new phenomena. This might appear to be a contradiction, but it is the plain truth. I prepare the reader to appreciate how this came about by discussing some important facts about classical electromagnetism in this chapter.

4.1 Fields and Potentials

The (classical) electromagnetic fields in vacuo that satisfy given boundary conditions can be calculated through Maxwell's equations. In the Gauss system they read as:

$$
\begin{aligned}
\vec{\nabla} \cdot \vec{E} &= 4\pi\rho, \\
\vec{\nabla} \cdot \vec{B} &= 0, \\
\vec{\nabla} \wedge \vec{E} &= -\frac{1}{c}\frac{\partial \vec{B}}{\partial t}, \\
\vec{\nabla} \wedge \vec{B} &= \frac{1}{c}\frac{\partial \vec{E}}{\partial t} + \frac{4\pi}{c}\vec{j},
\end{aligned}
\tag{4.1}
$$

where $\vec{j}$ and ρ are current density and charge density. In all, they are 4 functions of space and time. The fields can be computed and measured, however, it amazing that the Maxwell equations succeed in giving us 6 measurable quantities (3 components of $\vec{E}$ and 3 of $\vec{B}$) having only 4 quantities in input. This is a most remarkable property of the electromagnetic field. Moreover, we can obtain the same field more easily by working out 4 quantities, namely the scalar potential ϕ and the vector potential $\vec{A}$, such that

© Springer International Publishing AG, part of Springer Nature 2018
M. Cini, *Elements of Classical and Quantum Physics*,
UNITEXT for Physics, https://doi.org/10.1007/978-3-319-71330-4_4

$$B = \vec{\nabla} \wedge \vec{A},$$

$$\vec{E} = -\vec{\nabla}\phi - \frac{1}{c}\frac{\partial \vec{A}}{\partial t}.$$

From the potentials, one readily obtains the fields, but the potentials are unobservable and largely arbitrary. So, the 4 potentials not only contain all the information about the 6 field components, but also a disposable amount of irrelevant information. This property, called gauge invariance, is a key property of electromagnetism. The choice of the potentials is called *gauge*. In the Lorentz gauge $div\,\vec{A} + \frac{1}{c}\frac{\partial \phi}{\partial t} = 0$, the Maxwell equations give us:

$$\left[\nabla^2 - \frac{1}{c^2}\frac{\partial^2}{\partial t^2}\right]\vec{A} = -\frac{4\pi}{c}\vec{J},$$

$$\left[\nabla^2 - \frac{1}{c^2}\frac{\partial^2}{\partial t^2}\right]\phi = -4\pi\rho.$$

We shall obtain the potentials by solving these equations according to the Green's function method. For all $r \neq 0$, $\nabla^2(\frac{1}{r}) = 0$; indeed, for sphericosymmetric functions $\phi(r)$, one can write[1] $\nabla^2\phi(r) = \frac{1}{r}\frac{\partial^2}{\partial r^2}(r\phi(r))$, and since $\phi = \frac{1}{r}$, the result is 0. But this is not the end of the story. Considering a sphere around the origin,

$$\int d^3r\nabla^2\left(\frac{1}{r}\right) = \int d^3r\,div\,\,grad\frac{1}{r} = \int_S grad\frac{1}{r}\cdot\vec{n}\,dS$$

and since $grad(\frac{1}{r}) = -\frac{\vec{r}}{r^3}$, $\vec{n} = \frac{\vec{r}}{r}$, $dS = r^2d\Omega$, one is left with

$$\int d^3r\nabla^2\left(\frac{1}{r}\right) = -4\pi.$$

In conclusion, $\nabla^2(\frac{1}{r}) = -4\pi\delta(\vec{r})$. The function $g(r) = \frac{1}{r}$ la is called Green's function[2] of the Poisson equation

$$\nabla^2 V(\vec{x}) = -4\pi\rho(\vec{x}).$$

The solution is the sum of all the potentials of all the point charges that make up $\rho(\vec{x})$, namely,

$$V(\vec{x}) = \int d^3x'g(\vec{x} - \vec{x}')\rho(\vec{x}').$$

---

[1] An equivalent alternative which is in use is: $\nabla^2\phi(r) = \frac{1}{r^2}\frac{\partial}{\partial r}(r^2\frac{\partial}{\partial r}\phi(r))$.

[2] The Green's function are named after the Englishman George Green, a solitary amateur genius who working in his mill in the Midlands invented mathematical methods essential for the theory of electromagnetism and all of the modern Theoretical Physics, although at his time the Dirac's delta was not yet known.

4.2 Green's Function of the Wave Equation and Retarded Potentials

As a prototype calculation illustrating the main steps necessary to compute G for many systems of interest, I show in detail how the same reasoning leads to the Green's function of the wave equation; by definition,

$$\left[\nabla^2 - \frac{1}{c^2} \frac{\partial^2}{\partial t^2} \right] G(\overrightarrow{r}, t) = -4\pi \delta(\overrightarrow{r}) \delta(t).$$

Clearly, an instantaneous localized disturbance in ($\overrightarrow{r} = 0, t = 0$) stands as a source of a potential G that later arrives in ($\overrightarrow{r}, t$). It turns out that $G(\overrightarrow{r}, t) = \frac{\delta(t - \frac{r}{c})}{r}$. To see that, outside the origin, $G(\overrightarrow{r}, t)$ solves the wave equation, we use $\nabla^2 \phi(r) = \frac{1}{r} \frac{\partial^2}{\partial r^2}(r\phi(r))$. For $r \neq 0$, $\nabla^2 \left(\frac{\delta(t - \frac{r}{c})}{r} \right) = \frac{1}{r} \frac{\partial^2}{\partial r^2} \delta(t - \frac{r}{c})$; now, for each $f(t - \frac{r}{c})$, it holds that

$$\frac{\partial^2}{\partial r^2} f\left(t - \frac{r}{c}\right) = \frac{1}{c^2} \frac{\partial^2}{\partial t^2} f\left(t - \frac{r}{c}\right);$$

in particular, $\nabla^2 \left(\frac{\delta(t - \frac{r}{c})}{r} \right) = \frac{1}{c^2} \frac{\partial^2}{\partial t^2} \left(\frac{\delta(t - \frac{r}{c})}{r} \right)$, and

$$\left[\nabla^2 - \frac{1}{c^2} \frac{\partial^2}{\partial t^2} \right] G(\overrightarrow{r}, t) = 0, \quad r \neq 0.$$

To include the effect of the δ, integrate $[\nabla^2 - \frac{1}{c^2} \frac{\partial^2}{\partial t^2}] G(\overrightarrow{r}, t)$ over a sphere S with any radius R_0 centered on the origin. We denote the surface of the sphere by δS.

Contribution of the term in ∇^2

By the Gauss theorem:

$$\int_S \nabla^2 \left(\frac{\delta(t - \frac{r}{c})}{r} \right) d^3 r = \int_{\delta S} \text{grad} \left(\frac{\delta(t - \frac{r}{c})}{r} \right) \cdot \frac{\overrightarrow{r}}{r} dS.$$

In spherical coordinates,

$$\int_S \nabla^2 \left(\frac{\delta\left(t - \frac{r}{c}\right)}{r} \right) d^3 r = \int \left[\text{grad} \left(\frac{\delta\left(t - \frac{r}{c}\right)}{r} \right) \cdot \frac{\overrightarrow{r}}{r} \right]_{r=R_0} R_0^2 d\Omega. \quad (4.2)$$

Now, $\frac{\partial}{\partial x} r = \frac{x}{r}$, and so $\text{grad } r = \frac{\overrightarrow{r}}{r}$. Hence,

$$\text{grad } \delta\left(t - \frac{r}{c}\right) = -\frac{1}{c} \text{grad } r \, \delta'\left(t - \frac{r}{c}\right) = -\frac{\overrightarrow{r}}{rc} \delta'\left(t - \frac{r}{c}\right),$$

and

$$\operatorname{grad} \frac{\delta\left(t-\frac{r}{c}\right)}{r} = -\frac{\vec{r}}{r^2 c}\delta'\left(t-\frac{r}{c}\right) = -\frac{\vec{r}}{r^3}\delta\left(t-\frac{r}{c}\right).$$

Put into (4.2); the angular integral is 4π and

$$\int_S \nabla^2\left(\frac{\delta\left(t-\frac{r}{c}\right)}{r}\right)d^3r = -4\pi\left\{\delta\left(t-\frac{R_0}{c}\right) + \frac{R_0}{c}\delta'\left(t-\frac{R_0}{c}\right)\right\}. \qquad (4.3)$$

Term in $\dfrac{\partial^2}{\partial t^2}$

Again using $\frac{\partial^2}{\partial r^2}\delta(t-\frac{r}{c}) = \frac{1}{c^2}\frac{\partial^2}{\partial t^2}\delta(t-\frac{r}{c})$,

$$\int_S d^3r \frac{1}{c^2}\frac{\partial^2}{\partial t^2}\left(\frac{\delta(t-\frac{r}{c})}{r}\right) = \int_S d^3r \frac{1}{r}\frac{\partial^2}{\partial r^2}\delta\left(t-\frac{r}{c}\right)$$

$$= 4\pi\int_0^{R_0} dr\, r^2 \frac{1}{r}\frac{\partial^2}{\partial r^2}\delta\left(t-\frac{r}{c}\right).$$

Integrating by parts, we obtain

$$4\pi\int_0^{R_0} dr\left\{\frac{\partial}{\partial r}\left[r\frac{\partial}{\partial r}\delta\left(t-\frac{r}{c}\right)\right] - \frac{\partial}{\partial r}\delta\left(t-\frac{r}{c}\right)\right\}.$$

The integrand is a derivative, and one gets

$$= 4\pi\left[r\frac{\partial}{\partial r}\delta\left(t-\frac{r}{c}\right) - \delta\left(t-\frac{r}{c}\right)\right]_{r=0}^{r=R_0}.$$

So,

$$\int_S d^3r \frac{1}{c^2}\frac{\partial^2}{\partial t^2}\left(\frac{\delta\left(t-\frac{r}{c}\right)}{r}\right) 4\pi\left\{-\delta\left(t-\frac{R_0}{c}\right) + \frac{R_0}{c}\delta'\left(t-\frac{R_0}{c}\right) + \delta(t)\right\}. \qquad (4.4)$$

End result

In conclusion, putting together (4.3) and (4.4),

$$\int_S\left[\nabla^2 - \frac{1}{c^2}\frac{\partial^2}{\partial t^2}\right]G(\vec{r},t)d^3r = -4\pi\delta(t),$$

and so

$$\left[\nabla^2 - \frac{1}{c^2}\frac{\partial^2}{\partial t^2}\right]G(\vec{r},t) = -4\pi\delta(\vec{r})\delta(t).$$

Fig. 4.1 Potentials of a
moving point charge, which
moves along a trajectory
$\boldsymbol{R}(\tau)$ referring to the origin
O. The observation point is
$\boldsymbol{x}$, while $\boldsymbol{r}$ indicates the path
of the radiation emitted at
time τ towards the observer.
That radiation will be
detected in $\boldsymbol{x}$ at time
$t = \tau + \frac{r}{c}$

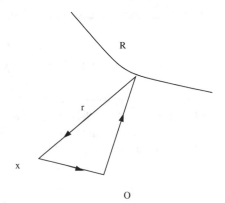

In summary, $G(\overrightarrow{r}, t) = \frac{\delta(t - \frac{r}{c})}{r}$ is the Green's function of the wave equation. Hence, from Fig. 4.1 we obtain the potential at $\overrightarrow{x}$ due to a distribution of charges, namely,

$$\phi(\overrightarrow{x}, t) = \int d^3x' \int dt' G(\overrightarrow{x} - \overrightarrow{x}', t - t') \rho(\overrightarrow{x}', t')$$

$$= \int d^3x' \int dt' \frac{\delta(t' - t + \frac{|\overrightarrow{x} - \overrightarrow{x}'|}{c})}{|\overrightarrow{x} - \overrightarrow{x}'|} \rho(\overrightarrow{x}', t'). \tag{4.5}$$

Integrating over source times t', one obtains the retarded potential

$$\phi(\overrightarrow{x}, t) = \int d^3x' \frac{\rho\left(\overrightarrow{x}', t - \frac{|\overrightarrow{x} - \overrightarrow{x}'|}{c}\right)}{|\overrightarrow{x} - \overrightarrow{x}'|}; \tag{4.6}$$

similarly,

$$\overrightarrow{A}(\overrightarrow{x}, t) = \int d^3x' \frac{\overrightarrow{j}\left(\overrightarrow{x}', t - \frac{|\overrightarrow{x} - \overrightarrow{x}'|}{c}\right)}{|\overrightarrow{x} - \overrightarrow{x}'|}. \tag{4.7}$$

4.3 Lienard–Wiechert Potentials

The use of the Dirac δ helps considerably in finding the potentials at the observation point $\overrightarrow{x}$ due to a point charge performing an arbitrarily assigned trajectory $\overrightarrow{R}(t)$. In the case of ϕ, from (4.5) with

$$\rho(\overrightarrow{x}', t') = e\delta(\overrightarrow{x}' - \overrightarrow{R}(t')),$$

one gets, integrating on $\vec{x}'$ first and writing τ in place of t',

$$\phi(\vec{x}, t) = e \int d\tau \frac{\delta\left(\tau - t + \frac{r(\tau)}{c}\right)}{r(\tau)}, \quad \vec{r}(\tau) = \vec{x} - \vec{R}(\tau).$$

We see that the potential is still of the Coulomb form, but is retarded: the light is received at time t, which is the emission time τ plus the propagation time $\frac{r(\tau)}{c}$; the latter depends on the position of the charge when the light is emitted.

Let τ^* denote the solution to $\tau^* = t - \frac{r(\tau^*)}{c}$.

Then,

$$\delta\left(\tau - t + \frac{r(\tau)}{c}\right) = \frac{\delta(\tau - \tau^*)}{\left|\frac{d}{d\tau}\left[\tau - t + \frac{r(\tau)}{c}\right]\right|} = \frac{\delta(\tau - \tau^*)}{\left|1 + \frac{1}{c}\frac{dr}{d\tau}\right|}.$$

Here, the absolute value is unnecessary, $|\frac{dr}{d\tau}| < c$. Since the charge is at $\vec{R}$, its velocity is $\vec{v} = \frac{d\vec{R}}{d\tau}$, while the time derivative of the distance covered by the light is

$$\frac{dr(\tau)}{d\tau} = \frac{d}{d\tau}\sqrt{\vec{r}(\tau) \cdot \vec{r}(\tau)} = \frac{1}{2\sqrt{\vec{r} \cdot \vec{r}}} 2\vec{r} \cdot \frac{d\vec{r}}{d\tau};$$

since $\frac{d\vec{r}}{d\tau} = -\vec{v}$,

$$\frac{dr(\tau)}{d\tau} = -\frac{\vec{r} \cdot \vec{v}}{r}.$$

Consequently,

$$\delta\left(\tau - t + \frac{r(\tau)}{c}\right) = \frac{\delta(\tau - \tau^*)}{1 - \frac{\vec{r} \cdot \vec{v}}{rc}},$$

and the potential is

$$\phi(\vec{x}, t) = \frac{e}{[r - \frac{\vec{r} \cdot \vec{v}}{c}]_{\tau^*}}. \tag{4.8}$$

Starting from

$$\vec{j} = e\vec{v}\,\delta(\vec{x} - \vec{R}),$$

the same reasoning leads to the vector potential

$$\vec{A}(\vec{x}, t) = \frac{e}{c}\left[\frac{\vec{v}}{r - \frac{\vec{r} \cdot \vec{v}}{c}}\right]_{\tau^*}. \tag{4.9}$$

4.4   Geometrical Optics and Fermat's Principle

Let u represent any component of the potential or of the field of a monochromatic electromagnetic wave with frequency ν in some region of space. Consider the propagation in a medium in which the velocity of light $v = \frac{c}{n}$ may differ from c and depend on position due to a spatially varying refraction index n; in addition, we may have complicated boundary conditions, like, e.g., in an optical instrument. In the wave equation $(\nabla^2 - \frac{1}{v^2}\frac{\partial^2}{\partial t^2})u = 0$ we may set $u(x, y, z, t) = \exp[i\omega(\frac{S(x,y,z)}{c} - t)]$. The new unknown S is known as the Eikonal. By substitution, one readily finds that it satisfies

$$c^2 - v^2[(grad(S))^2 - ic\nabla^2 S] = 0.$$

This nonlinear partial differential equation looks harder to solve than the wave equation. However, it simplifies in the limit $S \to \infty$ (that is, large phases and short wavelengths) when it reduces to the Eikonal equation

$$(grad\, S)^2 = n^2. \tag{4.10}$$

To explore the meaning of S, we observe that in one dimension, this becomes $\frac{dS}{dx} = \pm\frac{c}{v}$; integrating between x_A and x_B, one finds $S(x_B) - S(x_A) = \pm c \int_{x_A}^{x_B} \frac{dx}{v(x)}$. Apart from a factor $\pm c$, this is just the time the light takes to go directly from x_A to x_B. Even in 3 dimensions, the Eikonal has the same meaning, with the additional specification that the path allows for the shortest possible time. Indeed, Eq. (4.10) has the same structure as the Hamilton–Jacobi equation (2.83) where S is the action, $grad(S) = \boldsymbol{p}$ is the momentum of the material point and E is the energy. For a free particle, it reads $(\frac{\partial S}{\partial x})^2 + (\frac{\partial S}{\partial y})^2 + (\frac{\partial S}{\partial z})^2 = E$. The equation describes corpuscles that go along trajectories that make the action an extremum. These may be interpreted as the *light particles*[3] whose trajectories are the light rays. The wavelengths may be taken to be short in a given problem when all other lengths are much longer. In this limit there is no diffraction and the laws of Geometrical Optics are a good approximation. In the mechanical problem, S is the action and the path taken by the particle makes S an extremum. In the optical problem, S is an Eikonal, and the optical path grants the minimum time the light can take to go from A to B. The discovery that the light rays follow the quickest path is Fermat's principle, which is the oldest variational principle in Physics.[4] A light ray from $P = (-1, y_1)$ in a medium with refraction index n_1 reaches $Q = (1, y_2)$ a solid with refraction index n_2, following Snell's law[5] $n_1 sin(\theta_1) = n_2 sin(\theta_2)$. The time the light takes to go from P to Q is given by $ct = n_1\sqrt{1 - (y_1 - y_s)^2} + n_2\sqrt{1 - (y_2 - y_s)^2}$; looking for the minimum

---

[3]The quantum mechanical photons are emitted and adsorbed like particles but travel like waves, as we shall see below. However the description in terms of point particles works fine at short wavelengths.

[4]Pierre de Fermat stated it in 1662.

[5]This law is actually credited to the Arab mathematician Ibn Sahl in a manuscript of 984.

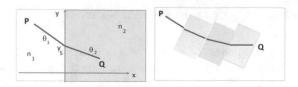

Fig. 4.2 Fermat's principle. Left panel: a light ray from a medium with refraction index n_1 to a solid with refraction index n_2. Right panel: an inhomogeneous medium behaves like a series of thin interfaces

as a function of y_S, we differentiate and find $n_1 \dfrac{y_1 - y_S}{\sqrt{1 - (y_1 - y_S)^2}} + n_2 \dfrac{y_2 - y_S}{\sqrt{1 - (y_2 - y_S)^2}} = 0$ which can be rewritten as Snell's law. A light ray in an inhomogeneous medium can be thought of as crossing a large number of interfaces following Snell's law, thereby minimizing the time it takes (Fig. 4.2).

4.5   Coherent Light

The plane wave with electric field $E(r, t) = E \exp(ikr - \omega t)$ is perfectly coherent light, meaning by that the field at one r and t allows to predict the field at all places and times.

In terms of the electric field $E(r, t)$, the correlation function

$$g^{(1)}(r_1, r_2, t) = \frac{\langle E(r_1, t) \cdot E(r_2, t + \tau) \rangle}{\langle E(r, t)^* E(r, t) \rangle} \tag{4.11}$$

measure the degree of coherence, which is low in the case of thermal light; it tends to be enhanced at small τ. For the intensity I the correlation function between two points at the same time

$$g^{(2)}(r_1, r_2) = \frac{\langle I(r_1, t) I(r_2, t) \rangle}{\langle E(r, t)^* E(r, t) \rangle} \tag{4.12}$$

also fluctuates in the case of natural light. A plane wave is perfectly coherent and the correlation functions do not fluctuate and never decay; the laser light is highly coherent.

4.5.1   The Measurement of Stellar Diameters

Disappointingly, even the nearby stars appear point-like in the most powerful telescopes. The Very Large Telescope under favourable conditions attains a resolution of

2×10^{-3} arc seconds; Sirius has a diameter 2.617×10^6 km and its distance obtained by parallax is 8.6 light years. A light year is about $9.46 \cdot 10^{12}$ km, This implies an angular diameter of about 5×10^{-6} radiants, which is below the detection limit. However, the diameter quoted above was obtained thanks to interesting interferometric techniques.

The light of a given frequency coming from a point source is a plane wave with a well-defined wave vector, so it is spatially coherent, but if the source is extended and round, the wave vectors of light rays coming from points that are one diameter apart make an angle $\Delta\alpha$; then, there is a bunch of wave vectors Δk in the light and we say that the transverse coherence is lost. In the Michelson interferometer, the monochromatized light impinging on two mirrors at points r_1 and r_2. The lines perpendicular to the mirrors make an angle ϕ. When $\phi = \Delta\alpha$ is the angle between the k vectors from opposite sides of the star, the light reflected by two mirrors and sent to a lens converges on a screen and interferes. There, the field is $E(r_1) + E(r_2) = E_{k1}(r_1) + E_{k2}(r_1) + E_{k1}(r_2) + E_2(r_2)$. The intensity is modulated by an interference term $g^{(1)}(r_1, r_2, t)$ (see (4.11)). For a monochromatic light of wave vector k, $g^{(1)}$ is modulated by a factor $\cos(\frac{kd\phi}{2})$ where $d = |r_1 - r_2|$. For larger ϕ the interference is lost. In this way, $\Delta\alpha$ can be measured.

A practical limitation of this method arises from the atmospheric turbulence that causes rapid fluctuations of the phase of the electric field. In 1956 Robert Hanbury Brown and Richard Twiss[6] demonstrated a method based on the measurement of intensity correlation at two different points. The stellar light collected at two different points. The two beams were collected by separate photomultipliers, where they produced fluctuating currents $I_1 + \Delta I_1$ and $I_2 + \Delta I_2$ (where ΔI_i denotes the fluctuations). In this way, the cross-correlation function $G = \langle \Delta I_1 \Delta I_2 \rangle$ was obtained and was converted in electronic signals which were used to measure the intensity correlation function (4.12) electronically. It is easily seen that the modulating factor $\cos(\frac{kd\phi}{2})$ again allows us to measure the stellar diameter without the disturbing influence of the amplitude fluctuations.

Recent Quantum versions of the Hanbury Brown and Twiss effect are deferred to Sect. 26.4.2.

---

[6]Correlation between photons in two coherent beams of light, R. Hanbury Brown and R. Twiss, Nature **4497**, 27 (1956).

Chapter 5
Thermal Physics

*Any macroscopic object in thermal equilibrium contains so
many particles in chaotic motion that the methods of Mechanics
are totally useless. This chapter is a simple introduction to
Thermodynamics and classical Statistical Mechanics.*

5.1 The Principles of Thermodynamics

Thermodynamics is an axiomatic part of Theoretical Physics, which is presented as
a set of phenomenological axioms or principles. Any investigation into the reasons
why the principles are true and how they are related to micro-Physics belongs to the
domain of Statistical Mechanics, while Thermodynamics gives guidelines that are
never contradicted by experiments, at least for macroscopic objects. Actually, the
principles are simple and part of the common wisdom by now, but Thermodynamics
deduces profound consequences.

Zero-th Principle

The *zeroth principle of Thermodynamics* states that two bodies can be in thermal
equilibrium and that such a state is transitive: if bodies A and B are equilibrium and
B is in equilibrium with C, then A is in equilibrium with C.

   This allows us to introduce the concept of temperature, which has a precise meaning only in equilibrium.

First Principle

Nowadays, the invoice of electricity reminds us that heat requires energy, but the
physical nature of heat has long remained unknown. It was really a scoop when Joule
in the 1840s measured the mechanical equivalent of the calorie, which is the amount
of heat required to raise the temperature of 1 g of water by 1 °C at the pressure of 1
atmosphere; the modern value is 4.1855 J. The *First Principle of Thermodynamics*
establishes the conservation of mechanical and thermal energy. If some amount δQ
of heat is supplied to any system, the effect is to increase its internal energy and/or

© Springer International Publishing AG, part of Springer Nature 2018   61
M. Cini, *Elements of Classical and Quantum Physics*,
UNITEXT for Physics, https://doi.org/10.1007/978-3-319-71330-4_5

Fig. 5.1 Carnot's cycle

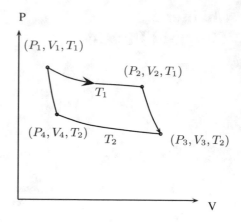

to cause some work to be performed. So,

$$\delta Q = dU + PdV, \tag{5.1}$$

where dU is the increase in internal energy, P the pressure, V the volume, and PdV is the work performed by the system. This result implies that a *perpetual motion machine of the first kind (producing work without a power supply)* cannot exist.

Second Principle

However, it is possible to make a motor, that is, a thermal machine that converts thermal energy into mechanical energy. To proceed, we need the notion of a *reversible* transformation. Such a process must take place so slowly and gradually that at all times, the system has pressure, temperature, and volume very close to the equilibrium values; therefore, it must be possible to do the same transformation in reverse, driving the system back to the same states of quasi-equilibrium.[1]

Aiming to build a motor, we need first to build a cycle, i.e., a process in which the fluid after producing some work, returns to its initial state ($\Delta U = 0$), ready to iterate the process. Consider a fluid such that the state is characterized by (P, V, T), that is, pressure, volume and temperature. Carnot's cycle is defined by the reversible transformations in Fig. 5.1, namely, two isotherms at temperatures T_1 and T_2 and two adiabatics. The machine is reversible and can work as a motor or as a refrigerator. In Fig. 5.1, we see its use as a thermal machine, which absorbs heat Q_1 at temperature T_1 and loses heat Q_2 at a lower temperature T_2. The first Principle grants that the work done in the cycle is $W = Q_1 - Q_2$; the efficiency is defined as

---

[1]A piece of diamond is out of equilibrium under normal conditions, because Graphite is the equilibrium form of Carbon; however since the kinetics of the transformation is extremely slow, one can do reversible transformations on diamond as well. Actually, the seemingly simple concepts on which Thermodynamics is based are far from obvious.

$$\eta = \frac{W}{Q_1} = 1 - \frac{Q_2}{Q_1}.$$

The possibility of trading heat for work implies that the heat content and the work done by a fluid depend on the previous history of the machine. In mathematical terms, δQ is not the differential of a function of the parameters that specify the state, since in a cycle, $\Delta Q \equiv \oint \delta Q$ might be non-vanishing, and the fluid can have undergone many cycles before returning to a given state.

The first principle would allow for $\eta = 1$ (total conversion of the heat from a source to work, equivalent to unlimited energy for free).

The situation changes radically when we introduce the Second Principle. The *Second Principle* (stated by Kelvin and Planck) says that it is impossible to build a *perpetual motion machine of the second kind*. This would consist of a device that makes mechanical work at the expense of the heat taken from one source. Clausius states the same principle in an even simpler and more striking way: heat goes spontaneously from hot to cold bodies, and work is needed to force it go from cold to hot. Actually, the two statements are equivalent. If it were possible to extract work from the cold source (in violation of the Kelvin–Planck statement), then we could use the work to heat the hot source, thus violating the Clausius statement. If it were possible to violate the Clausius statement, we could produce and increase the temperature difference between two bodies at the expense of the energy of the cold body; then we could action a Carnot cycle and obtain work, still at the expense of the cold body, violating the Kelvin–Planck statement.

The second Principle puts severe limits on what can be done by the Carnot cycle. The efficiency of a Carnot cycle depends only on the temperature of the heat sources. Indeed, if the Carnot cycles C_1 and C_2 between the same temperatures had efficiencies η_1 and η_2 with $\eta_1 > \eta_2$ we could use the first cycle to fool the second principle in the second cycle.

Problem 10 How?

Solution 10 C_1 would be used as a motor to produce the amount W_1 of work. Using work $W_2 < W_1$, one could use C_2 as a freezer, giving back to the hot source the heat taken by C_1. We should be left with an amount $W_1 - W_2$ of work, produced at the expenses of the cold source.

For the same reason, no other cycle can beat Carnot's cycle in efficiency.

Now we are in position to define the thermodynamic temperature, which is one of the basic achievements of thermal Physics. It is clear that any definition based on Mercury, ideal gases or any other (real or ideal) material can be used in a limited interval of temperatures and has no universal significance. But since η depends only on the temperatures, one can use measurements of η to define a universal thermometric scale. More precisely, we can define the absolute temperature T such that

$$\frac{T_2}{T_1} = \frac{Q_2}{Q_1}; \tag{5.2}$$

this law fixes all temperatures, including the extreme ones, in terms of a referential one, without having to rely on the properties of any substance. We need to be able to measure amounts of heat. Having done that, we have the extra bonus that in any cycle (and in any combination of cycles),

$$\frac{Q_2}{T_2} - \frac{Q_1}{T_1} = 0 = \oint \frac{\delta Q}{T}.$$

Thus, the *second principle of Thermodynamics* grants that in any reversible transformation at temperature T,

$$\frac{\delta Q}{T} = dS \tag{5.3}$$

is the exact differential of a function of temperature, pressure and any other relevant parameter, which is called Entropy. In any reversible cycle,

$$\oint dS.$$

The temperature is dimensionless, while S is energy.

Thermodynamic Potentials

Putting together the first and the second principles, we find that two equilibrium states close to each other are related by:

$$TdS = dU + PdV. \tag{5.4}$$

No doubt, the internal energy U is a function of state, with a well-defined value independent of the history; this result shows that it is precisely $U = U(S, V)$, that is, it depends on S e V. However, it is not easy to work with it, since S is extensive (i.e., proportional to the size of the system) like U. As a result, it is not so easy to measure: there is no such instrument as a *entropometer*.

From (5.4), we obtain the temperature,

$$T = \left(\frac{\partial U}{\partial S}\right)_V, \tag{5.5}$$

which can be measured easily with a thermometer. To change the independent variable, we introduce chemical potentials by means of Legendre transformations. The free energy is suitable for us;

$$F = U - TS. \tag{5.6}$$

The differential is, using (5.4),

$$dF = dU - TdS - SdT = -PdV - SdT. \tag{5.7}$$

Hence,

$$\left(\frac{\partial F}{\partial V}\right)_T = -P, \quad \left(\frac{\partial F}{\partial T}\right)_V = -S. \tag{5.8}$$

$F = F(V, T)$ is particularly well-suited for processes at constant volume, because only the T dependence is left. However, keeping V constant during a transformation may be difficult. Usually, one prefers to let V vary while keeping the pressure P constant. A possible choice is the enthalpy $H(S, P)$ defined by

$$H \equiv U + PV. \tag{5.9}$$

One finds that

$$dH = TdS + V dP.$$

This contains the differential of entropy, which has the drawbacks we have just discussed. The natural choice for most experimental conditions is the Gibbs free energy

$$G = U + PV - TS. \tag{5.10}$$

Its differential reads as

$$dG = -SdT + V dP.$$

Maxwell Relations

Comparing (5.4) with

$$dU = \left(\frac{\partial U}{\partial S}\right)_V dS + \left(\frac{\partial U}{\partial V}\right)_S dV.$$

one finds that

$$T = \left(\frac{\partial U}{\partial S}\right)_V, \qquad P = -\left(\frac{\partial U}{\partial V}\right)_S.$$

Since the mixed derivatives of $U(S, V)$ must agree, one obtains the Maxwell relation

$$\left(\frac{\partial T}{\partial V}\right)_S = -\left(\frac{\partial P}{\partial S}\right)_V.$$

Other Maxwell relations can be obtained from the other potentials. For example, from the differential of Enthalpy, one finds

$$T = \left(\frac{\partial H}{\partial S}\right)_P, \qquad V = -\left(\frac{\partial H}{\partial P}\right)_S.$$

Problem 11 What Maxwell relation follows?

Solution 11

$$\left(\frac{\partial T}{\partial P}\right)_S = -\left(\frac{\partial V}{\partial S}\right)_P.$$

Often, it is preferable to work with a thermodynamic system embedded in an environment (or thermal bath) with which it can exchange not only heat but particles, too. When the number of particles is variable, one introduces the chemical potential μ by extending (5.4):

$$dU = TdS - PdV + \mu dN.$$

This is familiar, since we ground our electrical apparatus to avoid shock hazards; the electric potential is the μ that we must control.

Perfect Classical Gas

For a perfect gas, the thermodynamic properties depend on the number f of degrees of freedom of the molecules; $f = 3$ for monoatomic gases to which only translation contributes, but diatomic molecules have $f = 5$ due to rotational freedom; extra degrees of freedom at relatively high temperatures (several hundreds of Kelvin degrees) result from molecular vibrations. The internal energy is:

$$U = \frac{f}{2}NK_BT. \tag{5.11}$$

The equation of state is the Clapeyron law

$$PV = NK_BT = nRT, \tag{5.12}$$

where $K_B = 1.381 \times 10^{-16}\frac{\text{erg}}{°\text{K}}$ is Boltzmann's constant, n the number of moles, and $R = N_AK_B$ is called the gas constant. For an adiabatic expansion ($\delta Q = 0$), inserting (5.11) into $dU + PdV = 0$, one finds $(1 + \frac{f}{2})PdV + \frac{f}{2}VdP = 0$. Integrating, one finds PV^γ =constant, where $\gamma = \frac{f+2}{f}$. The specific heat is by definition the thermal capacity of a mole ($N = N_A$). The specific heat at constant volume is

$$C_V = \left(\frac{\delta Q}{dT}\right)_V$$

(the notation means that the derivative is taken by an infinitesimal change of temperature keeping constant volume V). One obtains

$$C_V = T\left(\frac{\partial S}{dT}\right)_V.$$

The specific heat at constant pressure, which is generally easier to measure, is $C_P = T\left(\frac{dS}{dT}\right)_P$.

Problem 12 Calculate $\frac{C_P}{C_V}$ for a perfect gas.

Solution 12 One finds $C_V = \frac{f}{2}N_A K_B$, $C_P = (1 + \frac{f}{2})N_A K_B$. At constant volume, the term PdV is discarded, while at constant pressure, V can be eliminated using the Clapeyron equation. From this result, one finds that $\frac{C_P}{C_V} = \gamma$.

The Entropy of the Universe is Growing

Entropy is maximum in equilibrium. Take a container A having volume V_A that is part of a larger container B having volume V_B; A is separated from the rest of B by a partition and is filled with some perfect gas. Suppose that the walls are thermal insulators. When the gas is in equilibrium at temperature T_A, we suddenly remove the partition. I think that you will easily agree that the gas will occupy all the volume B, and after some time, it will be in equilibrium again at some temperature T_B.

Actually, since in the process, $\int \delta Q = 0$ and the work done on the gas or by the gas is nothing, $\Delta U = 0$. For a perfect gas, $U = U(T)$, so $T_B = T_A$. The process is a typical *irreversible process* and the second principle implies that

$$S_B - S_A > \int_A^B \frac{\delta Q}{T} = 0.$$

To compute $S_B - S_A$, consider a reversible isothermal expansion. Since $U = U(T)$, the expansion is isothermal if $dU = 0$. To expand the gas, we slowly move a piston, but the gas makes work on the piston when the volume increases by dV. Now, $dU = 0 \rightarrow \delta Q = PdV$, and we must supply this heat in the process. The change i entropy is

$$S_B - S_A = \int_A^B \frac{PdV}{T},$$

and inserting P from (5.12), one obtains

$$S_B - S_A = NK_B \log\left(\frac{V_B}{V_A}\right). \tag{5.13}$$

The motion of each molecule is mechanically reversible, but the time reversed process never occurs. The future can be distinguished from the past because of the sign of $S_B - S_A$. The spontaneous evolution of any system increases the entropy. This is the *arrow of time*.

More generally, we can consider a reversible transformation in which, besides changing the volume we also exchange heat with n moles of perfect gas. This can be done by adding an adiabatic transformation. Then, $N = nN_A$ and $dS = nC_V dT + PdV$, and inserting Clapeyron's law,

$$dS = nC_V dT + nR\frac{dV}{V};$$

integrating,

$$S_B - S_A = nC_V \ln\left(\frac{T_B}{T_A}\right) + nR \ln\left(\frac{V_B}{V_A}\right). \tag{5.14}$$

It is important to note that the entropy is an extensive quantity, that is, each molecule contributes a molecular entropy $\frac{S_B - S_A}{N}$, and the set of two independent tanks of gas has an entropy that is the sum of the two entropies. This is like the kinetic energy of molecules, that sum up to give the internal energy of the gas; however, the observation is more subtle, since the molecular entropy depends on T, which is not a property of the single molecule.

Even the Universe as a whole (Sect. 8.12) is growing old, the star formation rate is decreasing and the synthesis of heavy elements inside stars is clearly an irreversible process. The cosmological models that proposes cyclic or pulsating universes must face the basic problem of 'resetting' the entropy at the start of each cosmic cycle.

Third Principle

The *third principle of Thermodynamics* was formulated by Walther Nernst in 1906–1912. It states that $S \to 0$ for $T \to 0$. This removes the arbitrariness that was left by the second principle. One can show that this requires that the specific heat C of every perfect crystal tends to 0 at absolute zero, provided that the lowest state of the system is unique. More generally, for any system $\frac{\delta Q}{T} = \frac{C \delta T}{T}$ must stay small for any process with $T \to 0$, making it increasingly difficult to extract heat from a substance when the temperature gets lower. The most important implication is that it is impossible to reach the absolute zero in a finite number of operations. Today, there are people working with ultra-cold atoms at micro-Kelvin temperatures and manipulating them to make studies on the quantum behavior of matter under such conditions. One could wonder what the significance of the third principle is when a disordered mixture is brought to ultra-low temperatures; how is it possible that the measure of disorder S gets small by cooling? But at absolute zero, the system should get ordered; the paradox comes from the fact that the kinetics for approaching the equilibrium $S = 0$ state gets extremely slow at low T.

5.2 Black Body

The objects that we observe in everyday life are colored according of their nature. Viewed in white light, a body turns blue if it absorbs mainly red light and red if it better absorbs blue radiation; it looks white if it reflects all that is visible. Any body reflects some light, even without being an ideal mirror. A mirror also preserves direction and images. Some bodies absorb little light and reflect little; they look transparent and colorless. It is the light that the body reflects that we see and gives it a color. If a body in daylight reflects little visible light, it appears black; but coal also reflects a bit. An ideal black body absorbs all radiation completely. Such a body is an idealization of great importance in the historical development of Theoretical Physics. It is due to Kirchhoff.[2] The starting point was a focus on equilibrium conditions. Only a deep

---

[2]Gustav Kirchhoff was a German physicist (Königsberg (now Kaliningrad, Russia), 1824- Berlin, 1887). He established the well-known laws on linear electric circuits, and, together with Bunsen,

reflection allowed for the discovery that this variability, which is so necessary to our life, is due not only to the nature of the body but primarily to the color of light, and especially to the lack of equilibrium between thermal radiation and matter. (Life is itself a phenomenon that is far from equilibrium.) According to the first principle of Thermodynamics, energy must be conserved. For a body in equilibrium, the energy that is absorbed is re-emitted. This leads to the idea of an equilibrium distribution in frequency of the radiation absorbed and emitted by all bodies.

The ancient philosophers did not even remotely imagine such things. Everybody knew that a piece of iron when heated starts to emit light that changes from red brown to red to orange to yellow to with increasing temperature, and become much more bright. They also knew that all hot bodies behave in essentially the same way. Of course, the problem is complicated by the fact that our eye sees only wavelengths between 4000 and 8000 Å. In the nineteenth century, experiments led to Wien's law $\nu_{max} \propto T$, saying that the frequency ν where maximum intensity is issued is proportional to the temperature. The power emitted increases as ν^4, as we shall see.

A closed cavity in equilibrium at temperature T is filled with thermal radiation, which is emitted from the walls. In equilibrium, as much has to be absorbed as is emitted. Through a small hole in the walls, one can observe this radiation. This is an experimental realization of the black body, since every radiation entering the hole from outside will be partially reflected by the walls many times and eventually will be absorbed.

Kirchhoff Laws

Consider a cavity in equilibrium at temperature T. It is full of isotropic radiation in equilibrium with the walls. Let $u(\nu, T)$ denote the energy per unit volume and frequency ($[u(\nu, T)] = \frac{\text{erg.s}}{\text{cm}^3}$). The energy density is $u(T) = \int_0^\infty u(\nu, T)d\nu$. Gustav Kirchhoff, in 1859, has shown that $u(\nu, T)$ does not depend on the nature of the walls of the cavity. Suppose two different cavities A and B at the same temperature had different densities $u_A(\nu, T)$ and $u_B(\nu, T)$. We could let them exchange radiation and, using a color filter, we could allow the exchange of energy at frequencies such that $u_A(\nu, T) > u_B(\nu, T)$ but not at frequencies such that $u_A(\nu, T) < u_B(\nu, T)$. In this way, B should become hotter than A. The perpetual motion machine of the second kind would be operating in contrast to the Second Principle. Kirchhoff concluded that $u(\nu, T)$ must be some universal function. One can put objects of different colors, bright or opaque, but at equilibrium the same $u(\nu, T)$ is obtained.

To explain that, one must consider the energy balance of a surface element dS of the walls in the unit time. The energy emitted in an frequency interval $d\nu$ centered around ν is $e(\nu, T)dS$, where e is the emissive power which depends on the nature of the body. On the other hand, when isotropic radiation with frequency ν falls on the same surface element, the nature of the body also determines the absorption power $a(\nu, T)$ which is the ratio of absorbed energy over incidence energy. An ideal body with $a \equiv 1$ is called black. Remarkably, Kirchhoff found that the ratio $\frac{e(\nu,T)}{a(\nu,T)}$ does

---

invented the spectroscope; he also discovered Cesium and Rubidium. He wrote "Vorlesungen über mathematische Physik".

not depend on the body. Indeed, the same amount of energy in the interval $d\nu$ falls in the unit time over each dS of the walls and of the surfaces of all the bodies therein; a fraction, $a(\nu, T) \cdot dE$ is absorbed, but an equal amount, $e(\nu, T) \cdot dS$ must be emitted. There is a separate equilibrium condition at each frequency. For two bodies 1 and 2, the equilibrium conditions are:

$$\begin{cases} e_1(\nu, T)dS = a_1(\nu, T)dE(\nu, T), \\ e_2(\nu, T)dS = a_2(\nu, T)dE(\nu, T), \end{cases}$$

and we may conclude that

$$\frac{e_1(\nu, T)}{a_1(\nu, T)} = \frac{e_2(\nu, T)}{a_2(\nu, T)} = \frac{dE(\nu, T)}{dS}.$$

The universal function $\frac{dE(\nu,T)}{dS}$ is the *emitting power* of the black body; it is the energy that is emitted for unit time and frequency by the unit surface of the black body. Experimentally, one can make a good black body by producing a small hole on the surface of a cavity kept at temperature T. If the temperature is several hundred degrees Celsius the hole appears to an external observer as a bright spot. Its absorbing power is high (nearly all the incident radiation bounces on the walls many times and is absorbed). The emissive power is very high.

$$\frac{dE(\nu, T)}{dS} = \frac{c}{4}u(\nu, T). \tag{5.15}$$

To see the relation between dE and u let us take the surface element dS as the origin of spherical coordinates; a volume element of the cavity at (r, θ, ϕ) is

$$dV = r^2 dr d\Omega = r^2 dr d\phi \sin(\theta)d\theta;$$

at time $t = 0$, dV contains the energy $u(\nu, T)d\nu dV$ in the frequency interval $d\nu$. Such radiation is isotropic, and a fraction will arrive at dS at time $t = \frac{r}{c}$. Seen from dV, the surface element is reduced by a factor of $\cos(\theta)$ and occupies a fraction $fdS = \frac{dS \cos(\theta)}{4\pi r^2}$ of the spherical surface of radius r. In the time between $t = 0$ and $t = \frac{r}{c}$, dS receives the contributions of the elements dV up to a distance r;

$$dE = dS \int_0^{ct} dr r^2 \int_0^{2\pi} d\phi \int_0^{\frac{\pi}{2}} d\theta \sin(\theta)f,$$

the integral over θ covers the half space "above the surface". Then, we must divide by t to get the energy per unit time. The result is:

$$dE(\nu, T) = c\frac{u(\nu, T)dS}{4\pi} 2\pi \int_0^1 d\cos(\theta)\cos(\theta) = \frac{c}{4}u(\nu, T)dS.$$

Stefan–Boltzmann Law

Consider an empty cavity with volume V in thermal equilibrium at temperature T. We know the cavity contains electromagnetic radiation, the black body radiation. According to the Kirchhoff laws, the spectral components $u(\nu, T)$ obey a universal law (which will turn out to be Planck's law in later chapters); the total internal energy is

$$U = Vu(T). \tag{5.16}$$

$U = U(V, T)$ does not depend purely on T, as for the classical ideal gas, but also on Volume. There is a pressure on the walls. It is known from Electromagnetism that the radiation pressure P is related to the energy density u by

$$P = \frac{u}{3}. \tag{5.17}$$

A thought experiment is needed here. The cavity is closed by a (frictionless) piston in contact with the heat source at temperature T and the radiation is used as the fluid to make a reversible Carnot cycle. The four phases are:

A isothermal expansion $V \to V + \Delta V$ at temperature T. The internal energy grows by $u\Delta V$ and the work $P\Delta V$ is performed; therefore, from the First Principle, we know that from the source at temperature T, an amount of heath $Q = (P+u)\Delta V$ is extracted;

B adiabatic expansion (cavity and piston are now thermally isolated) $V + \Delta V \to V + \Delta V + dV$ with infinitesimal dV. The system makes work at the expense of its internal energy; the temperature decreases to $T - dT$, while the pressure drops to $P - dP$ with $dP = \frac{dP}{dT}dT$;

C isothermal compression with the cavity in contact with the source at temperature $T - dT$. The volume decreases from $V + \Delta V + dV$ and crosses the adiabatic through the initial state. The volume decreases by ΔV (neglecting higher order corrections). The heat given to the cold source at temperature $T - dT$ is $Q = [(P - dP) + u]\Delta V$;

D an adiabatic compression closes the cycle.

Up to higher order infinitesimals, the work done n the cycle is $dW = dP\Delta V$; the efficiency is

$$\eta = \frac{dW}{Q} = \frac{dP\Delta V}{(P+u)\Delta V} = \frac{dP}{dT}\frac{dT}{P+u}.$$

But the Second Principle says that the efficiency must be $\frac{dT}{T}$. So,

$$\frac{dP}{dT} = \frac{P+u}{T}.$$

Substituting (5.17) and integrating, one finds

$$u = aT^4.$$

This is the *Stefan–Boltzmann* law. A hot black body is in equilibrium with an energy density, which grows with the fourth power of Temperature; this is also the trend of the power radiated by a black body in the absence of equilibrium, when the body cools down. The stars have internal heat sources and emit by irradiation in agreement with the *Stefan–Boltzmann* law, since they are black to a good approximation. Stefan obtained the result experimentally. The theory was formulated by Boltzmann[3] in 1884; the constant a remained an empirical parameter for a long time, until it was determined with the advent of Quantum Mechanics.

Problem 13 The black body radiation carries entropy S. Consider the phases A and B of the above Carnot cycle; calculate $(\frac{\partial S}{\partial V})_T$, $(\frac{\partial S}{\partial T})_V$ and the second derivatives $(\frac{\partial^2 S}{\partial T \partial V})$.

Solution 13 The variation of S is

$$dS = \frac{\delta Q}{T} = \frac{dU + PdV}{T}.$$

From (5.16),

$$dU = u(T)dV + V\frac{\partial u}{\partial T}dT.$$

Also inserting (5.17), one finds

$$dS = \frac{\frac{4}{3}udV + V\frac{\partial u}{\partial T}dT}{T}. \tag{5.18}$$

From (5.18), $dS = (\frac{\partial S}{\partial V})dV + (\frac{\partial S}{\partial T})dT$, where

$$\left(\frac{\partial S}{\partial V}\right) = \frac{4u}{3T},$$

while

$$\left(\frac{\partial S}{\partial T}\right) = \frac{V}{T}\frac{\partial u}{\partial T}.$$

According to the *second Principle, dS* must be exact; hence, the mixed derivatives must be equal:

$$\frac{\partial}{\partial T}\left(\frac{\partial S}{\partial V}\right) = \frac{\partial}{\partial V}\left(\frac{\partial S}{\partial T}\right).$$

Therefore,

$$\frac{4}{3T}\frac{\partial u}{\partial T} - \frac{4}{3}\frac{u}{T^2} = \frac{1}{T}\frac{\partial u}{\partial T},$$

---

[3]Ludwig Boltzmann (Vienna 1844- Duino (near Trieste) 1906, Austria (now Italy(suicide))) pioneered Statistical mechanics with Gibbs.

that is,

$$\frac{\partial u}{\partial T} = 4\frac{u}{T}.$$

One can again derive the Stefan–Boltzmann law from this result.

Problem 14 The temperature of the solar photosphere is $T_S \sim 5800\,°\mathrm{K}$ and the size of the solar disk seen from here is $\frac{1}{4}°$. Evaluate the mean temperature of the Earth assuming that the Earth and the Sun are black bodies. What would the temperature of the Earth become if the distance from the sun were doubled?

Solution 14 The angle is $\frac{\pi}{4\times180}$ rad, therefore, the fraction of the solid angle occupied by the Sun is $\Delta\Omega = \int d\phi\,d\cos(\theta) = \frac{2\pi(1-\cos(\alpha))}{4\pi} = 4.76\,10^{-6}$. By the Stefan–Boltzmann law, the power goes with T^4. Therefore the ratio of temperatures is $\frac{T_E}{T_S} = 0.0467$. We obtain $T_E = 270.1\,°\mathrm{K}$, which is realistic (only slightly low, mainly because we neglected the greenhouse effect). Dividing α by 2 we should find $191.6\,°\mathrm{K}$.

5.3 Statistical Mechanics

The results of many experiments performed on macroscopic bodies depend on the thermal state. The theoretical interpretation of these experiments in microscopic terms is the subject of statistical mechanics, which, unlike Thermodynamics, assumes prior knowledge of the interactions between the microscopic components of the system, i.e., the Hamiltonian $H(p, q)$. Moreover it is supposed to know certain parameters that determine the system from the macroscopic point of view, such as pressure, temperature, number of molecules for each species, and so on. All kinds of measurements may be considered. Observables can be macroscopic, like the pressure of a gas as a function of the temperature or its specific heat, but all spectroscopies, including those designed to determine the microscopic properties of solids, are influenced by temperature effects and must be interpreted through statistical mechanics.

For example, Zartmann, in a famous experiment, measured the speed distribution of the molecules of a gas. The gas is contained in a tube; a small hole in the bottom of the tube is opened for a very short time and the molecules come out with the speed they had within the gas. The molecular beam so obtained impinges on a quickly spinning wheel, and the molecules arrive in different points depending on the time of flight. The velocity distribution, that one can convert into an energy distribution; is an asymmetric bell-shaped distribution whose maximum moves to higher energies by increasing temperature. Thus Zartmann verified the law established theoretically by Maxwell[4] (in obvious notation)

---

[4]Note that $\int_0^\infty \sqrt{x}e^{-x}dx = \frac{\sqrt{\pi}}{2}$, and so $\int dN = N$.

$$dN = N \frac{2}{\sqrt{\pi}} \frac{1}{(K_B T)^{3/2}} e^{-\frac{\epsilon}{K_B T}} \sqrt{\epsilon} d\epsilon. \qquad (5.19)$$

We shall derive this law in Sect. 5.7.1.

Most of the $N_A \sim 6 \times 10^{23}$ molecules in a mole have a kinetic energy of order $K_B T$ but a few are much slower or much faster than that. Every macroscopic measurement of some magnitude $f(p, q)$ which depends on the same arguments as $H(p, q)$ returns its average $\bar{f}$; the microscopic state cannot be measured and is therefore irrelevant. One cannot evaluate the evolution starting with an integral of the canonical equations. The proper use of the Hamiltonian was found by Gibbs.[5]

5.4   Gibbs Averages

Suppose our macroscopic system is a mole of some substance that we wish to treat as a complicated mechanical system; a point γ of the *phase space* or Γ *space* has $2s$ coordinates (s are space coordinates and s the conjugated momenta). So, the coordinates

$$q_1, q_2, \ldots, q_s, p_1, p_2, \ldots, p_s$$

of γ are just the arguments of the Hamiltonian. This huge list of variables specifies the microscopic state (or simply *microstate*) of the system. The evolution of the system is described by the chaotic motion of $\gamma(t)$.

The outcome of the macroscopic experiment of some magnitude $f(\gamma)$ does not depend on any details of $\gamma(t)$, and many microstates would give the same outcome. There are only small fluctuations, which scarcely perturb a regular behavior. If the measurement is done in equilibrium, and the system is perturbed in any way, the equilibrium can be reached again after a characteristic waiting time. If the measurement is then repeated, the measurement yields the equilibrium results again, without any memory of the twists of $\gamma(t)$ in Γ space. Therefore, obtaining $\gamma(t)$ from $\gamma(0)$ is not only impossible, but also irrelevant. What one measures is better represented by a time average,

$$\overline{f(p, q)} = \frac{1}{T} \int_0^T f(p, q) dp dq, \quad T \to \infty,$$

and the above discussion strongly suggests that $\bar{f}$ must be independent of the initial condition $\gamma(0)$. The time average would wash out all the information gained by the integration of the equations of motion and any distinction between the many microstates that would give the same results.

---

[5]The great theoretician Josiah Willard Gibbs (New Haven, Connecticut, U.S.A., 1839 - Yale 1903) was the first professor of Mathematical Physics in U.S.A.

We must do without detailed initial conditions, since the system evolves spontaneously towards equilibrium[6] within some characteristic times.

The fundamental approach of Statistical Mechanics is credited to J.W. Gibbs, and
the basic principle is very simple:

*The fundamental assumption of Statistical Mechanics is: if a system in equilibrium has
energy E, all the microscopic states of energy E are equally likely.*

Consider a system with Hamiltonian $H(p, q)$ in a state specified by all the relevant
macroscopic parameters, like pressure, volume, number of particles, total energy.
What can we say about its *microscopic* state? Gibbs considered the probability dP
that it be in a given bunch $d\Gamma$ of microstates; more precisely,

$$dP = \rho(p, q)d\Gamma, \tag{5.20}$$

where $d\Gamma = \prod_i^s dp_i dq_i$ is the volume element of *phase space* Γ of the system. ρ is
the *distribution function* . It must be normalized, that is,

$$\int_{\text{spazio } \Gamma} \rho(p, q)d\Gamma = 1.$$

Thus,

$$\bar{f} = \int f(p, q)\rho(p, q)d\Gamma.$$

A priori all the microstates are equally likely; therefore, all the portions of phase
space compatible with the macroscopic parameters must be treated in the same way.
$H(p, q)$ determines $\rho(p, q)$ and thus the probability of any given element $d\Gamma$. Put
another way, one considers an *ensemble* of systems with the same H and with given
macroscopic parameters; they are distributed in Γ space according to $\rho(p, q)$; we
perform a statistical average on this *ensemble*. Is the *ensemble* average really equal
to the time average of the corresponding quantity that one would compute from
the microscopic evolution? This is a very good question, which is best known as
the ergodic problem. Physicists use this as a sort of postulate, but from the formal
mathematical viewpoint, it is a difficult issue, which has been settled by Fermi and
others, but only under certain assumptions. However one fact is true: based on this
assumption, Statistical Mechanics works.

Liouville Theorems

The Gibbs theory, of course, is mathematically sound and does not break down with
a change of Lagrangian coordinates (punctual transformation $(p, q) \rightarrow (P, Q)$). In
other words,

$$\bar{f} = \int f(P, Q)\rho(P, Q) \prod_i^s dP_i dQ_i = \int f(p, q)\rho(p, q) \prod_i^s dp_i dq_i.$$

---

[6]If the system is the Universe, it is quite clear that it is not in thermal equilibrium, however, who
knows if it evolves towards an equilibrium of some kind?

This requires that the Jacobian be 1, that is,

$$\frac{\partial(Q, P)}{\partial(q, p)} = 1 = \frac{\partial(q, p)}{\partial(Q, P)}.$$

In the case of a single degree of freedom, one finds

$$\frac{\partial(Q, P)}{\partial(q, p)} = Det\begin{pmatrix} \frac{\partial Q}{\partial q} & \frac{\partial Q}{\partial p} \\ \frac{\partial P}{\partial q} & \frac{\partial P}{\partial p} \end{pmatrix} = \{Q, P\} = 1$$

because the Poisson brackets are invariant, as shown in Chap. 2. Liouville has shown that this remains true in the presence of many degrees of freedom, and therefore

$$\int dq\, dp = \int dQ\, dP,$$

that is, if you cut out a volume inside Γ, its extension is conserved during the evolution.

A second Liouville theorem states that

$$\frac{d\rho}{dt} = 0. \tag{5.21}$$

This is tantamount to saying that ρ depends only on the constants of the motion, and an *ensemble* evolves like an incompressible fluid. Indeed,

$$\frac{d\rho}{dt} = \frac{\partial \rho}{\partial t} + \{\rho, H\},$$

and while $\frac{\partial \rho}{\partial t} = 0$ since ρ cannot have any explicit time dependence.

5.5 Microcanonical Ensemble

The Microcanonical Ensemble is a collection of a huge number of isolated systems; they must be systems with the same number of particles and about the same energy E_0; the energy uncertainty δE must be so small that adding it to the system does not change the experiment appreciably, while it may well be large compared to a molecular energy.

The microstates are random. All the magnitudes, including the conserved energy E_0, can only be specified with macroscopic accuracy. It is not necessary to specify δE further. Experimentally, it is very hard to work with isolated systems on equilibrium, but conceptually, this is a good starting point. We need to find the distribution function $\rho(p, q)$. This is the probability density in Γ space, proportional to the number of systems in the ensemble that have their position in an element $d\Gamma$ around the point (p, q). In a macroscopic description, $\delta E \to 0$ and ρ_{micro} must be proportional to $\delta(H(p, q) - E_0)$ and normalized to 1. Letting $\omega(E_0) = \int d\Gamma \delta(H(p, q) - E_0)$ we arrive at the microcanonical distribution

$$\rho_{micro}(p, q) = \frac{\delta(H(p, q) - E_0)}{\omega(E_0)}, \quad \omega(E_0) = \int d\Gamma \delta(H(p, q) - E_0).$$

Here, $\omega(E_0)$ is the measure of the constant energy surface.

We must perform statistical averages over an *ensemble*, specifically a *microcanonical ensemble* of systems that are represented in Γ-space by points that are uniformly scattered in the gap $d\Omega$ between the hypersurfaces $H = E_0$ and $H = E_0 + \delta E$. Consider a perfect gas in equilibrium and suppose the volume of the container divided into two parts A and B with volumes V_A and V_B. In phase space, there is a set Γ of points that correspond to unbelievable situations in which all particles in A. These points are treated on the same footing as the others. All ways of distributing the molecules are represented in the ensemble and the energy is the same for all of them. Imagine the molecules are numbered. The probability that number 1 is in A is $\frac{V_A}{V_B}$, and the same holds true for the others; the probability of having all in A is $(\frac{V_A}{V_B})^N$. Therefore, if $\frac{V_A}{V_B} \frac{1}{2}$ and $N = 10^{23}$, the probability of having all in A is 10 to a power -3×10^{22}. This is practically 0 in Physics. The probability of having k molecules on one side and $N - k$ on the other side is $P(k, N) = 2^{-N} \binom{N}{k}$, and is shown as a function of k in the next figure for $N = 20$ and $N = 80$. While fluctuations around the most probable value always exist, very large deviations are impossible. This is why the evolution of the system has a preferred direction, from unlikely to likely. This explains the arrow of time (Fig. 5.2).

Boltzmann generalized this concept and conceived a Γ space divided into small cells (the criterion of smallness will be made more precise below). Let $\delta\Gamma$ denote the volume of each cell, which corresponds to a microstate of the system, with all the q and p specified with a small uncertainty. The cells are finite, but so small that the uncertainty of q and p is much smaller than any characteristic length or momentum

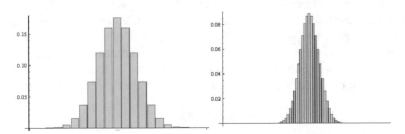

Fig. 5.2 Histograms of the probability distributions of N molecules in two equal boxes. Left: N = 20 Right: N = 80. The distribution becomes very sharp when N grows

of the problem. Since $q \times p$ is an *action*, let the uncertainty of pq be a small action[7]
h; then the volume of the cell is h^s, where s in the number of degrees of freedom of
the system.

$\hbar$ is kept finite. In this way, each volume of Γ space corresponds to a very large
number of cells. Let $\Omega_{tot}(E)$ be the volume of the region of Γ where the energy of
the system is in the interval between 0 and E and the volume, pressure and other
macroscopic parameters are also fixed; dividing the volume by per h^s one obtains a
very large number of microstates

$$W_{tot}(E) = \frac{\Omega_{tot}(E)}{h^s}.$$

This number is a measure of $\Omega_{tot}(E)$ for the given macroscopic conditions.

A small increment δE corresponds to a gap $\delta\Omega_{tot}(E_0)$ of phase space that contains
some number $W(E_0)$ of cells.

*W is the number of microstates that appear macroscopically as the same state of the system,
with energy E.*

Note that W is given by

$$W(E_0)h^s = \delta\Omega_{tot} = \left(\frac{\partial\Omega_{tot}}{\partial E}\right)\delta E = \omega(E_0)\delta E. \tag{5.22}$$

W is proportional to the measure of the constant-energy hypersurface

$$W = \frac{\partial W_{tot}}{\partial E}\delta E. \tag{5.23}$$

The argument is very ingenious. It appears to *stress* the mathematics, but it is
perfectly sound. Since δE is not infinitesimal, W is a large number. This appears to
be a difficulty, since then W is defined only up to a multiplicative constant. If we can
vary h, we must vary W accordingly. Nature actually defines a scale for h, which is
Planck's constant, and that does not fit in the classical scheme.

Now we can compare two hypothetical states of a macroscopic system, both
compatible with the macroscopic parameters. Compare for example, the state with
the gas that occupies all the available volume and the state in which the gas fills only
one part of the partition. We have seen that the number W of ways in which the latter
situation can occur is negligible. States with larger W are closer to equilibrium, which
requires maximum W. The equilibrium states are those with maximum entropy S.
This reasoning motivated Boltzmann to look for a functional link of the form

$$S = f(W).$$

---

[7]Plack's quantum h had not yet been discovered.

S is extensive: the entropy of a system formed by two parts is the sum of the entropies $S = S_1 + S_2$, while by the laws of probability, $W = W_1 W_2$. Therefore, f must satisfy

$$f(xy) = f(x) + f(y).$$

That is enough to identify f with the logarithm: so Boltzmann proposed the fundamental equation

$$\boxed{S = K_B \log W,} \tag{5.24}$$

that prompts a *statistical meaning* of entropy. This is the microcanonical definition of S. Remarkably, any change of h or δE, $\log W$ only changes an additive constant like in thermodynamics, but S is a state function. According to the enlightening equation (5.24), S is a measure of our ignorance of the microstate of the system. The system evolves by increasing the entropy; in a sense, it is shy and tries to hide the microstate by minimizing our knowledge of it. S increases with the degree of disorder. Only at absolute zero is the microstate unveiled because the system is compelled to reach the minimum energy state, and then $S = 0$ in agreement with the third principle of thermodynamics. However, the prudish system does not allow this, since one cannot reach absolute zero!

5.5.1 Entropy of the Perfect Gas

Consider a perfect gas of N atoms with mass m, and Hamiltonian

$$H = \sum_i^N \frac{p_i^2}{2m}.$$

in a volume V. To obtain S by (5.24), we need the number of cells of volume h^{3N} in the set of Γ space where the energy is $\leq E$. The projection of this domain on the space of coordinates $q_1 \ldots q_{3N}$ is a hypercube with side $V^{1/3}$; its projection on the space with coordinates $p_1 \ldots p_{3N}$ is a hypersphere, and therefore,

$$W_{tot}(E) = \frac{1}{h^{3N}} \int \prod_j^N d^3 p_j d^3 q_j \theta \left(2mE - \sum_i^N p_i^2 \right).$$

Integrating over q yields V^N, and we must work out

$$W_{tot}(E) = \left(\frac{V}{h^3} \right)^N \Omega_{3N}(\sqrt{2mE}),$$

where $\Omega_{3N}(R)$ is the volume of the hypersphere of radius R in 3 N dimensions, see (3.28). Consequently,

$$W_{tot}(E) \equiv \frac{\Omega_{tot}(E)}{h^{3N}} \frac{\pi^{\frac{3N}{2}}}{(\frac{3N}{2})!} \left(\frac{V}{h^3}\right)^N (2mE)^{\frac{3N}{2}} = \frac{1}{(\frac{3}{2}N)!} \left[\frac{V}{h^3}(2\pi mE)^{3/2}\right]^N. \quad (5.25)$$

So,

$$\log(W_{tot}(E)) = N \log\left[\frac{V}{h^3}(2\pi mE)^{3/2}\right] - \log\left[\left(\frac{3N}{2}\right)!\right].$$

Using Stirling's formula, one obtains

$$\ln W_{tot} = N \log\left[V\left(\frac{4\pi m}{3h^2}\frac{E}{N}\right)^{\frac{3}{2}}\right] + \frac{3}{2}N. \quad (5.26)$$

According to (5.24), (5.22), (5.23) in order to calculate S, we must take the logarithm of $W = \frac{\partial W_{tot}}{\partial E}$. By (5.25),

$$W_{tot}(E) = \text{constant} \times E^{\frac{3N}{2}} \Rightarrow W(E)\frac{\partial W_{tot}}{\partial E} = \frac{3N}{2E}W_{tot}.$$

So,

$$\log W(E) = \log W_{tot} + \log N + \text{costante}.$$

Recalling (5.22) and (5.23), this result has a geometrical content. In terms of the volume $\Omega(E)$ of the hypersphere, it gives the measure $\omega(E)$ of its hypersurface. By Eq. (5.26), the logarithm of $W_{tot}(E)$ for large N is dominated by the terms in $N \ln N$ plus terms linear in N; the factor that allows us to transform $\Omega(E)$ into $\omega(E)$ is a logarithm, which diverges for $N \to \infty$, but much more slowly than the dominant terms. For this reason, the logarithm is negligible in the thermodynamic limit. The constant δE is even more negligible. To sum up, $\log \omega(E)$ and $\log \Omega_{tot}$ are equal in the thermodynamic limit $N \to \infty$. Therefore, we take $\Omega(E)$ instead of $\omega(E)\delta E$. Thus,

$$\frac{S}{K_B} = \log W_{tot}(E) = N \log\left[V\left(\frac{4\pi m}{3h^2}\frac{E}{N}\right)^{\frac{3}{2}}\right] + \frac{3}{2}N. \quad (5.27)$$

To test the validity of the present statistical approach, we can compare it with the result (5.14), from thermodynamics. Does it agree? No!!

The Paradox

Gibbs noted that this *honest* calculation (done applying the known rules) leads to a paradox. The entropy is an *extensive* quantity, that is, it is proportional to N. The expression (5.27) for S is surely wrong since it is not extensive. So the rules must be changed, but how? To fix this severe problem, Gibbs proposed a new and very far-reaching physical idea. He suggested that the phase space for a system of N identical molecules is not $\Omega(E)$, but $\frac{\Omega(E)}{N!}$ (in the usual action units). Now, $N! = N(N-1)(n-2)\ldots3.2.1$ is the number of permutations on N objects. All the micro-states that differ by permutations of the particles are identified by Gibbs, that is, they are taken as the same configuration. Classically the exchange of two like atoms should lead to a different microstate, since it is possible (in principle, not in reality) to observe them continuously as they move and exchange place; at least, it is possible to mark them somehow, in the same way as billiard balls are marked by chalk without a visible effect on their trajectories. According to Gibbs, atoms cannot be treated like tiny billiard balls. Then, $S \to S - \log N! \sim S - (N \log N - N) = S - N \log(\frac{N}{e}) = S + N \log(\frac{e}{N})$, and we get the additive result

$$\frac{S}{K_B} = \log \Omega(E) = N \log \left[\frac{eV}{N} \left(\frac{4\pi m}{3h^2} \frac{E}{N} \right)^{\frac{3}{2}} \right] + \frac{3}{2} N \qquad (5.28)$$

in agreement with Thermodynamics. Gibb's intuition anticipated the quantum indistinguishability of identical particles!

5.6 Canonical Ensemble

The basic principle of Statistical Mechanics affirms the equal probability of the microstates; this idea is realised most directly in the microcanonical ensemble. However, in order to be able to assign a sharp energy to the microstates, we must assume that the system does not exchange heat or work with anything. This is a problem, since experiments are never done on isolated systems. Keeping a system effectively isolated is very difficult, if not impossible, and interesting problems usually deal with samples that are in equilibrium with a thermal bath at some temperature.[8] In the Canonical Ensemble the system defined by the Hamiltonian $H(p, q)$ is in equilibrium at temperature T. The system can be large or small, even a single molecule. Some interaction with a thermal bath is needed to fix T, but the coupling must be so small that we can continue to think in terms of $H(p, q)$. These requirements appear to be conflicting, but they are not really, since the coupling to the external bath can be very small. In practice it will take a very long time to establish equilibrium by a tiny interaction, but this is equilibrium Statistical Mechanics and there are no time constraints.

---

[8] A system at a fixed temperature exchanges energy with a heat bath, therefore its energy fluctuates; in the microcanonical ensemble the energy is fixed, and therefore the temperature fluctuates. However for a macroscopic sample such fluctuations are relatively unimportant. Therefore both schemes should lead to the same results for large systems.

5.6.1   Canonical Temperature

When a body is in equilibrium with a thermal bath, its internal energy fluctuates. There are exchanges of thermal energy $dE = \delta Q$. Even the entropy fluctuates. In thermodynamics, $dS = \frac{\delta Q}{T}$. Up to now, we have a statistical definition of S, but not yet of temperature T. In the microcanonical ensemble, a microstate has fixed E, but T is not sharp. A canonical definition of temperature is inherent in the above remarks. Let us introduce the inverse canonical temperature

$$\frac{1}{T} = \left(\frac{\partial S}{\partial E}\right); \tag{5.29}$$

this links energy fluctuations to entropy fluctuations. Now let us look at the equilibrium condition. Suppose an isolated system has two weakly interacting subsystems:

$$H \sim H_1 + H_2, \quad d\Gamma = d\Gamma_1 d\Gamma_2.$$

They cannot have a sharp energy content, since there are energy fluctuations dE_1, dE_2, but, since the mutual interaction is negligible, $dE_1 + dE_2 = 0$. The equilibrium condition is that S must be maximum. Therefore,

$$dS = \left(\frac{\partial S_1}{\partial E_1}\right) dE_1 + \left(\frac{\partial S_2}{\partial E_2}\right) dE_2 = 0.$$

Thus, the equilibrium is given by

$$\left(\frac{\partial S_1}{\partial E_1}\right) = \left(\frac{\partial S_2}{\partial E_2}\right).$$

This is in line with the definition (5.29) of the temperature.

5.6.2   Information Entropy, Irreversible Gates and Toffoli Gate

In 1854 at the University College of Cork, George Boole developed a new algebra to embody the logic of propositions. This is at the foundation of modern computing. The basic unit of information is the bit, which can be $0 \equiv$ FALSE or $1 \equiv$ TRUE. Logic gates are devices that transform some input bits into output bits according to some prescribed rule. For instance, NOT is a gate with one input bit and one output bit; it changes TRUE into FALSE and FALSE into TRUE. Some gates have two inputs and one output. A AND B (written as $A \wedge B$ gives 1 if both inputs are 1, while

A OR B (written as $A \vee B$) gives 1 if at least one of the inputs is 1. NOR gives 1 if OR gives 0 and 0 if OR gives 1. Similarly, NAND gives the opposite reply to AND.

NAND and NOR are two examples of the so-called universal gates. This means that by combining enough NAND gates (or NOR gates) one can perform any logical operation and eventually carry on any calculation in Boolean logic algebra. One could set up a computer that only operates with combinations of one kind of gate, such as NAND or NOR. However, such gates have the property of wasting some information. For instance, AND gives 0 if the input is 00, 01 or 10, and NAND gives 1 in such cases; in this way, the output does not allow for the unique specification of the input. One classifies such gates as irreversible. In 1961, the German Rolf Landauer worked at IBM and realized that computation must be subject to the laws of Thermodynamics like any other macroscopic process. In particular, this implies that a minimum entropy amount equal to $K_B T \log 2$ is produced when a bit of information is erased. To avoid this, one can use reversible gates, which, in order to avoid information loss, must have as many output as input channels. NOT is a trivial example. Many of them perform permutations of the input bits, and therefore can be represented by permutation matrices. For example, with three inputs, one has the 8 combinations $000, 001, 010, 011, 100, 101, 110, 111$. Taking these in the above order, the Toffoli gate, proposed by Toffoli in 1980, is represented by the permutation matrix

$$
\begin{pmatrix}
1 & 0 & 0 & 0 & 0 & 0 & 0 & 0 \\
0 & 1 & 0 & 0 & 0 & 0 & 0 & 0 \\
0 & 0 & 1 & 0 & 0 & 0 & 0 & 0 \\
0 & 0 & 0 & 1 & 0 & 0 & 0 & 0 \\
0 & 0 & 0 & 0 & 1 & 0 & 0 & 0 \\
0 & 0 & 0 & 0 & 0 & 1 & 0 & 0 \\
0 & 0 & 0 & 0 & 0 & 0 & 0 & 1 \\
0 & 0 & 0 & 0 & 0 & 0 & 1 & 0
\end{pmatrix}
$$

This gate gives an output identical to the input in the following cases: $(0, 0, 0)$, $(0, 0, 1)$, $(0, 1, 0)$, $(0, 1, 1)$, $(1, 0, 0)$, $(1, 0, 1)$. Instead, $(1, 1, 0)$ gives $(1, 1, 1)$ and $(1, 1, 1)$ yields $(1, 1, 0)$. The Toffoli gate is clearly reversible and can be shown[9] to be universal.

5.7 Canonical Distribution

In the Canonical Ensemble, the energy is not fixed, so we must replace the distribution function with a different one $\rho(p, q)$. It is evident that possible interactions of other systems with the thermal bath are not relevant if the bath is very large. The system S and the bath B make up an isolated system C with Hamiltonian $H_C \sim H_S + H_B$; we may consider C an isolated system having a fixed energy E_C. We may deal with

---

[9] C.P. Williams, Explorations in Quantum Computing, Springer (2011).

$\mathcal{C}$ in the microcanonical scheme. The element of the phase space is $d\Gamma_C = d\Gamma_S d\Gamma_B$ and the probability of finding $\mathcal{C}$ in $d\Gamma_C$ is

$$dP_C = \frac{\delta(H_B + H_S - E_C)}{\int \delta(H_B + H_S - E_C)d\Gamma_C} d\Gamma_B d\Gamma_S.$$

The denominator depends only on the fixed constant E_C; in this calculation, we wish to compute the entropy of the bath, and since the constant produces an additive constant to the entropy, we drop it. For every microstate of $\mathcal{S}$, there are many microstates of $\mathcal{B}$ such that the total energy is E_C. The probability of finding $\mathcal{S}$ in $d\Gamma_S$ independently of the position of $\mathcal{B}$ in its Γ space can be obtained by *integrating out the bath*, that is, by integrating on $d\Gamma_B$: up to a constant,

$$dP_S = d\Gamma_S \int \delta(H_B + H_S - E_C)d\Gamma_B. \tag{5.30}$$

The integral is the measure of the hypersurface of phase space available to $\mathcal{B}$ at a fixed value of the energy of $\mathcal{C}$. Its calculation appears to be impossible.

However, a remark saves us. The same integral appears in the definition of the microcanonical entropy of the isolated $\mathcal{B}$:

$$S_B(E) = K_B \log(W_B(E)) = K_B \log\left[\delta E \int \delta(H_B - E)\frac{d\Gamma_B}{h^{s(\mathcal{B})}}\right], \tag{5.31}$$

where $s(\mathcal{B})$ is the number of degrees of freedom of the bath and $E = E_C - H_S$. Here, δE adds a constant to the entropy and can be ignored. Exponentiating,

$$e^{\frac{S_B(E)}{K_B}} = \int \delta(H_B - E)\frac{d\Gamma_B}{h^s(\mathcal{B})}.$$

In this case, however, $E = E_C - H_S$. Therefore, we need

$$\int \delta(H_B - E)\frac{d\Gamma_B}{h^{s(\mathcal{B})}} = e^{\frac{S_B(E_C - H_S)}{K_B}}.$$

Since the bath is very large, we can expand,

$$S_B(E_C - H_S) \sim S_B(E_C) - \frac{\partial S_B}{\partial E}H_S$$

and in view of (5.29),

$$S_B(E_C - H_S) \sim S_B(E_C) - \frac{H_S}{T}.$$

Therefore,

$$dP_S = Ce^{-\frac{H_S(p,q)}{K_BT}}d\Gamma_S \equiv \rho d\Gamma_S.$$

The new constant C will be eliminated by the normalization condition.

The *canonical distribution* for S, that is, for the molecule, is the dimensionless $\rho(p,q) = Ce^{-\frac{H_S(p,q)}{K_BT}}$; we simplify the notation writing $d\Gamma$ for $d\Gamma_S$ (since $\mathcal{B}$ is eliminated) with

$$\beta = \frac{1}{K_BT}.$$

So,

$$\boxed{\rho = \frac{e^{-\beta H_S(p,q)}}{\int e^{-\beta H_S(p,q)}d\Gamma}}. \tag{5.32}$$

The denominator is the *partition function* (which ensures that the integral of ρ over Γ is 1),

$$Z = \int e^{-\beta H_S(p,q)}d\Gamma, \tag{5.33}$$

which is a central quantity, as we shall see. The canonical average of any magnitude $A(p,q)$ is

$$\overline{A} = \frac{1}{Z}\int e^{-\beta H_S(p,q)}A(p,q)d\Gamma.$$

If the system is formed by two independent parts, $H_S = H_1 + H_2$, and thus $\rho = \rho_1\rho_2$; the parts 1 and 2 have their own canonical distribution.

5.7.1 Maxwell Distribution

In the case of a perfect gas in a volume V, the gas is C while an atom is S; then, $H = \frac{p^2}{2m}$, $d\Gamma = d^3xd^3p$, and so,

$$dP = C\exp\left[-\frac{p^2}{2mK_BT}\right]d^3xd^3p.$$

Since $\int d^3x$ is the volume V the normalization condition is

$$CV\int \exp\left[-\frac{p^2}{2mK_BT}\right]d^3p = 1.$$

Inserting $\int_{-\infty}^{\infty}dxe^{-\alpha x^2}\sqrt{\frac{\pi}{\alpha}}$,

$$CV = \frac{1}{(2\pi m K_B T)^{3/2}} \Leftrightarrow Z = V(2\pi m K_B T)^{3/2}.$$

The velocity distribution is obtained by $(p_i \rightarrow mv_i)$ and reads as:

$$dP(\mathbf{v}) = \left(\frac{m}{2\pi K_B T}\right)^{2/3} \exp\left(\frac{1}{2}\frac{mv^2}{K_B T}\right) dv_x dv_y dv_z.$$

One sets $d^3v = v^2 dv \sin\theta d\theta d\phi$ and integrates over the angles. The result is:

$$dP(v) = 4\pi \left(\frac{m}{2\pi K_B T}\right)^{2/3} \exp\left(\frac{1}{2}\frac{mv^2}{K_B T}\right) v^2 dv.$$

The Maxwell distribution (5.19) can be obtained by inserting $\frac{1}{2}mv^2 = \epsilon$, with $dv = \sqrt{\frac{1}{2m\epsilon}}d\epsilon$, and the number of molecules N. Indeed, we find:

$$dN = N\frac{2}{\sqrt{\pi}}\frac{1}{(K_B T)^{3/2}}e^{-\frac{\epsilon}{K_B T}}\sqrt{\epsilon}d\epsilon.$$

5.7.2 Perfect Gas in the Canonical Ensemble and Boltzmann Statistics

Above, we derived the Maxwell distribution from the canonical distribution. In this section, we check that it arises directly from Gibb's principle that all the micro-states have the same probability. The perfect gas consists of $N >> 1$ independent material points with total energy E in equilibrium. The gas is the thermal bath $\mathcal{B}$ and $\mathcal{S}$ is just the molecule. The the whole system $\mathcal{C} = \mathcal{B} + \mathcal{S}$ is treated as in the microcanonical ensemble, with fixed energy end particle number. All the micro-states of energy E are taken to be equally likely. We work out the distribution $n(\epsilon)$ of the single molecule energy ϵ that grants the maximum of the number W of microstates of $\mathcal{C}$. In short, the equilibrium distribution is the one that can be realized in more ways.

Since the molecules are independent, instead of the Γ space we can work in the single molecule μ space, with coordinates x, y, z, p_x, p_y, p_z. We divide μ space in cells of volume $\Delta x \Delta y \Delta z \Delta p_x \Delta p_y \Delta p_z$. In this way we can discretize. The energy axis ϵ of a molecule will be sliced in small intervals of microscopic width $\Delta\epsilon$. The exact method of slicing is irrelevant as long as the cells are macroscopically small and the energy $\Delta\epsilon$ is sufficiently small. However the cells must still be so large that each cells contains several molecules.

Each way to assign the individual molecules to the cells is a microstate. We shall treat the molecules as distinguishable as if each could be labelled. This is a natural assumption in classical physics, that we shall drop in Quantum Mechanics. Therefore, classically the exchange of the μ space coordinates (p, q) of two molecules takes to a distinct microstate. The equilibrium distribution

$$\bar{n}_i = N\rho(\epsilon_i), \tag{5.34}$$

must be normalized according to

$$\sum_i \rho(\epsilon_i) = 1. \tag{5.35}$$

The probability of any distribution $n_1, n_2, n_3 \ldots$, with n_i molecules in ith cell, is proportional to the number of ways it can be realized, compatibly with the restrictions

$$\sum n_i = N$$

and

$$\sum \epsilon_i n_i = E.$$

The number of ways[10] to choose n_1 particles to put in cell 1 is

$$\binom{N}{n_1};$$

when this is done, one is left with $N - n_1$ molecules from which one can choose n_2 for the second cell. A distribution $n_1, n_2, n_3, \ldots$ is realized in

$$W = \binom{N}{n_1}\binom{N-n_1}{n_2}\binom{N-n_1-n_2}{n_3}\ldots \tag{5.36}$$

different ways. Here a colossal simplification occurs, since

$$\binom{N}{n_1}\binom{N-n_1}{n_2}\binom{N-n_1-n_2}{n_3}$$

$$= \frac{N!}{n_1!(N-n_1)!}\frac{(N-n_1)!}{n_2!(N-n_1-n_2)!}\frac{(N-n_1-n_2)!}{n_3!(N-n_1-n_2-n_3)!}\ldots$$

$$= \frac{N!}{n_1!n_2!n_3!\ldots},$$

and so,

$$W = \frac{N!}{\prod_r n_r!}. \tag{5.37}$$

---

[10] $\binom{N}{n} = \frac{N!}{n!(N-n)!}$ is the number of different choices of n objects from N, regardless the order of the selected objects.

Since the logarithm is everywhere an increasing function of its argument, we maximize $\log(W)$; by the Stirling formula

$$\ln W \approx N \ln N - \underbrace{N} - \sum n_i \ln n_i + \underbrace{\sum n_i} = N \ln N - \sum n_i \ln n_i. \quad (5.38)$$

This is the entropy of the gas

$$\frac{S}{K_B} = \ln W \approx N \ln N - \underbrace{N} - \sum n_i \ln n_i + \underbrace{\sum n_i} = N \ln N - \sum n_i \ln n_i \quad (5.39)$$

in terms of the (yet unknown) cell populations. According to the Lagrange method, we impose

$$\frac{\partial}{\partial n_r} \left[-\sum n_i \ln n_i - \gamma \sum n_i - \beta \sum \epsilon_i n_i \right] = 0,$$

where β, γ are multipliers. We get

$$-\ln n_r - 1 - \gamma - \beta \epsilon_r = 0,$$

that is,

$$\bar{n}_r = e^{-\alpha} e^{-\beta \epsilon_r}.$$

with $\alpha = 1 + \gamma$, $\beta = \frac{1}{K_B T}$. Now we can grasp better the meaning of the Boltzmann distribution. The parameter α is fixed by (5.35). To obtain the entropy, we cast (5.34), (5.35) into (5.39):

$$\frac{S}{K_B} = N \log N - N \sum_r \rho(\epsilon_r) \log(N \rho(\epsilon_r)) - N \sum_r \rho(\epsilon_r) \log(\rho(\epsilon_r)).$$

The molecular distribution function reads as:

$$\rho(\epsilon_r) = \frac{e^{-\beta \epsilon_r}}{Z}.$$

The canonical entropy per molecule of the gas is:

$$\frac{S_m}{K_B} = - \sum_r^{cells} \rho(\epsilon_r) \log(\rho(\epsilon_r)). \quad (5.40)$$

A more general statement, valid for any system, reads as

$$\frac{S_m}{K_B} = - \sum_i p_i \log(p_i)), \quad (5.41)$$

where p_i is the probability of finding the system in the microstate i. The reader could wonder why the canonical entropy, which is taken as a definition in Feynman's book *Statistical Mechanics*, is different from the microcanonical result (5.28). The reason is that in the microcanonical calculation, the system is the gas, while here, the system is the molecule. In an isolated system, all the microstates are equally likely, $p_i = \frac{1}{W}$ and (5.41) returns $S = K_B \log(W)$. Below, we shall compute the gas entropy via the canonical partition functions and the result will agree with Thermodynamics again.

The canonical partition function reads as:

$$Z = \sum_r e^{-\beta \epsilon_r}, \tag{5.42}$$

and the canonical mean of any quantity A is:

$$\overline{A} = \frac{1}{Z} \sum_r^{cells} e^{-\beta \epsilon_r} A(r).$$

Problem 15 A *mini-gas* of 10 molecules has 4 states (or cells) available. The cell energies are $0, \epsilon, 2.\epsilon, 3.\epsilon$. The total energy is $E = 14\epsilon$. Let n_k denote the number of molecules in the kth cell of energy $k\epsilon$. Find the most probable value of n_0.

Solution 15 There are 18 possibilities. Below I report the cell occupations:
$((0, 6, 4, 0), (0, 7, 2, 1), (0, 8, 0, 2), (1, 4, 5, 0),$
$(1, 5, 3, 1), (1, 6, 1, 2), (2, 2, 6, 0), (2, 3, 4, 1),$
$(2, 4, 2, 2), (2, 5, 0, 3), (3, 0, 7, 0), (3, 1, 5, 1),$
$(3, 2, 3, 2), (3, 3, 1, 3), (4, 0, 4, 2), (4, 1, 2, 3),$
$(4, 2, 0, 4), (5, 0, 1, 4)).$
The maximum probability occurs for $n_0 = 2$ and for $n_0 = 3$.

5.7.3 Thermodynamic Magnitudes in the Canonical Ensemble

The thermodynamical internal energy U is identified with the average energy

$$U = N\overline{H_S},$$

where the factor N arises from the sum over molecules; however U is obtained from Z:

$$\overline{H} = -\frac{1}{Z}\frac{\partial Z}{\partial \beta} = -\frac{\partial \ln(Z)}{\partial \beta}. \tag{5.43}$$

The perfect gas example suggests to define the entropy per molecule

$$\frac{S_m}{K_B} = - \sum_r^{\text{cells}} \rho(\epsilon_r) \log(\rho(\epsilon_r)).$$

The canonical distribution (5.32) yields

$$\frac{S_m}{K_B} = - \sum_r \rho(\epsilon_r)[-\beta\epsilon_r + \log(Z)] = \frac{U}{NK_BT} + \log(Z). \qquad (5.44)$$

From (5.6) $F = U - TS$, the molecular free energy reads

$$F = -K_BT \ln(Z).$$

Moreover, combining (5.44) with (5.43), we find

$$S_m = -\frac{1}{T}\frac{\partial}{\partial\beta}\log(Z) + K_B \log(Z),$$

and since

$$\frac{\partial}{\partial\beta} = -\frac{1}{K_B\beta^2}\frac{\partial}{\partial T},$$

the molecular entropy is

$$S_m = K_BT\frac{\partial}{\partial T}\log(Z) + K_B \log(Z)$$

or, equivalently,

$$S_m = \frac{\partial}{\partial T}[K_BT \ln(Z)] = -\left(\frac{\partial F(V,T)}{\partial T}\right)_V, \qquad (5.45)$$

in accord with (5.7); recalling (5.8), it is easy to obtain the pressure

$$P = -\left(\frac{\partial F}{\partial V}\right)_T. \qquad (5.46)$$

So, from the partition function all the thermodynamic quantities are readily obtained.

Problem 16 For the monoatomic perfect gas, calculate the partition function Z, the internal energy U, entropy S and pressure P.

Solution 16

$$Z = \frac{1}{h^3} \int d^3p d^3 exp\left[-\frac{p^2}{2mK_BT}\right] = \frac{1}{h^3} V(2\pi mK_BT)^{\frac{3}{2}}.$$

Hence,

$$U = -N\frac{\partial}{\partial \beta} \log Z = \frac{3}{2}NK_BT.$$

From (5.44),

$$\frac{S_m}{K_B} = \log\left(\frac{V(2\pi mK_bT)^{\frac{3}{2}}}{h^3}\right),$$

up to an additive constant.

$$F = -K_BT\log(\frac{V(2\pi mK_BT)^{\frac{3}{2}}}{h^3}.$$

Finally, from (5.46),

$$P = \frac{K_BT}{V}.$$

5.7.4 Theorem of Equipartition

Consider the harmonic oscillator

$$H(p,q) = \frac{p^2}{2m} + V(q), \quad V(q) = \frac{1}{2}m\omega^2 q^2.$$

Since

$$V(q) = \frac{1}{2}q\frac{\partial H}{\partial q},$$

the thermal average of V at temperature T is given by

$$\overline{V} = \frac{1}{2Z}\int dqdp q\frac{\partial H}{\partial q}e^{-\beta H} = \frac{1}{2Z}\int dqdp\frac{q}{(-\beta)}\frac{\partial}{\partial q}e^{-\beta H};$$

and by an integration by parts one finds

$$\overline{V} = \frac{-K_BT}{2Z}\int dqdp\left\{\frac{\partial}{\partial q}(qe^{-\beta H}) - e^{-\beta H}\right\}.$$

This can be simplified, since $H \to \infty$ for $q \to \pm\infty$ and the first contribution vanishes; we are left with

$$\overline{V} = \frac{K_BT}{2}.$$

Similarly, one can obtain:

$$\overline{T} = \frac{K_BT}{2}.$$

The traditional statement is that for any system of oscillators, there is a contribution $\frac{K_B T}{2}$ for each degree of freedom, i.e. from each q and each p. Systems of oscillators include the vibrations of molecules and solids and the normal modes of the electromagnetic field in a cavity. The failure of this theorem for the electromagnetic field is called ultraviolet catastrophe, and it led Planck to introduce the quanta (see below). But here, we go ahead with the classical theory.

Example: Monoatomic Perfect Gas

Since

$$H = \sum_i^N \frac{p_{xi}^2 + p_{yi}^2 + p_{zi}^2}{2m},$$

we rederive the internal energy is

$$U = \frac{3}{2} N K_B T$$

and the specific heat at constant volume $C_V = (\frac{\partial U}{\partial T})_V = \frac{3}{2} N K_B$. For one mole of gas, $N \sim 6.022 \times 10^{23}$ is the Avogadro number and $C_V = \frac{3}{2} R$, where $R = N K_B$ is the gas constant. These well known results are all wrong at low temperatures: in particular, the specific heats vanish for $T \to 0$ for the third principle of Thermodynamics.

Example: Biatomic Gas

For biatomic molecules, if one considers only the translational degrees of freedom, the result is $C_V = \frac{3}{2} N K_B$ as above. But one should add to the kinetic energy the rotational Hamiltonian, which can be roughly represented by a rigid rotator

$$H_{\rm rot} = \frac{1}{2I} \left(P_\theta^2 + \frac{1}{\sin^2 \theta} P_\phi^2 \right).$$

Due to the quadratic form, $H_{\rm rot}$ adds another R per mole. This is not all. The rotator is not rigid, therefore one should add a (roughly) harmonic potential and another R. Experimentally, e.g., for H_2, well above the boiling point of liquid Hydrogen, $C_V = \frac{3}{2} N K_B$ and the only degrees of freedom that contribute are translational, while the others are frozen. They start to contribute at higher temperatures. The explanation is possible only in the quantum theory.

Dulong and Petit Law

Einstein modeled the vibrations of an elemental solid as an harmonic oscillator associated to each of the N atoms. Thus, H is the sum of N oscillator terms and classically $C_V = 3R$. This agrees with a law by Dulong e Petit, which is about right around room temperature. However, at low temperatures, C_V drops, and this can be understood by Quantum Mechanics. Actually, Einstein had proposed a qualitatively correct explanation in terms of vibration quanta already in 1907 (see Chap. 25).

Black Body and Ultraviolet Catastrophe

The spectrum, that is the dependence of energy density on frequency $u(\nu)$, was measured through its link (5.15) with the emissive power. At each temperature, the data show a bell-shaped $u(\nu)$ with a characteristic asymmetry that grows from 0 to a maximum and then decreases to zero more slowly at high frequencies. The maximum shifts to a higher frequency with increasing the temperature T and the maximum frequency is proportional to T (Wien's law). Rayleigh and Jeans, in 1900, treated the problem as follows. In a large cubic cavity of volume V the number of right polarized modes of the electromagnetic field with wave vector in d^3k is

$$d\mathcal{N} = \frac{V d^3k}{(2\pi)^3};$$

this must be multiplied by 2 to include left polarization. Integrating over the angles of $\vec{k}$ s by $\int_\Omega d^3k = 4\pi k^2 dk$ and setting $|k| = \frac{\omega}{c}$, one finds

$$\int_\Omega d^3k = \frac{4\pi\omega^2 d\omega}{c^3};$$

finally,

$$d\mathcal{N} = 2\frac{V k^2 dk}{\pi^2} = V\frac{\omega^2 d\omega}{\pi^2 c^3}.$$

Each mode is a harmonic oscillator and has energy $K_B T$ according to the equipartition theorem. The energy per unit volume is:

$$u(\omega)d\omega = \frac{K_B T \omega^2}{\pi^2 c^3}d\omega. \tag{5.47}$$

This result is in excellent agreement with experiments in the region of low frequencies (well below the maximum), and this represents a great success of the above theory. Alas, when ω increases, things go so disastrously that each physicist refers to this problem as the *the ultraviolet catastrophe*. The calculation of the number of modes is simple geometry and cannot be wrong. It is the equipartition theorem that would cause the absurd result $\int_0^\infty u(\omega)d\omega = \infty$, and each body should emit infinite energy. This can be traced back to the fact that the Boltzmann statistics does not apply to the states of the field and those of energy above $K_B T$ are effectively quenched. Again, the problem is deep and the solution requires Quantum Mechanics, which explains in detail the law of Wien and fixes the constant of the Stefan–Boltzmann law.

5.8   Grand-Canonical Ensemble

In most problems, a more general ensemble is preferable. It is the Grand-canonical Ensemble in which the system S can exchange with the bath B besides energy, even particles. Instead of fixing the energy U = E as in the microcanonical ensemble, we fix the temperature T as in the canonical one; instead of fixing the number N of particles as one does in both the microcanonical and canonical scheme, one keeps the chemical potential μ fixed. Now all N values are possible, and the G space becomes the union of all those who describe the system with N defined. It is this monster space that must be divided into cells. In the search for the most probable distribution, we must to maximize W at constant N, and this requires a new Lagrange multiplier that turns out to be μ. Ultimately, the formalism works as in the canonical scheme, but with the Hamiltonian $H(p, q)$ replaced with $H(p, q) - \mu N$. Using the discrete notation, the grand canonical partition function is

$$Z_G = \sum_i e^{-\beta[\epsilon_i - \mu N]}.$$

The grand canonical average of any A becomes

$$\bar{A} = \frac{1}{Z_G} \sum e^{-\beta[\epsilon_i - \mu N]} A(i).$$

The internal energy is given by

$$U = -\frac{\partial}{\partial \beta} \log Z_G$$

in analogy with the canonical case. The only new quantity is the average number of molecules,

$$N = \frac{\partial}{\partial \mu \beta} \log Z_G. \tag{5.48}$$

5.8.1   Monte Carlo Methods

Many activities require a draw or the intervention of chance. Truly random numbers can be generated experimentally, using, for example, the nuclear decay of radioactive substances, but it is usually preferable to generate pseudo-random sequences (that are deterministic but are started by a seed, as the clock of a computer, whose results are practically unpredictable. The most commonly used is the one by Lehmer. One starts by choosing three large secret integers a, b, m. Let the integer r1 be the random seed. More random numbers are then generated by the formula

$$ri + 1 = (ari + b) \qquad\qquad mod \quad m.$$

That is, $ri + 1$ is the remainder of $(ari + b)/m$; in this way, the ration ri/m are pseudo-random numbers between 0 and 1.

Apart from lotteries and some situations in which a draw is prescribed by laws, a random number generation has important applications in Science, often dealing with the calculation of integrals. Given a square of side a we can inscribe in it a circle of radius a/2. Suppose one uses a random number generator to produce pairs (x, y) with $-a < x < a$, $-a < y < a$. The dot with coordinates (x, y) is inside the circle with probability $\pi/4$. This is actually one way to approximate π. In a one-dimensional integral $\int_0^L dxf(x)$, the error due to a finite mesh goes like $\epsilon \sim \frac{\Delta L}{L}$. With $n \sim \frac{1}{\epsilon}$ points the integral starts to converge, unless f has narrow structures. In d dimensions, the number of times one must compute f goes like $n \sim \epsilon^{\frac{1}{d}}$. If $d = 100$ and we want $\epsilon < \frac{1}{1000}$, we need 10^{300} computations of f. Assuming that one computation of f takes 10^{-6} s, the calculation lasts about 10^{294} s. Comparing this figure with the age of the universe, which is less than 10^{18} s, we may conclude that the waiting time is excessive and a faster method must be sought.

Actually, the variance of the Monte Carlo method decreases like $\frac{1}{\sqrt{N}}$ with increasing the number N of points, regardless of d. Moreover, if one knows which regions of the integration field contribute most, one can improve the convergence by an *importance sampling*. Molecular dynamics simulations can be based on the numerical solution of the equations of motion of samples of molecules; the samples are prepared using a number of realizations of the phenomenon under study, weighted by the probability of occurrence (the Metropolis method). In this way, one can compute mean values of the thermodynamic potentials.

The Monte Carlo approach is also applied with success to quantum mechanical simulations. The Car–Parrinello method is often used for this purpose. It is based on quantum mechanical *ab initio calculations* using the Density Functional approach, which is an efficient method to find the ground state density and energy in many-electron systems. Due to its computational complexity, it is applied on model systems that are much smaller than those that can be considered in classical molecular simulations. Car–Parrinello calculations have also been applied successfully to systems of interest for biology.

Chapter 6
Special Relativity

Everyone is entitled to use their own measurements to describe physical reality, and the descriptions of observers who are in different inertial frames are related by simple transformation laws. This was well understood by Galileo, but the finite speed of light required Einstein to fit electromagnetism into the scheme, with paradoxical and astounding consequences.

6.1 Galileo's Ship

Galileo stated the *Principle of Relativity* using a series of thought experiments to be performed inside a ship. This was revolutionary in a time dominated by Aristotelian misconceptions, when the common wisdom was that there is an absolute rest system and a force is needed to keep objects moving.[1] The worst difficulty was that the 2000-years-old teaching of Aristotle was upheld by the Church.

It was the first time that thought experiments were introduced and their power demonstrated, and in many ways, that was the beginning of Science in the modern conception of the term. Galileo's *Dialogo dei Massimi Sistemi* is also a masterpiece of Italian Literature, written in the colorful but still perfectly clear Italian of 1600. He demonstrated that no experiment conducted inside the ship without looking outside could reveal the state of motion of the ship, as long as it is uniform. Let me summarize three novel principles that were clearly demonstrated by Galileo in that work and in other publications:

1. There are *inertial frame of reference*, such that a free body moves along a straight line with constant speed once the friction is removed;
2. Given an inertial frame, all the frames that move with constant speed in it (without rotations) are also inertial;
3. No experiment can distinguish one inertial frame from another, and the laws of Physics are the same for all.

---

[1]However, the times were ripe and some of the works by Giordano Bruno (1548–1600) imply correct ideas about the relativity of motion.

© Springer International Publishing AG, part of Springer Nature 2018
M. Cini, *Elements of Classical and Quantum Physics*,
UNITEXT for Physics, https://doi.org/10.1007/978-3-319-71330-4_6

Remarkably, he arrived at that without ever having access to any even approximately
inertial system (apart from a freely falling one), since the ship actually is in a field
of gravity. Let us discuss (in modern terms) the implications.

Consider two inertial references $\mathcal{K}$ and $\mathcal{K}'$ and a material point P; let the coordinates of P be $\vec{r} = (x, y, z)$ in $\mathcal{K}$ and $\vec{r}' = (x', y', z')$ in $\mathcal{K}'$. If the origin of $\mathcal{K}'$
travels in $\mathcal{K}$ with speed $\vec{u}$, $\vec{r}'$ is related to $\vec{r}$ by

$$\vec{r}' = \vec{r} - \vec{u}\,t. \qquad (6.1)$$

The time is just a parameter, the same in both references, and Newton's equations
are trivially the same; a change of variables in the Lagrangian makes the change.
This is the Galilean Relativity. In the nineteenth century everyone believed in Classical Mechanics, but the crisis arrived with Maxwell's equations, which predict that
light moves with velocity c. Velocity relative to what reference? For a long time, it
was thought that the privileged reference in which Maxwell's equations hold must
be sought by seeking variations of the speed of light with direction. A celebrated
experiment used the Michelson interferometer of Fig. 6.1 but failed to find any effect.
Nowadays, interferometric experiments using laser light measure the speed of light to
a part in 10^9 or better, and any discrepancy would be a celebrated discovery, yet none
has been reported. So we start to discuss Einstein's Special Relativity, which is based

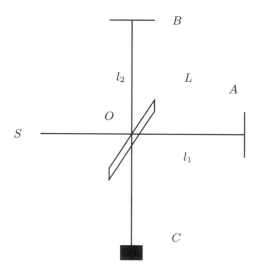

Fig. 6.1 Scheme of the Michelson interferometer used to seek a dependence of c on the direction.
Light coming from a source S is split in O by a half-silvered mirror L making an angle of 45°; the
rays OA and OB are reflected by mirrors that send them back to L. Here, the ray BO is split in OC
and OS, while AO is split in OS and OC. In C there is an eyepiece, and one can see the interference
between the rays SOAOC and SOBOC. Any difference of the light speed in the two arms should
produce a phase shift. No dependence on the direction was found, despite the fact that the Earth
travels at 29 km/s around the Sun

on the simple assumption that Galileo's Relativity is not restricted to Mechanics, but all the Laws of Physics must hold in the same form in any inertial system. In General Relativity, all reference systems are included, but this is deferred to Chap. 8.

6.2 From Thought Experiments to the Lorentz Transformation

Einstein[2] started from the idea that the Maxwell equations were right and the laws of Electromagnetism must hold in any inertial system. In particular, any measurement of c yields the same c, with no dependence on the speed of the source of the light and no privileged reference system. Einstein in his Special Relativity Principle made the stronger statement that all the laws of Physics must have the same form in all inertial reference systems. This was in line with Galileo, but the Principle of General Relativity states that all the laws of Physics must have the same form in all reference frames, including the accelerated ones. In this chapter we develop the special theory.

An event, like the emission or the absorption of a light signal at some point in some instant, is something real, something that will receive different space-time coordinates by different observers, but is something objective in itself. All phenomena consist of a succession of events, each characterized in a given frame by 4 coordinates $(\overrightarrow{x}, t)$. We must give precise meanings to these coordinates. We need to examine critically the physical (i.e., operational) meaning of the notions of space separation and time interval between two events. The theory must refer to real facts and measurable quantities that likewise will be transformable from one reference to the other. As we shall see shortly, all such objects will be described mathematically as tensors.

Meters, Clocks and Frames

By a meter, one can measure the distance between two points provided that they are fixed; a clock measures the time between two events occurring in proximity to the clock. For the rest, an observer is assumed to set a Cartesian reference and to be able to set meters and clocks wherever they are needed. Moreover all the clocks must be synchronized. This can be achieved by sending a signal from the origin; each clock is set at the time $t = \frac{r}{c}$, where r is the distance from the origin. Then, the observer is ready to measure the space-time coordinates (x, y, z, t) of any event. The length of moving rod is measured as the distance between two simultaneous events in which the ends are observed at the same t.

Some quantities are invariant in a trivial way since they are defined in terms of measurements done in special reference systems. A stationary rod can be measured

---

[2] Albert Einstein (Ulm 1879 - Princeton 1955) is so famous that it is hardly necessary to mention his merits. Definitely, the most important physicist since the time of Newton, besides inventing and developing Special and General Relativity he gave a formidable contribution to Quantum Mechanics as well. Being of Jewish origin, he was forced to flee from Germany already in 1933. He spent the rest of his life in Princeton (USA).

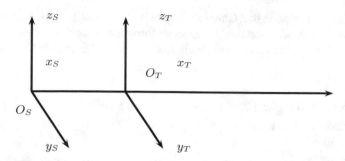

Fig. 6.2 The thought experiments comparing the results of measurements performed on the train with those performed in the station. The train reference $\mathcal{K}_T (O_T x_T y_T z_T)$ travels with speed u along the x axis relative to the station reference $\mathcal{K}_S(O_S x_S y_S z_S)$; the train crosses the station at time $t_T = t_S = 0$

using a meter; such is a *proper length*, since all observers rely on the measurement done in a particular reference. Likewise, when two events in a given system occur at the same point, the time separation is a *proper time*. In the same way, the *rest mass* of a body is measured in a reference where it is motionless.

We wish to give a meaning to the length of moving rods (which cannot be measured by simply using a meter) and to the time interval between events that occur at distant locations (although a clock is not enough in this case). The meaning of such quantities is fully defined when we agree on the way in which they can be measured (Fig. 6.2).

Instead of the ship used by Galileo, we shall use the example of a train that moves with constant speed $\vec{u}$ relative to a station. The train represents the reference (x_T, y_T, x_T, t_T), and the station (x_S, y_S, z_S, t_S).

The train driver and the station manager agree to make some measurements and compare the results. They agree on the relative velocity, although each of them pretends to be motionless while the other is moving. Often they disagree on the results of the measurements. Each observer will interpret the experiments based solely on his own measurements. The relation between the results of both observers is fixed by the Principle of Relativity. The station manager can measure the distance between the rods of the rail using a meter. This is a proper length for him. But the train driver is also able to do the same, with his own meter, because although the iron of the rods is moving, the rods always appear to be the same for him. The conclusion is that the lengths in transverse directions (orthogonal to the velocity) are the same for both.

Relativity of Simultaneity

The train driver switches a lamp at O and measures the time the light takes to reach mirrors A and B placed on the train at equal distances from O. To do so, he fixes a clock close to each mirror. The train driver finds that the light hits both mirrors at the same time (Fig. 6.3). The two hits are simultaneous events, taking place at different places but at the same time.

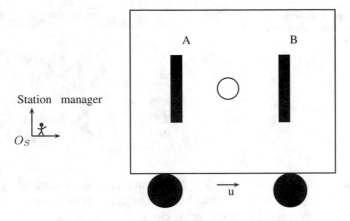

Fig. 6.3 Experiment to show the Relativity of Simultaneity

The line AOB is parallel to the velocity of the train and B is ahead. If the station manager wants to measure the arrival times at A and B, the experimental set-up must include two clocks fixed to the ground and close to the mirrors A and B at the times when each receives the light, since the clocks must be in the reference of the station. This arrangement is more complicated than the measurement on the train, but can be done. He finds that the light hits A first. This is in line with the fact that the speed of light is c, since during the propagation of the light, A approaches the point from which the light was emitted while B goes farther. Who is right? Both! In fact, an experiment is always right! Moreover, the simultaneity becomes absolute if the A-B distance vanishes. Finally, if the train driver sets mirrors at the same distance from O in a direction orthogonal to the rail, then the station manager agrees with him that the light reaches both mirrors at the same time; for both observers, the setup is symmetrical.

Experiment on the Relativity of Time

The train driver installs a lamp fixed to the ceiling at height h over the ground. When he switches the light on, how long does it take before it reaches the ground?

$$t_T = \frac{h}{c},$$

of course. According to the station manager, however, the light must cover the distance $\sqrt{h^2 + (ut_s)^2}$, where u is the speed of the train; hence, the time is given by $t_s = \frac{\sqrt{h^2+(ut_s)^2}}{c}$. This implies

$$t_s = \frac{h}{c} \frac{1}{\sqrt{1 - \left(\frac{u}{c}\right)^2}}. \tag{6.2}$$

Fig. 6.4 Arrangement of the thought experiment on the relativity of lengths

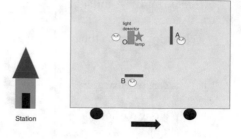

For the station manager, the time is longer, because the distance covered by the light is longer. Actually, the time measured by the train driver is the shortest possible. It is a proper time, measured solely by using clocks.

Relativity of Lengths

In the next experiment (Fig. 6.4) the train driver puts a lamp, a light detector and a clock in a proper position, where he sets the origin of the coordinates on the train, a mirror A with a detector and a clock at a distance L from the lamp along the x axis parallel to the rail and a second mirror B with a detector and a clock at a distance L from the lamp along the y axis. At time $t = 0$, the lamp emits a flash of light; the light reaches both mirrors simultaneously at time $t_1 = \frac{L}{c}$ and both reflected signals return to D_0 at time $t_2 = \frac{2L}{c}$. The simultaneity of the two returns to the origin is absolute. The experiment can be arranged such that the first flash starts from the origin at time 0 also in the reference of the station. The reflection from A has space-time coordinates (x_T, y_T, z_T, t_T) with $t_T = (l, 0, 0, \frac{L}{c})$ and the reflection from B coordinates $(0, L, 0, \frac{L}{c})$; at time $(t_{OBO})_T = (t_{OAO})_T = 2\frac{L}{c}$, the two flashes return simultaneously in O. All observers must agree on the simultaneity, since both events take place at the same point.

The station manager agrees that mirror B is at $y = L$ since this is a transverse length, but considers this experiment a way to measure the length of OA not by observing O and A simultaneously but by observing them at different times. Since OB moves with velocity u, the ray OBO becomes the two equal sides of a triangle having as its basis the length of the distance covered by O during the time τ_{OBO}. Letting O' and O'' denote the initial and final positions of O, $O'O'' = u\tau_{OBO}$, and the length covered by the ray is

$$O'BO'' = 2\sqrt{l_2^2 + \left(\frac{u\tau_{OBO}}{2}\right)^2} = \sqrt{4l_2^2 + (u\tau_{OBO})^2}.$$

The time measured in the reference of the station is longer, since the light makes a longer trip; actually, $\tau_{OBO} = \frac{2l_y}{c}\frac{1}{\sqrt{1-\frac{u^2}{c^2}}}$, but this is just the time dilation (6.2) noted above. On the other hand, the station manager cannot agree on the length of OA. Let us find $L_S = L_{OA}$ in the reference of the station. During the time τ_{OA} of the

OA trip, the mirror A initially at L_{OA} has moved forward by a distance $u\tau_{OA}$. The actual distance covered is $L_{OA} + u\tau_{OA}$. Since $\frac{L_{OA}+u\tau_{OA}}{\tau_{OA}} = c, \tau_{OA} = \frac{L}{c-u}$. Similarly, the reverse trip takes the time $\tau_{AO} = \frac{L}{c+u}$. To sum up, the time taken to go to A and back is

$$\tau_{OAO} = \frac{2L}{c}\frac{1}{1-\frac{u^2}{c^2}}.$$

The trip along x should take longer if the distance were the same. Since it is an absolute fact that $\tau_{OAO} = \tau_{OBO}$, it follows that

$$L_S = L\sqrt{1-\frac{u^2}{c^2}}. \tag{6.3}$$

The proper length (measured by a meter at rest) is the longest. A shortening by 1% requires $u = 0.15c$, and the effect is too small to be noted in everyday life. Of course, the descriptions of both observers are equally right.

Note that the shortening of moving bars and the time dilation are *real* effects. Such are the results of any measurement. The time dilation

$$\tau = \frac{\tau_0}{\sqrt{1-\frac{u^2}{c^2}}}.$$

where τ represents a proper time, is the reason why a relativistic muon lives much longer in the laboratory than in the rest system.

But one should not think that "in reality the experiment of the train driver lasts $2\frac{L}{c}$ and it *appears* to longer to the station manager." There is no privileged reference, and the duration is really longer in the station; for instance, the station manager has a longer time to intervene in the experiment.

The station manager sees the clock in the train showing the time τ_0 while the station clock shows τ, and concludes that the train clock runs slow. Many people are tempted to think that the train driver then must see that the station clock runs too fast. By the principle of Relativity the train driver must see the station clock going slow. He says: I am motionless, while the station runs with a speed $-u$, so there is a time dilation by a factor of $\sqrt{1-\frac{u^2}{c^2}}$. Is that contradictory? Not at all. Both observers in 1 second can see what the other does in $\sqrt{1-\frac{u^2}{c^2}}$ seconds.

It is time to define the often-used symbol

$$\gamma = \frac{1}{\sqrt{1-\frac{u^2}{c^2}}} \tag{6.4}$$

and express the time dilation by $\tau = \tau_0\gamma$.

It is common practice also to introduce also the symbol $\beta = \frac{u}{c}$.

The Famous Twin Paradox

Identical twins, Anthony and Bartholomew, were born in an inertial system S. Bartholomew starts at time t_1 on a spacecraft, reaches a star 12 light years away traveling at 0.6c and soon returns home with the same speed. Anthony stays at home and celebrates the return of his brother at time t_2 after $t_2 - t_1 = 2 \times \frac{12}{0.6} = 40$ years. Meanwhile Bartholomew is aged $t_2' - t_1$ prime $= \int_{t_1}^{t_2} dt \sqrt{1 - (0.6)^2} = 40 \times 0.8 = 32$ years, and is younger than his twin.

Bartholomew says: *I never traveled, it was Anthony with all the Earth who has taken a curved path with speed 0.6c. Also, during the trip, I have often observed that terrestrial clocks were slow. Really, then I would have expected to find Anthony younger than myself.* This is the paradox. The resolution is that Bartholomew must have felt inertial forces during acceleration at the start, when he bent to return home and during the braking at arrival. So he has no right to consider himself stationary in an inertial system. The frame of Antony, instead, was inertial (or rather, almost inertial, because of gravity).

This experiment was conducted in 1972 comparing two Cesium clocks with each other, one of which had toured the world in an airplane. Despite the smallness of the velocity compared to c, the relativistic effect was found as Einstein had predicted.

Relativity of Synchronization

Two clocks at rest in the same inertial reference can be synchronized in the following way. The distance L is a proper distance and is measured by using a meter, after which one of the clocks, say clock A, emits a signal at time t_A and the other is set at time $t_A + \frac{L}{c}$ when the signal arrives. The observers in different inertial systems, however, do not agree that the clocks are synchronized. An observer who travels on a spacecraft from A to B with speed u relative to the clocks and is at A at time t_A finds that the distance is actually $\frac{L}{\gamma}$, so he reaches B at the time $t_A + \frac{L}{u\gamma}$; this is his proper time. Instead, clock B indicates $t_A + \frac{L}{u}$. So, in the system of the spacecraft the clocks are not synchronized.[3] The synchronization is relative to the reference.

The Lorentz Transformation

We were able to deduce some of the most characteristic effects of Special Relativity through thought experiments, using only the fact that c is the same in all the inertial systems. To make progress, we must derive the Lorentz transformation. This can be done in a formal, abstract way, but we are ready to determine the result very simply through a new thought experiment, again with the help of the station manager and the train driver. We arrange that their origins O_S and O_T coincide at time $t_S = t_T = 0$. The station manager prepares a traffic light (which is fixed in the station reference

---

[3] If the astronaut is unaware of Relativity, he will find a still stronger disagreement. He notes that the clocks in the Earth system go slow by a factor of $\frac{1}{\gamma}$; so, if he keeps the length $\frac{L}{\gamma}$ that he measures as the true length, then he expects that on the Earth, the elapsed time should be $\frac{L}{u\gamma^2}$, while in reality it is $\frac{L}{u}$, as we know.

$\mathcal{K}_S$) at position x_S along the rail; a timer turns the light on at some station time t_S. This is an event E with coordinates (x_S, t_S).

In the reference $\mathcal{K}_T$ running at speed u, it is observed that the event E has coordinates $(x_T^{(E)}, y_T^{(E)}, z_T^{(E)}, t_T^{(E)})$. As we know, the y and z coordinates are the same for both references and we can forget about them. The train driver measures the distance $(O_S E)_T$ of E from the station at in O_S. The distance along the x axis is of E from O_T is just $x_T^{(E)}$. In order to find the distance from O_S, he adds the distance $O_S O_T = u t_T$ from the station, still measured on the train by a meter and a clock; the result is evidently

$$(O_S E)_T = x_T^{(E)} + u t_T^{(E)}.$$

Instead, the station manager measures $x_S = (O_S E)_S$ by a meter, as a proper length. Then, taking into account the relativistic contraction,

$$(O_S E)_T = x_S \sqrt{1 - \left(\frac{u}{c}\right)^2} \equiv \frac{x_S}{\gamma};$$

therefore, (understanding $^{(E)}$),

$$x_S = \gamma[x_T + u t_T]. \tag{6.5}$$

By the principle of Relativity, it follows that

$$x_T = \frac{x_S - u t_S}{\sqrt{1 - \frac{u^2}{c^2}}} \equiv \gamma[x_S - u t_S]. \tag{6.6}$$

Eliminating x_T from Eqs. (10.14) and (6.6), one obtains $x_S = \gamma^2(x_s - u t_S) + \gamma u t_T$; but $1 - \gamma^2 = -\beta^2 \gamma^2$. In this way one can find t_T in terms of measurements done by the station manager:

$$t_T = \gamma[t_S - \frac{u}{c^2} x_S]. \tag{6.7}$$

Having completed the Lorentz transformation, (6.6) and (6.7) we realize that it is symmetric in x and ct. The Galileo transformation (6.1) is the limit for $c \to \infty$.

The inverse transformation is

$$x_S = \gamma[x_T + u t_T], \qquad c t_S = \gamma\left[c t_T + \frac{u x_T}{c}\right]. \tag{6.8}$$

with $y_S = y_T$, $z_S = z_T$. Usually, one replaces the time with a zeroth space dimension $x_0 = ct$ and the velocity by the pure number $\beta = \frac{u}{c}$; then the transformation becomes

$$\begin{cases} x_T = \gamma(x_S - \beta x_{0S}), \\ x_{0T} = \gamma(x_{0S} - \beta x_S), \end{cases} \tag{6.9}$$

still with $y_T = y_S$ and $z_T = z_S$. More generally, the velocity is not parallel to the position vector (that, is, the station is not along the railway). Then we may decompose the position vector on the train by $r_T = r_\perp + r_\parallel$ and the transformation becomes

$$r_S = (r_\perp)_T + \gamma((r_\parallel)_T + vt_T)$$

$$ct_S = \gamma \left(ct_T + \frac{r_T.u}{c} \right);$$

$$\begin{cases} r_S = (r_\perp)_T + \gamma((r_\parallel)_T + vt_T), \\ ct_S = \gamma \left(ct_T + \frac{r_T.u}{c} \right); \end{cases} \tag{6.10}$$

the inverse transformation is done by exchanging the frames and reversing the velocity.

Check that the Lorentz Transformation Yields the Contraction

The Lorentz transformation leads to the phenomenon of the Lorentz contraction, whereby a bar moving parallel to its length is shorter than its rest length. Suppose the station manager wishes to measure the length of a bar on the train parallel to the velocity of the train. The train driver can measure L using a meter. If one uses (6.8) for the ends of the bar setting the same t_T for both, the result is that the bar length increases by a factor γ. This is wrong, because the positions of both ends A and B must be taken simultaneously in the station, not in the train. In order to find the correct condition, we apply (6.9) to both ends of the bar, then write the difference, $\Delta x_T = \gamma \Delta x_S - \beta \Delta x_{0S}$, $\Delta x_{0T} = \gamma(\Delta x_S - \Delta x_{0S})$; now we set $\Delta x_{0S} = 0$. Since $\Delta x_S = \gamma \Delta x_S$, one finds that $L_T = \gamma L_S$. This is the contraction $x_B^{(S)} - x_A^{(S)} = L\sqrt{1 - \frac{u^2}{c^2}}$.

Problem 17 The space-time coordinates of two events as measured by an observer O are $x_1 = 6 * 10^4$ m, $y_1 = z_1 = 0$, $t_1 = 2 * 10^{-4}$ s, $x_2 = 12 * 10^4$ m, $y_2 = z_2 = 0$, $t_2 = 1 * 10^{-4}$ s. The observer O' travels with a speed v relative to O along the x axis, and finds that the two events are simultaneous. Find v using $c = 3 \times 10^8 \frac{m}{s}$.

Solution 17 Since $ct_1 - \frac{u}{c}x_1 = ct_2 - \frac{u}{c}x_2$, one finds that $\frac{v}{c} = -\frac{1}{2}$.

Problem 18 The rocket O' travels with speed v relative to a space station O. The astronaut looks at the station through a 4 m porthole and says that the size of the porthole is the same as the size of the window in the station. The station manager using a meter finds that the window is 5 m long. Evaluate v.

Solution 18 From $5 = 4\gamma$ one finds $\frac{v}{c} = \frac{3}{5}$.

6.2.1 Relativistic Addition of Velocities

The train driver observes a body moving with a speed

$$W = \frac{dx_T}{dt_T}.$$

For the station manager, the speed is $V = \frac{dx}{dt}$, where $dx = \gamma[dx_T + udt_T]$, $cdt = \gamma[\frac{u}{c}dx_T + cdt_T]$. So,

$$V = \frac{W + u}{1 + \frac{uW}{c^2}}.$$

For W and u less than c, V is always $< c$.

Problem 19 A rocket R flies away from Earth E along a straight line, while a UFO is seen from E and from R. Both objects proceed in the same direction. An observer on E finds that the UFO is going at $0.5c$, while the pilot of the rocket finds that it goes at $-0.5c$. What is the speed of R relative to E?

Solution 19 $\frac{v}{c} = 0.8$.

Problem 20 In the reference (x_S, t_S), the station manager sees a train and a rocket travelling in the same direction (say, the x direction), and finds that the speed of the train is u. The train driver in his reference (x_T, t_T) finds that the speed of the rocket is v. Write the Lorentz transformations from the station to the train and from the train to the rocket. Then, eliminating (x_T, t_T) obtain the transformation from the station to the rocket. Show that this is of the Lorentz form and verify the theorem of the sum of velocities.

Solution 20 The required transformation reads as:

$$x_S = \gamma(u)\gamma(v)[(1 + uv)x_R + (u + v)t_R], \; t_S = \gamma(u)\gamma(v)[(1 + uv)t_R + (u + v)x_R].$$

This is a Lorentz transformation provided that $\gamma(u)\gamma(v)(1 + uv) = \gamma(w) = \frac{1}{\sqrt{1-w^2}}$, where w is the speed of the rocket in the reference of the station. Solving, one finds that $w = \frac{u+v}{1+uv}$.

6.3 The Geometry of Special Relativity: Minkowsky Chronotope

The Lorentz transformation mixes time and space coordinates. The transverse lengths are conserved, but the lengths along the velocity are not, so distances depend on the reference. However,

$$c^2t_T^2 - x_T^2 = \frac{c^2t^2 - 2uxt + \frac{u^2x^2}{c^2}}{1 - \frac{u^2}{c^2}} - \frac{x^2 - 2uxt + u^2t^2}{1 - \frac{u^2}{c^2}} = c^2t^2 - x^2.$$

The invariant quantity is the *interval* s, defined[4] by its square

$$s^2 = r^2 - c^2t^2. \tag{6.11}$$

The interval between two events, unlike a distance in a four-dimensional Euclidean space $\mathbb{R}^4$, can be positive, zero or negative; a vanishing interval $s^2 = 0$ is called *light-like*, because the light can reach one from the other. When $s^2 < 0$, the interval is *time-like* and imaginary, while if $s^2 > 0$ the interval is *space-like*. Space-time is a four-dimensional space, the Chronotope, whose coordinates $x_0 = ct, x_1 = x, x_2 = y, x_3 = z$ are lengths. A point in the Chronotope is called an event. In a four-dimensional Cartesian space $\mathbb{R}^4$, any linear transformation that leaves square distances $\sum_{i=1}^{4} x_i^2$ invariant is a rotation (improper, if it involves a reflection). In the Chronotope instead we can make Lorentz transformations with invariant s^2. Such a space is called pseudo-Euclidean. In $\mathbb{R}^4$, when the distance between two points vanishes, the two points coincide, while a light-like interval means that a signal can travel from one event to the other.

We can keep the Euclidean sum of the squares formula if we are ready to use the formal trick of replacing x_0 with $x_4 = ix_0 = ict$. The Minkowsky Chronotope is the pseudo-Euclidean space where a point has coordinates:

$$x_1 = x, x_2 = y, x_3 = z, x_4 = ict.$$

The transformation becomes

$$\begin{cases} x_1' = \gamma(x_1 + i\beta x_4), \\ x_2' = x_2, \\ x_3' = x_3, \\ x_4' = \gamma(x_4 - i\beta x_1). \end{cases} \tag{6.12}$$

Any object $v = (v_1, v_2, v_3, v_4)$ with 4 components that Lorentz-transform like x_1, x_2, x_3, x_4 is called a four-vector. We shall find several physical observables that behave as four-vectors. For instance, consider a particle moving with speed $v = \frac{dx}{dt}$ in some reference. The derivative $\frac{dx_\mu}{dt}$ has 4 components, but is no four-vector, because dt depends on the reference. One obtains a four-vector by differentiating the position four-vector x_μ with respect to invariant proper time of the particle[5] $d\tau = \frac{ds}{c}$. One finds the velocity four-vector $v_\mu = \frac{\delta x_\mu}{d\tau}$ with components $v_1 = \frac{dx}{d\tau}, v_2 = \frac{dy}{d\tau}, x_3 = \frac{dz}{d\tau}, i\frac{dt}{d\tau}$.

---

[4]About half of the authors define it as $s^2 = -r^2 + c^2t^2$.

[5]At first sight, it may seem odd that the components x_α measured in the system of the observer are differentiated with respect to the time measured in the reference of the moving body, but actually, $cd\tau = ds$ is an interval and does not depend on the reference system.

This is called four-velocity. Since $\frac{dt}{d\tau} = \gamma$ with $\gamma = \frac{1}{\sqrt{1-\frac{v^2}{c^2}}}$,

$$v_\mu = \gamma(\boldsymbol{v}, c),$$ (6.13)

where $\boldsymbol{v}$ is the Galilean speed.

The components are conventionally denoted by a Greek subscript, like x_α, with $\alpha = 1, 2, 3, 4$. The latin indices, like i in x_i, $i = 1, 2, 3$, are used for the first 3 (space) components of the four-vector.

The special transformation (6.12) can be put in the matrix form

$$x'_\mu = \sum_{\nu=1}^{4} \Lambda_{\mu\nu} x_\nu,$$ (6.14)

with

$$\Lambda = \begin{pmatrix} \gamma & 0 & 0 & i\gamma\beta \\ 0 & 1 & 0 & 0 \\ 0 & 0 & 1 & 0 \\ -i\beta\gamma & 0 & 0 & \gamma \end{pmatrix};$$ (6.15)

the transposed matrix has β replaced by $-\beta$ and is clearly Λ^{-1}.

Next, note that the Λ matrix has the property that all its columns are orthogonal vectors, that is, $\Lambda_{\mu\nu}\Lambda_{\mu\rho} = \delta_{\nu\rho}$. Hence,

$$\sum_\mu x'_\mu x'_\mu = \sum_{\mu,\nu} \Lambda_{\mu\nu}\Lambda_{\mu\chi} x_\nu x_\chi = \sum_\nu x_\nu x_\nu.$$

Since such sums occur quite often, Einstein introduced the convention of understanding the summation symbol when the same index occur twice: so, one simply writes $s^2 = x_\nu x_\nu$.

Let v_μ and w_μ be four-vectors; the scalar product is $v_\mu w_\mu$. A Lorentz transformation leads to

$$v'_\mu w'_\mu = \Lambda_{\mu\nu}\Lambda_{\mu\rho} v_\mu w_\mu,$$

but since $\Lambda_{\mu\nu}\Lambda_{\mu\rho} = \delta_{\nu\rho}$, we conclude that $v_\mu w_\mu$ is invariant, that is, it is a scalar. In particular we have checked that $x_\mu x_\mu$ is the invariant square interval. From the four-velocity one obtains the invariant

$$v_\alpha v_\alpha = \gamma^2(u^2 - c^2) = -c^2.$$

Differentiating again, we find $v_\alpha \frac{dv_\alpha}{d\tau} = 0$, so the four-acceleration is orthogonal to the four-velocity.

Other quantities of pre-relativistic Physics are recognized as scalars or four-vector components in Relativity. Consider an electromagnetic wave. The wave length is given by $\lambda = cT$, where T is the period; $\omega = ck$, where $k = \frac{2\pi}{\lambda}$ is the absolute value of the wave vector $\overrightarrow{k}$, pointing along the propagation direction $vers(\overrightarrow{E} \wedge \overrightarrow{H})$. Consider any component of the electromagnetic field of a wave in vacuo; its amplitude $f(\omega t - \overrightarrow{k} \cdot \overrightarrow{x})$ satisfies the wave equation $\frac{1}{c^2}\frac{\partial^2}{\partial t^2} f = \nabla^2 f$. Any observer in any reference can measure the phase difference between two space-time points by simply counting the maxima or the zeros of f between them. The phase difference is a relativistic invariant, because it is just a matter of counting. So, $\Phi = k_\mu x_\mu$ is invariant, and

$$k_\mu = \left(\overrightarrow{k}, k_0 = \frac{\omega}{c} \right)$$

is a four-vector (called the wave four-vector). Among the other relevant four vectors, there is the current $j_\mu = (\overrightarrow{J}, i\frac{\rho}{c})$, where ρ is the density, and the four-potential $A_\mu = (\overrightarrow{A}, i\frac{\phi}{c})$.

There are other benefits from the formal analogy between (6.12) and a rotation in the plane, that is (leaving aside the y and z axes), in ordinary $\mathbb{R}^2$, where the rotation matrix is

$$\begin{pmatrix} \cos(\theta) & \sin(\theta) \\ -\sin(\theta) & \cos(\theta) \end{pmatrix}.$$

Equation (6.12) corresponds to a rotation if we set $\cos(\theta) = \gamma$, $\sin(\theta) = i\beta\gamma \Rightarrow \tan(\theta) = i\beta$. Setting $\theta = i\xi$, since $\tan(i\xi) = i\tanh(\xi)$, we are left with $\tanh(\xi) = \beta$. The special Lorentz transformation is a rotation by $\theta = i\xi$ in in the $x_1 - x_4$ plane by an imaginary angle $i\xi$; this is such that $\tanh(\xi) = \beta$. The rotation is such that

$$x_1' = x_1 \cosh(\xi) + i x_4 \sinh(\xi) \quad x_4' = x_4 \cosh(\xi) - i x_1 \sinh(\xi).$$

The analogy with a rotation in an imaginary angle in $\mathbb{R}^4$ is not tremendously enlightening about the physical meaning, but it allows us to combine real rotations (which are also represented by 4×4 matrices) with special Lorentz transformations to make general Lorentz transformations in any directions,[6] which is the set of all possible Lorentz transformations and the rules to combine two of them. Moreover, this analogy helps us to find and classify the various kinds of invariant, which can be built in analogy for rotations and Lorentz transformations. These are the tensors.

6.3.1   Cartesian Tensors

Chronotopic Tensors are similar to those in $\mathbb{R}^4$; they represent physical quantities that are the same for all observers. The most obvious analogy is the one with vectors in ordinary space $\mathbb{R}^3$. An electric field E at a point in space may be written in a reference by specifying three components. Now, if one rotates the reference, the

---

[6] The infinite set of matrices so obtained are called a representation of the Lorentz Group.

components change in such a way that E remains the same, as it should. The fact that the components may vary with the frame does not spoil the objective character of E, but the components must transform in a specific way in any reasonable theory. Any physical, measurable object must be represented by a tensor. In Relativity, the base space is the Chronotope, so the tensors are functions of space and time.

Einstein mathematically formulated the Principle of Relativity as follows:

The laws of Physics are equations between tensors.

Thus we also gain a simple criterion to *relativize* the physical laws: they must be rewritten in covariant form, i.e. as equalities among tensors. One obtains laws that are good candidates since they (1) satisfy the principle of Relativity and (2) have the correct classical limits. In many cases this strategy is rewarding.

A relativistic invariant $\phi(x)$ is the simplest type of tensor. It is a single quantity that may depend on the point x_μ but is the same for all observers. It is also called a scalar or a zero-tank tensor (no indices). A four-vector $v_\mu(x)$, $\mu = 0, \ldots, 4$ Lorentz-transforms like x_μ and is a tensor with rank 1 and one index. Out of two four-vectors $v_\mu(x)$, $w_\nu(x)$ one can form an invariant $\phi(x) = v_\mu(x)w_\mu(x)$ which is their scalar product in analogy with scalar products in $\mathbb{R}^3$. We have just seen that $x_\mu x_\mu$ is invariant.

Now consider the gradient of a scalar ϕ. It is $d\phi = \frac{\partial \phi}{dx_\mu} dx_\mu$. Since $\phi(x)(x)$ is a scalar and dx_μ are the components of a vector (actually, for short, one says that dx_μ is a vector), it is clear that the gradient $\frac{\partial \phi}{dx_\mu}$ is a four-vector obtained by differentiation. A tensor of rank 2 $w_{\mu\nu}$, $(\mu = 1, \ldots, 4, \nu = 1, \ldots, 4)$ transforms like $x_\mu x_\nu$, that is, like the products of two four-vectors:

$$w'_{\mu\nu} = \Lambda_{\mu\rho} \Lambda_{\nu\sigma} w_{\rho\sigma}.$$

Higher rank tensors (i.e. ones with three or more indices) are defined similarly. It is fairly obvious that the sum of two tensors of the same rank is another tensor and so one can take linear combinations. A tensor T_{ik} such that $T_{ik} = T_{ki}$ is called a symmetric tensor; if instead $T_{ik} = -T_{ki}$ the tensor is antisymmetric[7]; the importance of such properties is evident if one considers that they are preserved by Lorentz transformations. So, for example, an antisymmetric tensor is such for all observers. By doing the derivatives, one can easily check that the electromagnetic field is an antisymmetric tensor:

$$F_{\mu,\nu} = \frac{\partial A_\nu}{\partial x_\mu} - \frac{\partial A_\mu}{\partial x_\nu}. \tag{6.16}$$

---

[7] In general, one can form tensors that belong to irreducible representations of the Group of permutations of the indices; such a property is invariant.

This can be written as an antisymmetric matrix:

$$F = \begin{pmatrix} 0 & B_z & -B_y & -i\,E_x \\ -B_z & 0 & B_x & -i\,E_y \\ B_y & -B_x & 0 & -i\,E_z \\ i\,E_x & i\,E_y & i\,E_z & 0 \end{pmatrix}. \tag{6.17}$$

Thus, we know how to transform the electromagnetic fields. For the special transformation (6.14), one finds:

$$\begin{aligned} B'_x &= B_x & E'_x &= E_x \\ B'_y &= \gamma(B_y + \beta E_z) & E'_y &= \gamma(E_y - \beta B_z) \\ B'_z &= \gamma(B_z - \beta E_y) & E'_z &= \gamma(E_z + \beta H_y). \end{aligned} \tag{6.18}$$

This is the fast way to transform the field: otherwise, one can Lorentz transform the charge-current four vector and then re-calculate the fields using Maxwell's equations. It is easy to realize that if in the unprimed system a charge is acted upon by an electric field, in the primed system a Lorentz force also arises. From the field tensor, one can form a third rank tensor

$$\frac{\partial}{\partial x_\lambda} F_{\mu\nu} = \frac{\partial^2 A_\nu}{\partial x_\lambda \partial x_\mu} - \frac{\partial^2 A_\mu}{\partial x_\lambda \partial x_\nu}.$$

Two more tensors can be obtained by permutation of indices:

$$\frac{\partial}{\partial x_\mu} F_{\nu\lambda} = \frac{\partial^2 A_\lambda}{\partial x_\mu \partial x_\nu} - \frac{\partial^2 A_\nu}{\partial x_\mu \partial x_\lambda}$$

and

$$\frac{\partial}{\partial x_\nu} F_{\lambda\mu} = \frac{\partial^2 A_\mu}{\partial x_\nu \partial x_\lambda} - \frac{\partial^2 A_\lambda}{\partial x_\nu \partial x_\mu}.$$

These are not independent, because their sum reads as:

$$\frac{\partial}{\partial x_\lambda} F_{\mu\nu} + \frac{\partial}{\partial x_\mu} F_{\nu\lambda} + \frac{\partial}{\partial x_\nu} F_{\lambda\mu} = 0.$$

To see the meaning, let us introduce the fields: what we got is actually $div\,\boldsymbol{B} = 0$. Moreover, $\frac{\partial F_{12}}{\partial x_4} + \frac{\partial F_{24}}{\partial x_1} + \frac{\partial F_{41}}{\partial x_2} = 0$ is nothing but $\frac{1}{c}\frac{\partial}{\partial t} B_3 + \frac{\partial E_2}{\partial x_1} - \frac{\partial E_1}{\partial x_2} = 0$, which is a component of the $\nabla \wedge \boldsymbol{E}$ Maxwell equation; the other components are obtained by permutation of the indices. The inhomogeneous Maxwell equations are rewritten as

$$\frac{\partial F_{\mu\nu}}{\partial x_\nu} = \frac{4\pi}{c} J_\mu.$$

We are in position to Lorentz transform the fields. Let $E_\parallel$, $B_\parallel$ denote the field components parallel to the speed v and $E_\perp$, $B_\perp$ the orthogonal components.

Problem 21 The relativistic law of sum of speeds must result from the sum of four-velocities. Verify that.

Solution 21 If in the train reference $\mathcal{K}_T$, a material point moves with three-velocity $\vec{v}^T = (v, 0, 0)$ the train driver finds that the four-velocity is

$$w_\mu^{(T)} = \frac{1}{\sqrt{1 - \frac{v^{T2}}{c^2}}}(\vec{v}^T, ic). \tag{6.19}$$

Lorentz transforming, we find the four-vector in $\mathcal{K}_S$; since the speed of the train is u,

$$w_1^{(S)} = \frac{w_1^{(T)} - \frac{iu}{c}w_4^{(T)}}{\sqrt{1 - \frac{u^2}{c^2}}},$$

and substituting,

$$w_1^{(S)} = \frac{1}{\sqrt{1 - \frac{v^{T2}}{c^2}}} \frac{1}{\sqrt{1 - \frac{u^2}{c^2}}}(v^T + u).$$

For coherence with (6.19) $w_1^{(S)} = \frac{v^S}{\sqrt{1 - \frac{v^{S2}}{c^2}}}$, where v^S is measured in the station. So,

$$\frac{(v^S)^2}{1 - \left(\frac{v^S}{c}\right)^2} = \varphi, \qquad \varphi = \frac{(u+v)^2}{\left(1 - \left(\frac{u}{c}\right)^2\right)\left(1 - \left(\frac{v}{c}\right)^2\right)},$$

which implies

$$(v^S)^2 = \frac{\varphi}{1 + \frac{\varphi}{c^2}} = \frac{(u+v)^2}{\left(1 - \left(\frac{u}{c}\right)^2\right)\left(1 - \left(\frac{v}{c}\right)^2\right) + \frac{(u+v)^2}{c^2}}.$$

The denominator is

$$1 - \left(\frac{u}{c}\right)^2 - \left(\frac{v}{c}\right)^2 + \left(\frac{uv}{c^2}\right)^2 + \left(\frac{u}{c}\right)^2 + \left(\frac{v}{c}\right)^2 + 2\frac{uv}{c^2} = \left(1 + \frac{uv}{c^2}\right)^2.$$

Finally, one finds the known result;

$$v^S = \frac{v^T + u}{1 + \frac{v^T u}{c^2}}.$$

Problem 22 A spaceship travels with speed v relative to a space station. The pilot puts a mirror A ahead and a mirror B on the tail at a distance $AB = 2l_0$; a light source S in the middle emits a signal that hits A and B simultaneously after a time τ_0 measured on the clock of the spaceship. An observer in the space station finds that the signal reaches B first and then A after a time τ_0. Find the speed of the space ship.

Solution 22 In the station, the space ship is long $l = l_0\gamma$; the time to reach A is $\tau_A = \frac{l}{c-v}$, while the time to reach B is $\tau_A = \frac{l}{c+v}$; So $\Delta t = 2\frac{l_0}{c}\frac{u}{\sqrt{c^2-u^2}}$ must be equated to $\frac{l_0}{c}$. So one finds that $\frac{v}{c} = \frac{1}{\sqrt{5}}$.

6.3.2   Action of the Free Field

The Lagrangian of the field can be taken to be proportional to $F_{\mu\nu}^2$; it is usually written in the form

$$L = \frac{i}{4\pi c}\left(\frac{\partial A_\nu}{\partial x_\mu} - \frac{\partial A_\mu}{\partial x_\nu}\right)\left(\frac{\partial A_\nu}{\partial x_\mu} - \frac{\partial A_\mu}{\partial x_\nu}\right). \tag{6.20}$$

Of course, the factor $\frac{i}{4\pi c}$ is arbitrary. The Lagrange equations are

$$\frac{\partial L}{\partial A_\mu} = \frac{\partial}{\partial x_\nu}\frac{\partial L}{\partial\left(\frac{\partial A_\mu}{\partial x_\nu}\right)}.$$

One finds that

$$\frac{\partial}{\partial x_\nu}\left(\frac{\partial A_\nu}{\partial x_\mu} - \frac{\partial A_\mu}{\partial x_\nu}\right) = \frac{\partial}{\partial x_\nu}F_{\mu\nu} = 0 \Leftrightarrow div B = 0, rot E + \frac{1}{c}\frac{\partial B}{\partial t} = 0.$$

6.3.3   Doppler–Fizeau Effect and Aberration of Light

Let us conduct a new thought experiment, with the help of the usual train running along the x axis with velocity u. The station is replaced by a monochromatic light source that emits a plane electromagnetic wave with pulsation $\omega_S = 2\pi\nu_S$ and wave vector $\vec{k}_S$. The index S stands for source. The modulus of $\vec{k}_S$ is $k_S = \frac{\omega_S}{c}$ and the direction makes an angle θ_S with the x axis, therefore the component along $\vec{u}$ is

$$(k_S)_\parallel = (\vec{k}_S)_x = k_S\cos\theta_S = \frac{\omega_S}{c}\cos\theta_S,$$

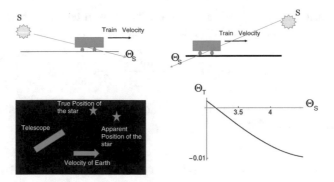

Fig. 6.5 Top Left: train away from source, $\theta_S < 0$; Top right: train towards source, $\theta_S > \pi$; bottom left: stellar aberration; bottom right: θ_T versus θ_S for $\beta = 0.1$ (exaggerated)

while the vertical component is

$$(k_S)_\perp = k_S \sin \theta_S = \frac{\omega_S}{c} \sin \theta_S.$$

One can see in Fig. 6.5 that when the train moves away from the source, θ_S is small, and, and when the train approaches the source, θ_S is around π.

In the reference of the train, the driver observes an electromagnetic wave, but we must find out which wave it is. Let us say, it has some angular frequency ω_T; its wave vector has some module $k_T = \frac{\omega_T}{c}$, and it makes some angle θ_T with the x axis, and therefore its components are:

$$(k_T)_\parallel = (\vec{k}_T)_x = \frac{\omega_T}{c} \cos \theta_T, \ (k_T)_\perp = \frac{\omega_T}{c} \sin \theta_T.$$

The train driver draws a line, makes two marks on it and -at least in principle- he can count the nodes (points with vanishing field) of the wave between marks at a given instant. The important remark is that the number of nodes between two marks cannot depend on the reference. Therefore, the phase $\Phi = k.r - \omega t$ must be invariant. Since x_μ is a four-vector, and is involved, the phase must be the scalar product with another four-vector, which generalises the pre-relativistic wave vector. This is evidently Therefore,

$$k_\mu = \left(\vec{k}, k_0 = \frac{\omega}{c} \right).$$

The Lorentz transformation written in real form (6.9), is

$$\begin{cases} (k_T)_\parallel = \gamma \left[(k_S)_\parallel - \beta(k_S)_0 \right] \implies \omega_T \cos \theta_T = \gamma \omega_S \left[\cos \theta_S - \beta \right], \\ (k_T)_\perp = (k_S)_\perp \qquad\qquad\quad \implies \omega_T \sin \theta_T = \omega_S \sin \theta_S, \\ (\vec{k}_T)_0 = \gamma((k_S)_0 - \beta(k_S)_\parallel) \implies \omega_T = \gamma \omega_S \left[1 - \beta \cos \theta_S \right]. \end{cases} \qquad (6.21)$$

The last of (6.21) gives the frequency on the train in terms of the angle in the station; while the first two yield the frequency on the train

$$\tan(\theta_T) = \frac{\sin(\theta_S)}{\gamma(\cos(\theta_S) - \beta)}. \tag{6.22}$$

By plugging $\cos(\theta_S)$ derived from the first and simplifying, one obtains the formula found by Einstein in 1905:

$$\omega_T = \frac{\omega_S}{\gamma[1 + \beta\cos(\theta_T)]}. \tag{6.23}$$

The source is seen in a direction that depends on the speed of the observer relative to the source. This phenomenon is known as the aberration of the light and has been well known to the astronomers since the eighteenth century. It was discovered by the Englishman James Bradley (1693–1762), later the Royal Astronomer. It is the apparent motion of all the stars due to the revolution of the Earth around the sun. The effect changes the apparent position of stars by 20 arc-seconds when the line of sight is perpendicular to the Earth's orbit, and is a direct evidence that our Planet rotates around the Sun.

The trend is illustrated in Fig. 6.5 bottom left; the reason is clear from the fact that the angle receives a negative correction illustrated in Fig. 6.5 bottom right, even if the parameter $\beta = 0.1$ used for illustration is too large.

For $\theta_S = 0$, the observer moves away from the source and experiences the longitudinal Doppler effect[8]:

$$\omega_T = \omega_S\sqrt{\frac{1 - \frac{u}{c}}{1 + \frac{u}{c}}}. \tag{6.24}$$

If the source is approaching, there is a blue shift which is a first-order effect for small speeds. Qualitatively, this is similar to the behavior of sound waves that reach us from moving objects. This effect allows for precise measurement of speeds and is of paramount importance in Astrophysics. Edwin Hubble, in 1929, discovered that the light from all the far Galaxies reddens with increasing distance and concluded that the Universe is expanding (see Sect. 8.12). The Doppler effect also allows the police to measure the speed of cars and fine the drivers who exceed the limits.

Equation (6.21) also shows that for $\cos(\theta_S) = \beta$, $\theta_T = \frac{\pi}{2}$, and

$$\omega_T = \gamma\omega_S\left[1 - \beta^2\right] = \omega_S\sqrt{1 - \beta^2}.$$

---

[8]Here u is the speed of the train; more usually the law is written in terms of the speed $-u$ of the source relative to the observer.

This *transverse Doppler effect* is second-order in β, and is not predicted by the pre-relativistic Physics. It can be understood as being due to the relativistic time dilation.

Accurate experimental verifications have been done using the recoilless emission discovered by Rudolf Mössbauer. The Mössbauer effect is a solid state effect. Typically, a free ^{57}Fe atom makes a nuclear decay within a crystal emitting a very narrow γ ray with a recoil energy of about 10^{-1} eV. The vibrations in solids are quantized harmonic oscillators, called phonons. Since the oscillator state with the recoiling atom is not orthogonal to the ground state, there is a probability that a nuclear decay emits a γ ray without exciting the oscillator at all. In this case, the whole crystal recoils. Since the crystal is macroscopic, in practice a recoil-less emission occurs. This effect allows to work with very well defined frequencies, since it avoids the recoil that is an important source of broadening. Doppler shifts due to velocities of the order of 1 cm per second can be measured.[9]

6.3.4 Relativistic Mechanics

Consider a free point mass with the action $S = \int_{t_1}^{t_2} \frac{1}{2} m v^2 dt$, where m is the mass. This is good in Newtonian mechanics, but does not comply with the criterion for a relativistic law, since it is not written in terms of real things like events and tensors. However, the principle of least action $\delta S = 0$ lends itself to a relativistically covariant reformulation. The law of motion must be invariant, so the simplest possibility is that the action S be a scalar. First, we replace the absolute times t_2 and t_2 with the proper times of two events a and b. So,

$$S = \int_{\tau_a}^{\tau_b} d\tau \; \times \; \text{something.}$$

The integrand must have the dimensions of an energy. The only scalar that characterizes the particle is the rest mass m_0, which is the mass measured in the rest frame of the particle; besides, we need a velocity to make an energy. Any velocity would be OK, since we know that if the action is multiplied by a constant, the equations of motion are not affected. Since c is one of the fundamental constants, we take the Lagrangian $L = -m_0 c^2$.

In the rest frame of the point mass, the differential of the interval is $ds = c d\tau$. The observer who sees the particle moving with velocity v finds that $ds^2 = c^2 dt^2 - dx^2 - dy^2 - dz^2$, and $d\tau = dt\sqrt{1 - \frac{v^2}{c^2}}$. So,

$$S = -m_0 c^2 \int_{t_a}^{t_b} \sqrt{1 - \left(\frac{v}{c}\right)^2} \, dt = \int_{t_a}^{t_b} dt \, L(q, \dot{q}, t);$$

---

[9]For the experimental confirmation of the transverse Doppler effect, see Hay et al., Phys. Rev. Letters 4, 165 (1960).

the constant $-m_0c^2$ is a good choice, since for $v \ll c$, L gives back the non-relativistic kinetic energy: $L \approx m_0c^2 + \frac{1}{2}m_0v^2 + \cdots$. The constant m_0c^2 does not change the equations of motion and can be ignored. The momentum conjugated with $\overrightarrow{x}$ is

$$\overrightarrow{p} = \frac{\partial L}{\partial \overrightarrow{v}} = \frac{m_0 \overrightarrow{v}}{\sqrt{1 - \left(\frac{v}{c}\right)^2}} \equiv m \overrightarrow{v}.$$

This reduces[10] to the Galilean momentum for small v. The change is that the mass m is velocity-dependent: $m \approx m_0$ for $v \to 0$, but m diverges for $v \to c$, So, c is a limiting velocity, unattainable for massive bodies. Nobody can overtake a light ray and look back to see what it looks like! The energy is given by the Hamiltonian $E = \overrightarrow{p} \cdot \overrightarrow{v} - L$. One finds

$$E = \frac{m_0c^2}{\sqrt{1 - \left(\frac{v}{c}\right)^2}} \equiv mc^2. \tag{6.25}$$

This formula is so famous that everyone knows it. It says that a fixed mass m has a *rest energy*

$$\boxed{E_0 = m_0c^2,}$$

while a moving mass also has kinetic energy; at low speed, this is

$$\boxed{E \approx m_0c^2 + \frac{1}{2}m_0v^2 + \cdots}$$

which coincides with the Galilean expression plus the rest energy. Like the momentum, the energy E blows up at c. But all that is really striking only when it is properly understood, because in Newtonian Mechanics, the origin of energy is arbitrary and only energy differences are meaningful. The real point is that in Relativity theory, there is no such arbitrariness. Indeed, the momentum

$$p_\mu = m_0 w_\mu = \frac{m_0}{\sqrt{1 - \left(\frac{v}{c}\right)^2}}(v, ic) = \left(\overrightarrow{p}, i\frac{E}{c}\right)$$

is a four-vector, and this property would be broken if one could add a constant to the fourth component. This implies that we can know the energy content of a system from its rest mass, and conversely, any amount of energy has inertia. That was a real change of paradigm in Science. The consequence is that the mass of bound systems (like, e.g., nuclei) is less than the sum of the masses of the constituents. For instance, this is observed in nuclides, since in the nucleo-synthesis, some energy was emitted as γ rays.

---

[10]Using $\frac{1}{\sqrt{1-x^2}} \approx 1 + \frac{x^2}{2} + \frac{3x^4}{8}$.

These findings are readily extended to deal with a charged particle in a field. The pre-relativistic Lagrangian is shown in (2.88). Once we modify the kinetic energy term as shown above, we get:

$$L(\vec{r}, \vec{v}, t) = -m_0 c^2 \sqrt{1 - \frac{v^2}{c^2}} - q\phi + q\vec{v} \cdot \vec{A}(\vec{r}, t). \qquad (6.26)$$

Now one can derive the relativistic equations of motion of a charge in a field

$$\frac{d}{dt}\vec{p} = \vec{F}, \quad \vec{p} = m_0 \gamma \vec{v}, \qquad (6.27)$$

where $\vec{F}$ is the Lorentz force (2.85).

Four-Force

Equation (6.27) must be put into a manifestly covariant form. The derivative must be $\frac{d}{d\tau}$, with respect to the proper time τ, and the force must become a four-vector. The equations of Dynamics become:

$$\frac{d}{d\tau} p_\mu = f_\mu. \qquad (6.28)$$

Multiplying by $v_\mu = \frac{1}{\sqrt{1-(\frac{v}{c})^2}}(\vec{v}, ic)$ and recalling that $v_\mu v_\mu = -c^2$, one finds

$$m_0 v_\mu \frac{d}{d\tau} v_\mu = \frac{m_0}{2} \frac{d}{d\tau}(v_\mu v_\mu) = 0 = f_\mu v_\mu.$$

So,

$$\vec{f} \cdot \vec{v} + i f_4 c = 0,$$

$$f_4 = \frac{i\vec{f} \cdot \vec{v}}{c}.$$

Since $dt = \gamma d\tau$ and $\frac{d}{d\tau} = \gamma \frac{d}{dt}$, the space part of (6.28) becomes

$$\frac{d}{d\tau}\vec{p} = \frac{1}{\sqrt{1-(\frac{v}{c})^2}} \frac{d}{dt}\vec{p} = \vec{f} = \frac{1}{\sqrt{1-(\frac{v}{c})^2}} \vec{F},$$

and this relates the modified three-vector $\vec{f}$ with the Lorentz force $\vec{F}$ (2.85). Moreover,

$$f_4 = \frac{i}{c} \frac{\vec{F} \cdot \vec{v}}{\sqrt{1-(\frac{v}{c})^2}},$$

and the four-force is $f_\mu = (\frac{\vec{F}}{\sqrt{1-(\frac{v}{c})^2}}, \frac{i}{c}\frac{\vec{F}\cdot\vec{v}}{\sqrt{1-(\frac{v}{c})^2}})$. The fourth component of the equations of motion is

$$\frac{d}{dt}\frac{m_0c^2}{\sqrt{1-\left(\frac{v}{c}\right)^2}} = \vec{F}\cdot\vec{v}.$$

In this way, the equations of Dynamics are relativized.

6.3.5 Field Lagrangian and Hamiltonian; Energy-Momentum Tensor

Consider a fluid having proper density (that is, measured in the rest reference) $\rho_0(x)$ in the chronotopic point x and four-velocity u_μ (also a function of space-time) with negligible pressure. We may take the *stress-energy tensor* in the form

$$T_{\mu\nu} = \rho_0(x)u^\mu u^\nu. \tag{6.29}$$

The divergence may be denoted by $T^{\mu\nu}_{,\mu} \equiv \frac{\partial T^{\mu\nu}}{\partial x^\mu} \equiv \partial_\mu T^{\mu\nu}$. At any rate,

$$T^{\mu\nu}_{,\mu} = \partial_\mu(\rho_0 u^\mu)u^\nu + \rho_0 u^\mu \partial_\mu u^\nu.$$

This can be simplified, since in the rest frame the velocity is zero and the second term vanishes; moreover, $\partial_\mu(\rho_0 u^\mu) = 0$ by the continuity equation. Thus, $T^{\mu\nu}_{,\mu} = 0$; in other words, the energy-momentum tensor is *divergenceless*.

If the fluid has proper internal energy U_0 and proper pressure P_0, extra contributions arise to the stress-energy tensor. One can show that a more complete, still divergenceless expression is:

$$T_{\mu\nu} = \delta_{\mu\nu}P_0 + \left(\rho_0(x) + \frac{\rho_0 U_0}{c^2} + \frac{P_0}{c^2}\right)u_\mu u_\nu. \tag{6.30}$$

This is a covariant way to represent the energy-matter contents of space-time and is crucially important for General Relativity. Indeed, a stress-energy tensor is naturally associated to any relativistic field. The formalisms of Classical Mechanics show their full power in relativistic field theory, where the coordinates q become the values of a field at each space-time point. At the present stage, we can stay generic about the precise nature of ϕ which is simply required to be a relativistic field that carries energy and momentum. In this section, we see how things go in the case of a one-component field $\phi(x)$. The action is the relativistic invariant

$$S = \int_{t1}^{t2} L dt = \int \mathcal{L}(\phi, \partial_\mu\phi)d^4x, \tag{6.31}$$

and $\mathcal{L}$ is called the Lagrangian density. One can carry on the argument in analogy with Classical Mechanics and find that the condition $\delta S = 0$ leads to the equations of motion

$$\frac{\partial \mathcal{L}}{\partial \phi} = \partial_\mu \left(\frac{\partial \mathcal{L}}{\partial(\partial_\mu \phi)} \right). \tag{6.32}$$

The analogy is complete. In terms of the momentum density

$$\pi(x) = \frac{\partial \mathcal{L}}{\partial \dot{\phi}(x)}, \tag{6.33}$$

one can define a Hamiltonian

$$H = \int d^3 x \mathcal{H}(x) = \int d^3 x [\pi(x)\dot{\phi}(x) - \mathcal{L}(x)], \tag{6.34}$$

where $\mathcal{H}$ is called Hamiltonian density. Conservation laws can be derived most simply by inserting into the derivative $\frac{\partial \mathcal{L}}{\partial x_\mu} = \frac{\partial \mathcal{L}}{\partial \phi}\frac{\partial \phi}{\partial x_\mu} + \frac{\partial \mathcal{L}}{\partial(\partial_\nu \phi)}\frac{\partial(\partial_\nu \phi)}{\partial x_\mu} = \frac{\partial \mathcal{L}}{\partial \phi}\frac{\partial \phi}{\partial x_\mu} + \frac{\partial \mathcal{L}}{\partial(\partial_\nu \phi)}\frac{\partial(\partial_\mu \phi)}{\partial x_\nu}$ the equation of motion $\partial_\mu \frac{\partial \mathcal{L}}{\partial(\partial_\mu \phi)} = \frac{\partial \mathcal{L}}{\partial \phi}$, getting

$$\frac{\partial \mathcal{L}}{\partial x_\mu} = \sum_\nu \left[\left(\partial_\nu \frac{\partial \mathcal{L}}{\partial(\partial_\nu \phi)} \right) \partial_\mu \phi + \frac{\partial \mathcal{L}}{\partial(\partial_\nu \phi)} \partial_\nu(\partial_\mu \phi) \right].$$

This can be rewritten as

$$\frac{\partial \mathcal{L}}{\partial x_\mu} = \sum_\nu \frac{\partial}{\partial x_\nu} \left[\frac{\partial \mathcal{L}}{\partial(\partial_\nu \phi)} \partial_\mu \phi) \right],$$

or

$$\sum_\nu \frac{\partial}{\partial x_\nu} \left[\frac{\partial \mathcal{L}}{\partial(\partial_\nu \phi)} \partial_\mu \phi) - \frac{\partial \mathcal{L}}{\partial x_\nu} \delta_{\mu\nu} \right] = 0. \tag{6.35}$$

We have obtained

$$\partial_\nu T_{\mu\nu} = 0, \tag{6.36}$$

where

$$T_{\mu\nu} = \partial_\nu \left[\frac{\partial \mathcal{L}}{\partial(\partial_\nu \phi)} \partial_\mu \phi \right] \tag{6.37}$$

is the *stress-energy tensor*, a.k.a. the *energy-momentum tensor*. Clearly, Eq. (6.36) is a conservation law; T_{00} represents the energy density and T_{0i} the components of the momentum density of the field.

Noether's theorem is even more powerful in field theory and provides conserved currents. The infinitesimal transformation analogous to the one in classical mechanics

$$\phi(x) \rightarrow \phi(x) + \alpha \Delta \phi(x) \tag{6.38}$$

must leave the equations of motion invariant if it has to be a symmetry. The change it introduces in the Lagrangian density is

$$\alpha \Delta \mathcal{L} = \frac{\partial \mathcal{L}}{\partial \phi}(\alpha \Delta \phi) + \left(\frac{\partial \mathcal{L}}{\partial(\partial_\mu \phi)}\right) \partial_\mu (\alpha \Delta \phi). \tag{6.39}$$

Introducing the equations of motion one can reduce this to

$$\Delta \mathcal{L} = \partial_\mu \left(\frac{\partial \mathcal{L}}{\partial(\partial_\mu \phi)}\right) \partial_\mu \Delta \phi. \tag{6.40}$$

The equations of motion remain unchanged if we add a four-divergence, that gives a contribution only on the boundary where, by assumption, $\delta \phi(x) = 0$. Therefore Noether's theorem states that

$$\partial_\mu j^\mu = 0, \tag{6.41}$$

where the conserved current is

$$j^\mu(x) = \frac{\partial \mathcal{L}}{\partial(\partial_\mu \phi)} \Delta \Phi - T^\mu. \tag{6.42}$$

An important example is provided by the infinitesimal translation $x^\mu \rightarrow x^\mu - a^\mu$, that is, $\phi(x) \rightarrow \phi(x) + a^\mu \partial_\mu \phi(x)$, $\mathcal{L} \rightarrow \mathcal{L} + a^\mu \partial_\mu \mathcal{L}$. Thus, the four-divergence is $a^\nu \partial_\mu (\delta_\nu^\mu \mathcal{L})$. There are four conserved currents:

$$T_\nu^\mu = \frac{\partial \mathcal{L}}{\partial(\partial_\mu \phi)} \partial_\nu \phi - \mathcal{L} \delta_\nu^\mu. \tag{6.43}$$

This is the *stress-energy tensor*, a.f.a the *energy-momentum tensor* such that $\int T^{0i} d^3 x = -\int \pi \partial_i \phi d_{3x}$ is the field momentum, while $\int T^{00} d^3 x$ represents the field energy.

Chapter 7
Curvilinear Coordinates and Curved Spaces

To proceed, we must develop some mathematical tools drawn from Differential Geometry.

Even in flat Euclidean space it may be useful to use curvilinear coordinates; for instance, in 3d problems having central symmetry, we obtain an important simplification when the line element $ds^2 = dx^2 + dy^2 + dz^2$ is replaced by $ds^2 = dr^2 + r^2 d\theta^2 + r^2 \sin^2(\theta) d\phi^2$. In a curved space we have no other choice, because Cartesian coordinates may exist only locally, that is, in an infinitesimal neighborhood. One example is the surface of a sphere of radius R. The spherical coordinates with r set equal to the radius R of the sphere do the job. In the latter case, we deal with a 2d curved subspace embedded in a 3d Euclidean space. The curved coordinates are intrinsic to the surface, and one can ignore the existence of a radial dimension.

The Minkowsky space is not general enough for our purposes. We need a Riemannian space, which has a curvature. Since we live on the curved surface of a more or less spherical planet, such ideas are not too far from common wisdom, but the mathematical apparatus that we need is far from trivial. The relevant theory was created by outstanding mathematicians: Karl Friederich Gauss (1777–1855), Berhard Riemann (1826–1866), E. Christoffel (1829–1900), G. Ricci (1853–1925) and T. Levi Civita (1873–1942). In order to introduce curvilinear coordinates, we can start from ordinary $\mathbb{R}^2$ since the extension to $\mathbb{R}^d$ for any larger dimension d is obtained for free. This is a good strategy: once we have developed the formalism, it fits our needs when working in curved spaces, where Cartesian coordinates do not exist.

In $\mathbb{R}^2$, we can choose an origin O and a Cartesian system such that any point may be assigned as $\mathbf{R} = (X^1, X^2)$. The square distance from a given point $\mathbf{R}$ to a near point

$$\mathbf{R} + d\mathbf{R} \tag{7.1}$$

© Springer International Publishing AG, part of Springer Nature 2018
M. Cini, *Elements of Classical and Quantum Physics*,
UNITEXT for Physics, https://doi.org/10.1007/978-3-319-71330-4_7

Fig. 7.1 When the coordinate lines are not orthogonal, the covariant basis vectors (red) are different from the contravariant ones (green) and do not have fixed length

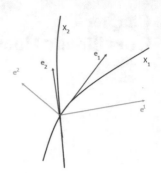

is

$$ds^2 = dX_1^2 + dX_2^2. \tag{7.2}$$

Let $x^i = x^i(X^1, X^2)$ be curvilinear coordinates in terms of the Cartesian $\mathbf{R} = (X^1, X^2)$. We assume that these functions are invertible, that is, $x^i = x^i(X^1, X^2) \Leftrightarrow X^i = X^i(x^1, x^2)$. If x^1 is varied while x^2 is kept fixed, $\vec{R}$ will trace out a curve, the x^1 curve. In this way one can think of a local coordinate system at each point with basis vectors

$$\vec{e}_i = \frac{\partial \vec{R}}{\partial x^i} = \left(\frac{\partial X^1}{\partial x^i}, \frac{\partial X^2}{\partial x^i} \right). \tag{7.3}$$

These are called *covariant* basis vectors; note that the component i is a subscript. Now the infinitesimal vectorial shift (7.2) becomes, understanding the sum over repeated indices,

$$ds^2 = \frac{\partial \mathbf{R}}{\partial x^i} \frac{\partial \mathbf{R}}{\partial x^j} dx^i dx^j = \mathbf{e}_i \mathbf{e}_j dx^i dx^j = g_{ij} dx^i dx^j,$$

where

$$g_{ik} = \vec{e}_i \vec{e}_k \tag{7.4}$$

is called the covariant component ik of the metric tensor. The Einstein convention now is that repeated indices are meant to be summed if they are a subscript and a superscript.

But there is the complication that, in general, the covariant basis vectors are not of unit length and not orthogonal; moreover, they do not coincide with the contravariant basis vectors which are defined by $\vec{e}^i \cdot \vec{e}_k = \delta_{ik}$ (see Fig. 7.1).

Let us choose a point and start from it along a curve $x^i = x^i(s)$; we can arrange that s is the curvilinear distance from the starting point: $ds^2 = d\vec{R} \cdot d\vec{R}$. Then, $\vec{t} = \frac{\partial \vec{R}}{\partial s}$ is the velocity, that is, is tangent to the curve and $\vec{t} \cdot \vec{t} = 1$. Note that

$$\vec{t} = \frac{\partial \vec{R}}{\partial s} = \frac{\partial \vec{R}}{\partial x^i} \frac{\partial x^i}{\partial s} = \vec{e}_i \frac{\partial x^i}{\partial s} \tag{7.5}$$

$$\vec{e}_i \vec{e}_k \frac{\partial x^i}{\partial s} \frac{\partial x^k}{\partial s} = 1 \Rightarrow ds^2 = \vec{e}_i \vec{e}_k dx^i dx^k. \tag{7.6}$$

As anticipated above, to proceed we need tensors. Using (7.3) the metric tensor components are:

$$g_{ij} = \left(\frac{\partial X^1}{\partial x^i}, \frac{\partial X^2}{\partial x^i}, \frac{\partial X^3}{\partial x^i} \right) \cdot \left(\frac{\partial X^1}{\partial x^j}, \frac{\partial X^2}{\partial x^j}, \frac{\partial X^3}{\partial x^j} \right)$$

$$= \frac{\partial X^1}{\partial x^i} \frac{\partial X^1}{\partial x^j} + \frac{\partial X^2}{\partial x^i} \frac{\partial X^2}{\partial x^j} + \frac{\partial X^3}{\partial x^i} \frac{\partial X^3}{\partial x^j}$$

$$= \frac{\partial X^k}{\partial x^i} \frac{\partial X^k}{\partial x^j}.$$

From $g_{ii} = \vec{e}_i^2$, the length of $\vec{e}_i$ is $|\vec{e}_i| = \sqrt{g_{ii}}$. From $g_{ik} = \vec{e}_i \vec{e}_k = |\vec{e}_i||\vec{e}_k|\cos(\theta_{ik})$, one finds $\cos(\theta_{ik}) = \frac{g_{ik}}{\sqrt{g_{ii} g_{kk}}}$. Any vector at the point O can be expanded: $\vec{A} = A^i \vec{e}_i$; in this case, A^i are the *contravariant* components and bring i as a suffix. Expanding the contravariant vectors on the covariant basis

$$\vec{e}^i = g^{ik} \vec{e}_k, \tag{7.7}$$

where g^{ik} are the contravariant components of the metric tensor. The theory uses scalars (i.e., functions of the point that do not depend on the coordinate system), vectors, that may be covariant or contravariant, and tensors that can have several high or contravariant indices and low or covariant ones. In Eq. (7.7) we succeeded to raise the covariant index k, which is a subscript, to a contravariant index i which is a superscript. To do so we have multiplied by the contravariant component g^{ik} and contracted, i.e. summed over k. This turns out to be a general rule. The converse rule allows us to lower a contravariant index k by multiplying for g_{ik} and contracting over k; the index has become a covariant index i. The general rule for raising and lowering indices involves the metric tensor. Each index can be raised (converting from covariant to contravariant) or vice versa, and back. To convert the covariant components A_i of a vector, one raises the index by setting:

$$A^i = g^{ij} A_j, \quad A_i = g_{ij} A^j. \tag{7.8}$$

This process of summing over repeated high and low indices is called contraction. A similar rule allows to raise and lower indices in tensors with any number of *legs* (that is, indices): for instance, $A^i_{jk} = g_{js} A^{is}_k$. To find g^{ik}, multiply by $\vec{e}_l$:

$$\vec{e}_l \vec{e}^i = \delta^i_l = g^{ik} g_{kl} \tag{7.9}$$

Thus, the contravariant components are the elements of the matrix inverse of the matrix of the covariant components:

$$\{g^{ij}\} = \{g_{ij}\}^{-1}.$$

Note that the matrix g^{ij} is the inverse of the g_{ij} matrix and consequently. $g^i_j = \delta^i_j = \delta_{i,j}$. As noted above, the components of the contravariant tensor are denoted by superscripts.

In d dimensions, we must be able to transform to new coordinates $x \equiv (x_1, \ldots, x_d)$ from old ones x'. The differentials transform according to

$$dx^i = \frac{\partial x^i}{\partial x'^j} dx'^j; \tag{7.10}$$

the Einstein convention (that repeated indices are meant to be summed over from 1 to d if they are a subscript and a superscript) is used. dx^i is a prototype contravariant vector, with the component labels are written as superscripts; the general contravariant transformation rule is $A^i = \frac{\partial x^i}{\partial x'^j} A'^j$.

On the other hand, consider a quantity $f(x^i)$ which is a scalar, i.e., does not change in the transformation of variables. The derivatives of a scalar transform according to the covariant rule

$$\frac{\partial f}{\partial x^i} = \frac{\partial f}{\partial x'^j} \frac{\partial x'^j}{\partial x^i}. \tag{7.11}$$

A set of d components A_i transforming this way are the covariant components of a vector. Note that a superscript in the denominator is counted as a subscript. A second-order contravariant tensor has d^2 components that transform according to

$$T^{mn} = \frac{\partial x^m}{\partial x'^i} \frac{\partial x^n}{\partial x'^j} T'^{ij}, \tag{7.12}$$

that is, like the product of two contravariant vectors. A second-order covariant tensor has d^2 components that transform like the product of two covariant vectors, $T_{mn} = \frac{\partial x'^i}{\partial x^m} \frac{\partial x'^j}{\partial x^n} T'_{ij}$. A mixed tensor with one superscript index and one subscript index transforms like the product of a covariant and a contravariant vectors, and so on. In general, tensors may have any number of indices.

7.1 Parallel Transport, Affine Connection and Covariant Derivative

In general, the covariant basis vectors at different point are not parallel and differ in length; however the change of $\vec{e}_i$ when shifting the point by an infinitesimal dx^j must be proportional to the shift. See Fig. 7.2, which presents a two-dimensional

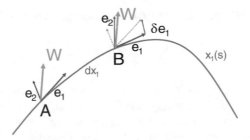

Fig. 7.2 Point A corresponds to the value x^1 of a coordinate line $x^1 = x^1(s)$ and is the origin of a local basis vectors $e_1 = \frac{dx^1}{ds}$ and $e_2 = \frac{dx^2}{ds}$. Point B on the coordinate line corresponds to $x^i + dx^i$ and a new local basis. For a generic line, this differs from the dashed basis that would result from a simple shift of the basis at A, since the coordinate line is not straight. The change in e_1 (shown in red) is of order dx

illustration of the parallel transport. The blue line represents a curve $x^i(s)$ and s is a parameter. The figure shows that when a vector W (green) at A is reproduced in B, its components undergo a change that is first-order in dx_1. The difference between the covariant basis vectors based at points that lie at infinitesimal distances is

$$d\vec{e}_i = \Gamma_{ij}^k dx^j \vec{e}_k, \tag{7.13}$$

with Γ_{ij}^k the Christoffel symbol of the second kind, also called affine connections. Consequently, the contravariant components of a constant vector W vary by

$$\delta W^i = -\Gamma_{kl}^i W^k dx^l. \tag{7.14}$$

Note that since $\frac{\partial \vec{e}_i}{\partial x^k} = \frac{\partial}{\partial x^k}\frac{\partial \vec{R}}{\partial x^i} = \frac{\partial}{\partial x^i}\frac{\partial \vec{R}}{\partial x^k} = \frac{\partial \vec{e}_k}{\partial x^i}$, it holds that

$$\Gamma_{ij}^k = \Gamma_{ji}^k.$$

Similarly, contravariant basis vectors at different points are not parallel, but the change of $\vec{e}^i$ when shifting the point by dx^j must be linear in the shift; when the change is expanded on the original basis, it is of the form

$$d\vec{e}^i = -\Gamma_{jk}^i dx^j \vec{e}^k; \tag{7.15}$$

the fact that the coefficients are indeed $-\Gamma_{jk}^i$ may be verified starting from $\vec{e}^i \cdot \vec{e}_k = \delta_{ik}$ which implies that $\frac{\partial}{\partial x_j}\vec{e}^i \cdot \vec{e}_k = 0 = (\frac{\partial \vec{e}^i}{\partial x^j}) \cdot \vec{e}_k + \vec{e}^i \cdot (\frac{\partial \vec{e}_k}{\partial x^j})$.

The covariant components of a constant vector W vary by

$$\delta W_i = \Gamma_{il}^k W_k dx^l. \tag{7.16}$$

This implies that the parallel transport on a closed circuit produces a change of a constant vector by

$$\Delta W_i = \oint \Gamma_{il}^k W_k dx^l. \tag{7.17}$$

Imagine a pendulum swinging at the North Pole along the meridian of Greenwich. If it is carefully taken to the Equator by parallel transport, it will keep swinging along the meridian of Greenwich; now, if it is taken along the Equator to the meridian of Rome and then taken back to the North Pole, it will be found to oscillate along the meridian of Rome. This is clearly a consequence of the curvature of the Earth. We shall soon find that actually Eq. (7.17) is the starting point for defining the curvature tensor.

It easy to generalize the above discussion about parallel transport to variable vectors.

The del operator

$$\nabla = \vec{e}^{\,i} \frac{\partial}{\partial x^i}$$

is an invariant, since by contracting vectors, one obtains scalars. This feature is useful when one wishes to differentiate tensors while keeping track of the tensorial character of the results. In particular,

$$\nabla \vec{A} = \vec{e}^{\,i} \frac{\partial}{\partial x^i} (A^j \vec{e}_j) = \vec{e}^{\,i} \vec{e}_j \frac{\partial A^j}{\partial x^i} + \vec{e}^{\,i} A^j \Gamma_{ji}^m \vec{e}_m$$

is a second-rank tensor, with components

$$A_{;i}^m = \frac{\partial A^m}{\partial x^i} + A^j \Gamma_{ji}^m; \tag{7.18}$$

the ; notation is used for the covariant differentiation, which takes the variation of the basis vectors into account. This means that one takes into account the effect of the curved coordinate lines on the differential by the substitution

$$dA^m = \frac{\partial A^m}{\partial x^s} dx^s \rightarrow DA^m = \left(\frac{\partial A^m}{\partial x^s} + \Gamma_{is}^m A^i \right) dx^s. \tag{7.19}$$

For a covariant vector, the rule is:

$$dA_m = \frac{\partial A_m}{\partial x^s} dx^s \rightarrow DA_m = \left(\frac{\partial A_m}{\partial x^s} - \Gamma_{ms}^i A_i \right) dx^s. \tag{7.20}$$

The rule extends to the tensors of any rank. If

$$T = T_{ik}\, \vec{e}^{\,i}\vec{e}^{\,k},$$

$$\nabla T = T_{ik;l}\, \vec{e}^{\,i}\vec{e}^{\,k}\vec{e}^{\,l}$$

with

$$T_{ik;l} = \frac{\partial T_{ik}}{\partial x^l} - T_{mk}\Gamma^m_{il} - T_{im}\Gamma^m_{lk}. \tag{7.21}$$

In addition, in agreement with the general rule for raising and lowering indices, $\Gamma_{i,kl} = g_{im}\Gamma^m_{kl}$ and $\Gamma^m_{kl} = g^{im}\Gamma_{m,kl}$. The derivatives of the metric tensor $g_{jm} = \mathbf{e}_j.\mathbf{e}_m$ with respect to the coordinates involve the affine connections: $\frac{\partial}{\partial x^k}g_{jm} = \vec{e}_j \frac{\partial}{\partial x^k}\vec{e}_m + \vec{e}_m \frac{\partial}{\partial x^k}\vec{e}_j$ and one finds that:

$$\frac{\partial}{\partial x^k}g_{jm} = \Gamma^p_{mk}g_{jp} + \Gamma^p_{jk}g_{pm}.$$

Hence one obtains Ricci's lemma

$$g_{ik;l} = 0. \tag{7.22}$$

However, if the metric tensor is given, such relations can be used to obtain the affine connections. With $(k, j, m) \to (j, m, k)$,

$$\frac{\partial}{\partial x^j}g_{mk} = \Gamma^p_{kj}g_{mp} + \Gamma^p_{mj}g_{pk}.$$

With $(k, j, m) \to (m, k, j)$, instead,

$$\frac{\partial}{\partial x^m}g_{kj} = \Gamma^p_{jm}g_{kp} + \Gamma^p_{km}g_{pj}.$$

The sum of the first two minus the third is:

$$\frac{\partial}{\partial x^k}g_{jm} + \frac{\partial}{\partial x^j}g_{mk} - \frac{\partial}{\partial x^m}g_{kj}$$

$$= \Gamma^p_{mk}g_{jp} + \Gamma^p_{jk}g_{pm} + \Gamma^p_{kj}g_{mp} + \Gamma^p_{mj}g_{pk} - (\Gamma^p_{jm}g_{kp} + \Gamma^p_{km}g_{pj})$$

$$= 2g_{pm}\Gamma^p_{jk}.$$

To obtain the connection, multiply by g^{im} and sum over m:

$$\Gamma^k_{ij} = \frac{1}{2}g^{im}\left(\frac{\partial g_{jm}}{\partial x^k} + \frac{\partial g_{mk}}{\partial x^j} - \frac{\partial g_{kj}}{\partial x^m}\right). \tag{7.23}$$

For a curve $x^i(s)$, where s is a parameter like a curvilinear length, the tangent at a point of coordinates x^i has components $\frac{dx^i}{ds}$, and so the tangent at a point of coordinates $x^i + dx^i$ has components $\frac{dx^i}{ds} + \frac{d^2 x^i}{ds}$, $i = 1, d$. There is an important special case to consider, i.e., the case of geodesic.

Problem 23 Find the non-vanishing Christoffel symbol for spherical geometry in 3d.

Solution 23

$$\Gamma_{22}^1 = -r, \; \Gamma_{12}^2 = r^{-1} = \Gamma_{13}^3 = \Gamma_{31}^3, \Gamma_{33}^1 = -r \sin^2 \theta, \; \Gamma_{23}^3 = \cot \theta = \Gamma_{32}^3,$$

$$\Gamma_{33}^2 = -\sin(\theta)\cos(\theta).$$

7.2 Geodesics

Let us start with a simple particular case. Suppose the space is $\mathbb{R}^2$, and the geodesic between two points is a straight line, which is the shortest line between any two points A and B, and its tangent is the same at all of its points and coincides with the straight line itself. Next, we decide to go to some set of curvilinear coordinates. Now the images of A and B are connected by the image of the straight line, the distance AB remains the same and the image of the straight line becomes the geodesic between A and B. We can search for it, requiring that the tangent t be the same at $x^i + dx^i$ as in x^i, except that its components change solely because of the change of the local reference according to (7.15). The components of the tangent vector are

$$t_i = \frac{dx_i}{ds}. \tag{7.24}$$

$$\frac{dt}{ds} = 0 = \frac{dt^i e_i}{ds} = \frac{dt^i}{ds} e_i + t^i \frac{de_i}{ds},$$

and inserting $\frac{de_i}{ds} = \Gamma_{ij}^k \frac{dx_j}{ds} e_k$,

$$0 = \frac{dt^i}{ds} e_i + t^i \Gamma_{ij}^k \frac{dx_j}{ds} e_k.$$

Now, reshuffling indices with $k \to i, i \to j, j \to k$ in the second term, we get:

$$0 = \left(\frac{dt^i}{ds} + t^j \Gamma_{jk}^i \frac{dx_k}{ds} \right) e_i.$$

Thus, using (7.24), we obtain the equation for the geodesic, namely,

$$\frac{d^2x^i}{ds^2} + \Gamma^i_{jk}\frac{dx^j}{ds}\frac{dx^k}{ds} = 0. \tag{7.25}$$

As noted above, a geodesic has another notable property: it is a length-minimizing curve. This property remains true, and also the above relations continue to hold, even if the base space is curved and has many dimensions. There is a pleasant and useful trick to generate geodesics, which is also a shortcut for evaluating the Γ connections. Consider the function

$$\mathcal{L}_{Geo}(x, \dot{x}) = \frac{1}{2}g_{mn}\dot{x}^m\dot{x}^n, \tag{7.26}$$

where the dots denote derivations with respect to s. Its dependence on coordinates and *velocities* suggests that it should be treated like a *Lagrangian*. I am going to show that the Euler-Lagrange equations $\frac{d}{ds}\frac{\partial\mathcal{L}_{Geo}}{\partial\dot{x}^p} - \frac{\partial\mathcal{L}_{Geo}}{\partial x^p} = 0$ coincide with the geodesic equations in covariant form. Indeed,

$$\frac{\partial\mathcal{L}_{Geo}}{\partial\dot{x}^p} = \frac{1}{2}(g_{mn}\delta^m_{np}\dot{x}^n + g_{mn}\delta^n_{np}\dot{x}^m) = g_{pn}\dot{x}^n$$

and

$$\frac{\partial\mathcal{L}_{Geo}}{\partial x^p} = \frac{1}{2}\partial_p g_{mn}\dot{x}^m\dot{x}^n.$$

Thus, one finds the equations of motion

$$\frac{d}{ds}g_{pn}\dot{x}^n = \frac{1}{2}(\partial_p g_{mn})\dot{x}^m\dot{x}^n.$$

Doing the derivative,

$$g_{pn}\ddot{x}^n + \partial_m g_{pn}\dot{x}^m\dot{x}^n - \frac{1}{2}\partial_p g_{mn}\dot{x}^m\dot{x}^n = 0.$$

Then, one symmetrizes $\partial_m g_{pn}\dot{x}^m\dot{x}^n$ with respect to the exchange of the dummy indices,

$$\partial_n g_{pm}\dot{x}^m\dot{x}^n \rightarrow \frac{1}{2}\partial_n g_{pm}\dot{x}^m\dot{x}^n + \frac{1}{2}\partial_m g_{pn}\dot{x}^m\dot{x}^n.$$

In this way, $\ddot{x}^m + \Gamma_{mpn}\dot{x}^p\dot{x}^n = 0$, which yields the geodesic equation by raising an index.

Curvature Tensor

The difference between the components of a vector at two nearby points is

$$d\vec{A} = A^k_{;j}dx^j\vec{e}_k, \tag{7.27}$$

where

$$A^k_{;j} = \left(\frac{dA^k}{dx^j} + A^i \Gamma^k_{ij}\right).\tag{7.28}$$

A vector such that $A^k_{;j} = 0$ does not change if $x_j \to x_j + dx_j$, and so it undergoes parallel transport.

The Riemann–Christoffel curvature tensor R can be obtained as follows. By applying the Stokes theorem to (7.17), one finds that $2\Delta W_k = R^i_{klm} W_i \Delta f^{lm}$, where Δf^{lm} is the surface element and

$$R^m_{ikl} = \Gamma^n_{il}\Gamma^m_{nk} - \frac{\partial}{\partial x^l}\Gamma^m_{ik} - \Gamma^n_{ik}\Gamma^m_{nl} + \frac{\partial}{\partial x^k}\Gamma^m_{il}.\tag{7.29}$$

In general, for a vector A, one verifies that

$$A_{i;k;l} - A_{i;l;k} = R^m_{ikl} A_m.\tag{7.30}$$

A nonzero R^m_{ikl} implies that the space is curved; It has d^4 components, where d is the dimensionality of the space. There are many symmetry relations among them. The Bianchi identity states that

$$R^m_{ikl;n} + R^m_{ink;l} + R^m_{iln;k} = 0.\tag{7.31}$$

The Ricci tensor is obtained by the following contraction:

$$R_{ab} = R^c_{abc}.\tag{7.32}$$

One finds the symmetric tensor:

$$R_{ik} = \frac{\partial \Gamma^l_{ik}}{\partial x^l} - \frac{\partial \Gamma^l_{il}}{\partial x^k} + \Gamma^l_{ik}\Gamma^m_{lm} - \Gamma^m_{il}\Gamma^l_{km}.\tag{7.33}$$

Moreover, the curvature scalar is defined by

$$R = R^a_a.\tag{7.34}$$

Spaces of constant curvature are of special interest. Let us consider a four-dimensional space with Cartesian coordinates x_i, $i = 1\ldots4$, in which we wish to embed a thee-dimensional constant curvature subspace. Immediately, one thinks about the 3-d surface of a 4-sphere. Let the radius be $a > 0$. The surface is defined by $\sum_i^4 x_i^2 = a^2$, and one can set up spherical coordinates in the 3-d surface with $r^2 = \sum_i^3 x_i^2$. Thus, on the surface of the hypersphere,

$$dx_4 = -\frac{x_1 dx_1 + x_2 dx_2 + x_3 dx_3}{\sqrt{a^2 - \sum_i^3 x_i^2}} = \frac{dr^2}{\sqrt{a^2 - r^2}}.$$

The metric in this space is:

$$dl^2 = \frac{dr^2}{1 - \frac{r^2}{a^2}} + r^2(d\theta^2 + \sin^2(\theta)d\phi^2). \tag{7.35}$$

This is an isotropic space that has a finite volume and constant curvature. It is customary to put $r = a\sin(\chi)$ with χ in the interval $(0, \pi)$ in order to simplify the first term and bring a in front of the expression; then,

$$dl^2 = a^2[d\chi^2 + \sin^2(\chi)(d\theta^2 + \sin^2(\theta)d\phi^2)]. \tag{7.36}$$

There is, however, another possible 3d constant curvature hypersurface, which is obtained by letting a to be imaginary, that is, replacing a^2 with $-a^2$ throughout. Then, the metric becomes

$$dl^2 = \frac{dr^2}{1 + \frac{r^2}{a^2}} + r^2(d\theta^2 + \sin^2(\theta)d\phi^2). \tag{7.37}$$

In this case, the first term is simplified if we put $r = a\sinh(\chi)$ and we may write

$$dl^2 = a^2[d\chi^2 + \sinh^2(\chi)(d\theta^2 + \sin^2(\theta)d\phi^2)]. \tag{7.38}$$

Chapter 8
Gravity

General Relativity Theory was admired by Lev Landau as the most beautiful, many years ago. Later, a series of striking experiments showed that it is extremely successful and far reaching, too. Einstein expected that many predictions could not be tested experimentally; now there is an impressive body of evidence with practical applications in Science, and also in everyday life. This chapter is not a substitute for a full course, but will introduce the reader to the main concepts and to recent developments.

8.1 Principle of Equivalence

Hidden in the Special Relativity, there are important hints that led Einstein to its generalization. The paradox named after Paul Eherenfest dates back to 1909. He considered a rigid cylinder rotating around its axis. By symmetry, the section of the cylinder must remain circular, and the radius R should not be affected by the motion, since it is always orthogonal to the velocity. But the circumference can be visualized as a polygon with many sides, and they move parallel to the velocity $v = \omega R$. So, Eherenfest concluded that the length of the circumference in the laboratory frame K should be $2\pi R\sqrt{1 - \frac{v^2}{c^2}}$, and this was a striking paradox. The problem was somewhat obscured by complications concerning the elastic response of the material constituting the cylinder, and by the practical impossibility of performing this experiment in the laboratory. However, Einstein pointed out the weak point of the above argument: it is not clear how Eherenfest would determine the length of the moving circle. The thought experiment must be done correctly. For example, if the cylinder could be measured when it is fixed and afterwards it could be set in motion; but in this case, Special Relativity cannot tell what the effects of the acceleration are. The safe procedure requires adopting the reference K′ which is rotating with the cylinder. For reasons of symmetry, a circumference in K is also a circumference in K′, but in K′ a length along the circumference is a proper length, and one can measure it using small rods. An observer in the inertial system K could count them, but would

© Springer International Publishing AG, part of Springer Nature 2018
M. Cini, *Elements of Classical and Quantum Physics*,
UNITEXT for Physics, https://doi.org/10.1007/978-3-319-71330-4_8

find that they are Lorentz contracted. Therefore, the solution of the paradox[1] is that in K' more rods are needed, and so the length is *increased* to $\frac{2\pi R}{\sqrt{1-\frac{v^2}{c^2}}}$, while in K the length is $2\pi R$, as it should be, according to the Euclidean geometry. The physical difference between K and K' is that K is inertial, while in K', there are inertial forces. The Euclidean rules do not apply in a curved space. A somewhat similar situation occurs in a straight route from the North Pole to Rome then along the 41.9th parallel; it would result that the parallel is shorter than 2π times the Rome-pole distance, because the Earth is almost spherical, and so plane Geometry does not apply. We met this argument already - recall Eqs. (7.17) and (7.29). This analogy suggests that the anomalous length of the circumference is the result of a curvature of space-time, and also of three-dimensional space, which is related to the accelerated path. Thus, Einstein's crucial point is that the Euclidean Geometry holds in inertial systems but not in accelerated ones. The observer in K' should feel inertial forces and note that the clocks that are further from the origin run slower than those that are nearer. We have already seen that Classical Mechanics allows us to choose any reference system and the inertial forces are automatically generated by the Lagrangian formalism in a simple way; therefore, the extension of the theory to include accelerated systems is a logical necessity in the first place. The above example reveals that the inertial forces are related to a more general geometry of space-time, and this is in line with the well-known fact that they produce accelerations (e.g., centrifugal and Coriolis) that are *mass-independent*. An elephant and a mosquito receive the same acceleration from a rotating platform. But this mass-independence of the acceleration has another time-honored, celebrated occurrence, namely, Gravity.

Throwing from the tower of Pisa two spheres of the same radius,[2] but different weight, Galileo allegedly demonstrated experimentally that all bodies have the same acceleration and touch the ground together. This statement is known as the weak equivalence principle. Aristotle had stated that the heavier mass must arrive before the light mass. Galileo arrived at a result that corrected a seemingly intuitive millennial error supported by a long indisputable authority, made unquestionable by the Church. Indeed, he guessed the result of experiments that physicists would continue to perform for centuries with increasing accuracy. Eötwös reached an accuracy of $5\,10^{-9}$ in 1922, and in 2012, Lunar Laser Ranging claimed an accuracy of 10^{-13}.

But Einstein realized that he needed a more general equivalence principle than that. He postulated that the outcome of any non-gravitational *local* experiment in a free-falling laboratory is independent of velocity and location in space. Local means in a region where fields are uniform.[3] This is the *Einstein principle of equivalence*. Einstein remarked that this principle implies the complete physical equivalence of a gravitational field and a corresponding acceleration of the reference system, and also

---

[1] A more formal derivation of this result will be found shortly.

[2] in order to eliminate the effects of the resistance of the air.

[3] The strong equivalence principle states that the outcome of any local experiment (involving Gravity or not) in a free-falling laboratory is independent of velocity and location. This requires that the gravitational constant G be the same at any location in space-time.

that no local experiment can distinguish between gravitational and inertial forces; all such forces have a geometrical nature.

The same principle is in operation on a satellite in circular orbit of radius R around some planet or star. If at the center of the circle, there is a star of mass M, the force felt by an Astronaut would be expected classically to be $F = m\frac{v^2}{R} - \frac{GM}{R^2}$. By suitably selecting v, the force F vanishes, so our Astronaut can be in free-fall. This is well known to happen in space laboratories in orbit. Moreover, the Astronaut could note that at smaller R, the attraction of the star prevails, while at larger R the centrifugal force wins. Therefore, the vanishing of F is only local, and extended objects are stretched by tidal forces. The fact remains that there is a trade-off between the field of gravity and an accelerated reference system. The equilibrium between forces of different origin is a familiar situation, but the striking fact is that the same situation holds regardless of the mass of the body, since the m that enters the centrifugal term (inertial mass) is the same as the m in the second term (gravitational mass). Therefore, inertial and gravitational forces are essentially the same sort of thing, and we may identify the two masses, although a priori they could be different.

We can already get interesting qualitative results through other thought experiments based on the Principle of equivalence. Consider an observer in a cabin without windows inside a spaceship; he drops a stone from a height of one meter, and finds that it falls downwards with acceleration g, directed, say, along the negative z axis. Suppose g is the same as the acceleration of gravity on Earth. The observer stands normally and does not float. He concludes that the ship probably stands on the ground, or maybe it flies like a plane, without a vertical acceleration. However, that conclusion is not granted. The ship could be in deep space, where the force of gravity is negligible, but subject to an upwards acceleration g. According to the Einstein principle of equivalence, no experiment made inside the cabin can distinguish between the two situations.

Now assume that a light ray with the Poynting vector along the x axis enters horizontally from a small hole in a wall of the cabin, at $z = z_0$. We know from Special Relativity that in any inertial system, independent of the vertical velocity of the cabin, the light propagates in a straight line, producing a spot on the opposite side of the cabin, still at $z = z_0$. If, instead, the ship were accelerated towards the high, the beam should sag and hit the opposite wall at a spot at $z < z_0$. In the time it takes for the light to cross the cabin, the latter has gained speed and advanced more than in the inertial situation. In addition, the light at the spot should be Doppler shifted towards the blue, because the Poynting vector there should have an additional negative component along z. But then, according to Einstein, the light *falls* in a field of gravity; although it has no mass, the ray bends and changes color. These effects are not contained in the Maxwell equations, which are implicity intended for inertial systems. The color change has been verified in the Laboratory, but is of special interest in Astrophysics as the *gravitational red shift* of the light coming from massive compact stars. To make the argument more quantitative, we can go back to the ship accelerated upwards with acceleration g. We set a light source at height h with respect to the detector; during the journey of the light, which lasts a time $\Delta t \sim \frac{h}{c}$, the speed of the spaceship changes by $\Delta v \sim \frac{gh}{c}$; the Doppler blue shift is $\frac{\Delta \nu}{\nu} \sim \frac{\Delta v}{c} \sim \frac{gh}{c^2}$. For any material body of mass m, this ratio could be rewritten as the ratio $\frac{mgh}{mc^2} = \frac{m\Delta\Phi}{mc^2}$ of the change in gravitational potential energy $mgh = m\Delta\Phi$ to the rest energy. $\Delta\Phi$ is the drop in gravitational potential. Here, the Principle of Equivalence enters. In a field of gravity, $\frac{\Delta \nu}{\nu} \sim \frac{\Delta\Phi}{c^2}$. The light falling toward the star shifts toward the blue, but the light reaching us from the star reddens.[4]

The ratio $\frac{m\Delta\Phi}{mc^2}$ of the gravitational energy to the rest energy is always small in laboratory experiments; it becomes of order unity in the case of black holes. Any mass M becomes a black hole if it is confined to a Schwarzschild radius $r_S = \frac{2GM}{c^2}$. This is very small compared to the actual size of most objects, except collapsed stars (see the following table).

Object	mass (Kg)	Radius R(m)	r_S	$\frac{r_S}{R}$
nucleus	10^{-26}	10^{-15}	10^{-53}	10^{-38}
atom	10^{-26}	10^{-10}	10^{-53}	10^{-43}
Earth	$6 \ 10^{24}$	$6 \ 10^6$	$9 \ 10^{-3}$	10^{-9}
white dwarf	$2 \ 10^{30}$	10^7	$3 \ 10^3$	$3 \ 10^{-4}$
neutron star	$2 \ 10^{30}$	10^4	$3 \ 10^3$	$3 \ 10^{-1}$
sun	$2 \ 10^{30}$	$7 \ 10^8$	$3 \ 10^3$	10^{-6}
galaxy	10^{41}	10^{21}	10^{14}	10^{-7}

But in order to achieve a coherent formulation of the Principle of Equivalence, another principle is needed.

8.2 The Principle of General Covariance and the Curved Space-Time

In accelerated frames and/or in the presence of gravity, the space-time is no longer flat, so we must be able to work with curvilinear coordinates x^μ, $\mu = 0, 1, 2, 3$ and to cope with general changes of frame to new coordinates x'^μ; the transformation needs to be invertible and differentiable, for physical reasons, but otherwise is general. The *principle of General Covariance* (also known as the *principle of General Relativity*) states that the laws of physics must be the same equations for *all* observers, inertial or not. Removing the inertial frames from their exclusive role in Special Relativity, makes things much more complicated but was logically necessary. We have noted that already the Lagrangian formulation of Classical Mechanics allows to work in any reference, and leads to a much deeper understanding of the theory.

---

[4]Again, this implies that clock go slower in a field of gravity, as they do under the action of a centrifugal force.

In Special Relativity, the infinitesimal interval between two events is

$$ds^2 = \eta_{\mu\nu} dx^\mu dx^\nu, \tag{8.1}$$

with $\eta_{\mu\nu} = 0$ for $\mu \neq \nu$ and $\eta_{00} = -1, \eta_{11} = \eta_{22} = \eta_{33} = 1$; this means that the geometry is pseudo-Euclidean and the space-time is flat.[5] A change of inertial reference may be done by combining the special Lorentz transformation

$$x_0' = \gamma(x_0 - \beta x_1), x_1' = \gamma(x_1 - \beta x_0), x_2' = x_2, x_3' = x_3$$

with translations and rotations. We must generalize Equation (8.1) by introducing a symmetric[6] *metric tensor* with covariant components $g_{\mu\nu}$ by setting

$$ds^2 = g_{\mu\nu} dx^\mu dx^\nu \tag{8.2}$$

From $dx'^\lambda = \frac{\partial x'^\lambda}{\partial x^\mu} dx^\mu, x'^\rho = \frac{\partial x'^\rho}{\partial x^\nu} dx^\nu$, one can deduce how the covariant tensor components must transform:

$$g_{\mu\nu} = g'_{\lambda\rho} \frac{\partial x'^\lambda}{\partial x^\mu} \frac{\partial x'^\rho}{\partial x^\nu}. \tag{8.3}$$

Equation (8.3) is the covariant transformation law for the components tensors with two indices. The contravariant components (see Chap. 7) transform according to the rule

$$g^{\mu\nu} = g'^{\lambda\rho} \frac{\partial x^\mu}{\partial x'^\lambda} \frac{\partial x^\nu}{\partial x'^\rho}. \tag{8.4}$$

As in Special Relativity, a central role is played by a variety of functions of the space-time point P that remain invariant under a change of frame. Collectively, these quantities (that may be scalars; other objects like 4-vectors and more general objects with many components) are called tensors. Generally the tensor components vary in such a way as to compensate for the frame change; some tensors generalize those of Special Relativity, but new tensors are also needed. We shall use the essential formalism summarised in the last chapter.

---

[5]I adopt the convention used by Landau-Lifschitz, Stephani and probably the majority of Authors. Moreover, the Greek indices run from 0 to 3 and the Latin indices from 1 to 3.

[6]g is a symmetric tensor, $g_{\mu\nu} = g_{\nu\mu}$, while the electromagnetic tensor F is antisymmetric. In the unified field theory proposed by Einstein and Schrödinger in the 50s the field is described by a tensor with a symmetric gravitational part and antisymmetric electromagnetic contribution.

8.2.1 Space Geometry in Stationary Problems

For two events that take place at the same point and almost at the same time, $dx^i = 0$ for i $= 1, 2, 3$ and the interval is $ds^2 = -c^2 d\tau^2 = g_{00} dx_0^2$, where τ is the proper time, and so

$$d\tau = \sqrt{-g_{00}} dx^0. \tag{8.5}$$

Clocks show their own proper time τ, not the coordinate time x_0. However from a calculated dx_0, one can obtain the physical $d\tau$ if g_{00} is known. In general, space and time are mixed in such a way that one cannot even define a space metric at a given time. In Special Relativity, in order to measure the distance to a far object, one can use the exchange of light signals and the fact that the proper time the light takes to cover any distance is zero. However, in the presence of gravity, this is not possible in principle. This is a fundamental complication when we try to define the distance of a galaxy. Apart from the obvious practical impossibility to exchange signals with remote objects like those considered in Astrophysics, the trouble is that the light can follow different routes while the Universe expands and the galaxies move.

However, some interesting problems are stationary, and we can (so to speak) separate time from space. Consider two nearby points A and B separated by an infinitesimal dx^μ. In principle, the exchange of signals can be done, although the light does not take the same coordinate time to go from A to B as it takes in the return trip. Nevertheless, imposing $ds^2 = 0$ (as appropriate to the propagation of light), one finds two roots corresponding to the coordinate times taken:

$$dx^0 = \frac{g_{0a} dx^k \pm \sqrt{(g_{0i} g_{0k} - g_{00} g_{ik}) dx^i dx^k}}{-g_{00}}, k = 1, 2, 3. \tag{8.6}$$

The difference between these times is the coordinate time taken by the exchange of signals. By using the above result, one finds the proper time $d\tau$ at A; multiplication by $c/2$ gives the distance AB. Thus, in the case of stationary problems, when $g_{\mu\nu}$ is time-independent, one can define the 3 × 3 metric tensor of space γ_{ij}, such that the element of distance is

$$dl^2 = \gamma_{ij} dx^i dx^j, \tag{8.7}$$

where i, j $= 1, 2, 3$. It turns out that

$$\gamma_{ij} = g_{ij} - \frac{g_{0a} g_{0b}}{g_{00}}. \tag{8.8}$$

8.2.2 Curved Space in a Rotating Frame

As a clear example of the mixing of space and time, the transformation (2.32) to a rotating reference takes from $-ds^2 = c^2 dt_0^2 - dx_0^2 - dy_0^2 - dz_0^2$ to

$$ds^2 = [c^2 - \omega^2(x^2 + y^2)]dt^2 - 2\omega(ydx - xdy)dt - dx^2 - dy^2 - dz^2. \quad (8.9)$$

This implies that $g_{\mu\nu} = \eta_{\mu\nu} + h_{\mu\nu}$, where $h_{\mu\nu}$ are the elements of the matrix

$$\begin{pmatrix} -\frac{\omega^2}{c^2}(x^2 + y^2) & -\frac{\omega y}{c} & \frac{\omega x}{c} & 0 \\ -\frac{\omega y}{c} & 0 & 0 & 0 \\ \frac{\omega x}{c} & 0 & 0 & 0 \\ 0 & 0 & 0 & 0 \end{pmatrix}.$$

Already, this simple example shows some features that we shall encounter later: the classical potential (centrifugal in this case) is in h_{00}, while h_{0i} shows the mixing of time with x and y that is related to the velocity-dependent Coriolis force. In this case, one finds the distance element

$$dl^2 = dr^2 + dz^2 + \frac{r^2 d\phi^2}{1 - \omega^2 \frac{r^2}{c^2}}. \quad (8.10)$$

This implies that an observer at rest in the rotating system would find that a circumference centered on the origin of the plane z = 0 is longer than $2\pi r$ by a factor $\sqrt{1 - \omega^2 \frac{r^2}{c^2}}$, as stated earlier by a thought experiment about the Eherenfest paradox.

8.2.3 Generalized Equation of Motion

In addition, we need extended versions of the equations of motion for a test mass, and also of the Maxwell equations. To satisfy the general covariance requirement, all quantities appearing in the fundamental equations must be tensors and all the derivatives must be replaced with covariant derivatives of the type (7.19), that is, $d \rightarrow D$. Therefore, the equation of motion of a test mass m subject to a force field of electromagnetic origin f must become

$$m \left(\frac{d^2 x^\mu}{d\tau^2} + \Gamma^\mu_{\nu\sigma} \frac{dx^\nu}{d\tau} \frac{dx^\sigma}{d\tau} \right) = f^\mu, \quad (8.11)$$

and in the absence of f, the mass follows a geodesic in space-time. Now it is obvious that the term involving the connection coefficients stands for effective inertial and gravitational forces. This allows us to figure out a physical and dynamical significance to the metric tensor and its derivatives, at least in problems in which the relativistic effects are small. As in the last subsection, one starts by writing

$$g_{\mu\nu} = \eta_{\mu\nu} + h_{\mu\nu}, \quad (8.12)$$

where $\eta_{\mu\nu}$ is the limit of Special Relativity and $h_{\mu\nu}$ is a correction that is considered small neglecting its powers. Now we take advantage of the observation made earlier that the non-relativistic potential resides in the g^{00} element; this observation leads quickly to the correct result. If in Eq. (7.23), one throws away the off-diagonal g elements, one finds

$$2\Gamma_{ij}^k \sim g^{00}\frac{\partial g^{00}}{\partial x^k}\delta_{i0}\delta_{j0} + g^{00}\frac{\partial g^{00}}{\partial x^j}\delta_{i0}\delta_{k0} - g^{ii}\frac{\partial g^{00}}{\partial x^i}\delta_{k0}\delta_{j0},$$

which yields

$$\Gamma_{00}^k \sim \frac{1}{2}g^{00}\frac{\partial g^{00}}{\partial x^k}, \quad \Gamma_{0j}^0 \sim \frac{1}{2}g^{00}\frac{\partial g^{00}}{\partial x^j}, \quad \Gamma_{i0}^0 \sim -\frac{1}{2}g^{ii}\frac{\partial g^{00}}{\partial x^i}.$$

Now we are interested in the equation of motion ($\mu = k = 1, 2, 3$), so we use Γ_{00}^k, which we may further approximate as $\Gamma_{00}^k \sim \frac{1}{2}\frac{\partial h^{00}}{\partial x^k}$. It turns out that the equations of motion are approximated by

$$m\frac{d^2x^i}{dt^2} = -m\frac{\partial}{\partial x_i}\frac{mc^2}{2}h_{00}. \tag{8.13}$$

This is the Newtonian result and confirms that the 00 element in the non-relativistic limit is

$$g_{00} = 1 + \frac{2V}{c^2}, \tag{8.14}$$

where V is the gravitational potential in the weak field case.

8.2.4 Generalized Maxwell Equations

The electromagnetic tensor

$$F_{\mu\nu} = A_{\nu;\mu} - A_{\mu;\nu} = \frac{\partial A_\nu}{\partial x_\mu} - \frac{\partial A_\mu}{\partial x_\nu} \tag{8.15}$$

keeps its form because the terms involving the Γ symbols cancel each other; the first couple of Maxwell's equations also keeps the same form as that in Special Relativity. However, the second pair of equations must be written as

$$F_{;\nu}^{\mu\nu} = \frac{4\pi}{c}J^\mu. \tag{8.16}$$

A light ray with wave vector k^μ goes along a straight line $dk^\mu = 0$ if there are no gravity fields; in a field, this generalizes to $Dk^\mu = 0$. In other words, the ray follows a geodesic,

$$\frac{dk^\mu}{dx^\beta} + \Gamma^\mu_{\alpha\beta} k^\alpha = 0. \qquad (8.17)$$

8.3 Einstein Field Equations

The non-relativistic equations of motion of a test particle in a gravitational field can be obtained through a gravitational potential V given by the Poisson equation

$$\nabla^2 V = 4\pi G \rho(\mathbf{r}, t). \qquad (8.18)$$

Here, $\rho(\mathbf{r}, t)$ is the matter density, which is a scalar in the non-relativistic theory, and $G = 6.67408^{-11} m^3 kg^{-1} s^{-2}$. This scheme allowed for an immense conceptual progress and an accurate description of the working of the solar system. However, it is not tenable in the light of the Principles of General Relativity and Equivalence. Before Relativity, Oliver Heavyside suggested that the Poisson equation should be supplemented by some gravito-magnetic field in order to allow for a finite speed of propagation of the gravitational interactions.

We must summarize the main results of a research that engaged Einstein for quite a long time, namely, the field equations that should replace the Newton equation (8.18). A general theory cannot be deduced from a special theory, so he had to choose the best conjecture. Besides the two Principles, Einstein considered criteria of simplicity and elegance; he had a unique sense of Physics.

The classical scalar ρ is not available as such, because it is not a relativistic invariant, and must be replaced with the energy-momentum tensor $T_{\mu\nu}$ of Eq. (6.38); then, at the left hand side, we must put a tensor with two indices, and initially the curvature was the favorite. But it had to be traceless. It can be verified through lengthy algebra that the Einstein tensor

$$G^{\mu\nu} = R^{\mu\nu} - \frac{1}{2} g^{\mu\nu} R \qquad (8.19)$$

has the property $G^{\mu\nu}_{;\nu} = 0$, so it can replace the l.h.s. of Equation (8.18). The Einstein equations are:

$$G^{\mu\nu} + \Lambda g^{\mu\nu} = 8\pi G_N T^{\mu\nu}. \qquad (8.20)$$

Here, Λ is the so-called cosmological constant, which is a source of gravitational field arising out of the vacuum. It can be interpreted as the energy density of the vacuum of space and produces a sort of antigravity if $\Lambda > 0$; it was introduced by Einstein in 1917, when it was realized that otherwise an infinite static Universe which seemed likely at the time, should collapse like a finite Newtonian cloud of particles. But in 1929 Hubble discovered that the Universe was actually expanding and Einstein repudiated the Λ term. For many years most people thought that the expansion should be slowed down by gravity, but at the turn of the century the dimming of far supernovae was taken as evidence that the expansion is actually accelerating. Further evidence arose from galaxy surveys and from the cosmic microwave background. Then, Λ was restored, even if nobody knows how this antigravity should arise. In order to agree with the data, Λ must be so small as to be completely immaterial when dealing with the Solar System and its surroundings, yet when the large scale Universe is considered, it should be a manifestation of the Dark Energy that is thought to overweight all matter (baryonic, dark or whatever else). However we shall put $\Lambda = 0$ since this is appropriate for most purposes.

Lowering an index in (8.20) with $\Lambda = 0$, one finds

$$R^{\mu}_{\nu} - \frac{1}{2}g^{\mu}_{\nu}R = 8\pi G_N T^{\mu}_{\nu};$$

since $g^{\mu}_{\nu} = \delta^{\mu}_{\nu}$, one finds $R = -8\pi G_N T$ where $T = T^{\mu}_{\mu}$. Hence the Einstein equations with $\Lambda = 0$ can be rewritten in the form

$$R^{\mu\nu} = 8\pi G_N \left(T^{\mu\nu} - \frac{1}{2}T g^{\mu\nu} \right) \tag{8.21}$$

where $T = T^{\mu}_{\mu}$. I stress that these equations are very general since they apply for any distribution of masses however large and fast moving, observed in a reference system that can rotate and be subject to forces while the observer is free to adopt any arbitrary curvilinear space-time set of coordinates.

8.3.1 Linearized Field Equations

The field equations are highly non-linear. It must be so, since the field is energy and is itself a source of the field; by contrast, the Maxwell equations are linear, since the field is neutral. But nonlinear equations are hard to solve. Only for a few, symmetric situations have analytic solutions been found. In most cases, one must resort to numerical methods. If the field is weak, $h_{\mu\nu}$ is small in Eq. (8.12), and it is possible to linearize and gain some physical insight in general. Linearizing, one finds that

$$\Gamma^{\sigma}_{\mu\nu} = \frac{1}{2}\eta^{\sigma\tau}\left(\frac{\partial h_{\mu\tau}}{\partial \nu} + \frac{\partial h_{\tau\nu}}{\partial \mu} - \frac{\partial h_{\mu\nu}}{\partial \tau} \right). \tag{8.22}$$

Since the h corrections to the metric become potentials in the weak field case, the Christoffel symbol is a combination of field components. Then, letting $h = h^{\alpha}_{\alpha}$, and using η instead of g to raise and lower indices, one finds through tedious algebra that the Ricci tensor is approximated by

$$R_{\mu\nu} = \frac{1}{2}\left(\frac{\partial^2 h_{\mu}^{\sigma}}{\partial x^{\nu}\partial x^{\sigma}} + \frac{\partial^2 h_{\nu}^{\sigma}}{\partial x^{\mu}\partial x^{\sigma}} - \Box h_{\mu\nu} - \frac{\partial^2 h}{\partial x^{\mu}\partial x^{\nu}} \right),\qquad(8.23)$$

where $\Box = \eta^{\mu\nu}\partial_{\mu}\partial_{nu}$. At this point, the reader could expect to find a wave equation yielding the recently detected gravitational waves. From the start, Einstein thought that the field should propagate with the speed of light, and the linearized equation in vacuo $R_{ik} = 0$ is not of the form of a wave equation, but is awfully complicated. This complexity essentially depends on our complete freedom to choose the coordinate system that might move about and rotate in an arbitrary way; in addition, there is a gauge invariance analogous to the electromagnetic one. In fact, one can show that by choosing the Fok gauge

$$\frac{\partial h_{\sigma}^{\nu}}{\partial \nu} - \frac{1}{2}\frac{\partial h}{\partial \sigma} = 0,\qquad(8.24)$$

we obtain simply

$$R_{\mu\nu} = -\frac{1}{2}\Box h_{\mu\nu}\qquad(8.25)$$

for the Ricci tensor in the limit of weak fields. Then, the Einstein equations reduce to

$$\Box h_{\mu\nu} = -\frac{16\pi G}{c^4}\left(T_{\mu\nu} - \frac{1}{2}T g_{\mu\nu} \right).\qquad(8.26)$$

Gravity waves will be discussed in Sect. 8.11. In the static case, only the only nonzero component is $T_0^0 = \rho c^2$, and one finds Poisson's equation for the 00 component and for the diagonal spatial components; consequently,

$$ds^2 = -\left(1 + \frac{2V}{c^2} \right)c^2 dt^2 + \left(1 - \frac{2V}{c^2} \right)(dr^2).\qquad(8.27)$$

8.4 Schwarzschild Solution and Black Holes

Already in the late eighteenth century, Pierre Simon, Marquis de Laplace, conjectured that stars having large mass and small radius could have an escape velocity larger than c, and therefore should be black and invisible. Let us see how black holes emerge from Relativity. The length $r_S = \frac{2MG}{c^2}$ is called the Schwarzschild radius for a body having mass M. In 1916, at the German-Russian front, Karl Schwarzschild found the static solution to the Einstein equations for the gravity field of a point-like spherical body in vacuum. In spherical coordinates, it reads as:

$$-ds^2 = \left(1 - \frac{r_S}{r} \right)c^2 dt^2 - \frac{dr^2}{1 - \frac{r_S}{r}} - r^2 d\theta^2 - r^2 \sin^2(\theta)d\phi^2.\qquad(8.28)$$

This is singular for $r = r_S$; therefore, the metric is valid for $r > r_S$. The meaning of the divergence is that an observer who is at rest at a great distances cannot observe what happens inside the so-called *horizon of events*. In order to understand the physical implications of the 4-d metric (8.28), we must use it to model thought experiments in which some observer measures times and lengths in terms of their own proper times and lengths. To start with, as in Special Relativity, $ds = cd\tau$, where τ is the proper time recorded by a clock placed at the chronotopic point specified as $(t = \frac{x_0}{c}, r, \theta, \phi)$. Then, what is the *coordinate time t*? The relation to the proper time is obtained by fixing the point, that is, by setting $dr = 0, d\theta = d\phi = 0$ in (8.28). One obtains

$$dt = \frac{d\tau}{\sqrt{1 - \frac{r_S}{r}}}. \tag{8.29}$$

Now it is clear that for $r \gg r_S$, the coordinate time t is close to the proper time τ, but in General Relativity, one is interested in the departures from the Newton theory and in situations when such departures may be substantial. In the same way, r cannot be identified with the distance from the center, since the metric blows up at $r = r_S$ and so the centre is off limits; moreover, setting in (8.28) $d\theta = d\phi = 0, dt = 0$ we get the equal-times interval, which is minus the square distance $dR^2 = ds^2 = \frac{dr^2}{1 - \frac{r_S}{r}}$; so, the infinitesimal proper length is $dR = \frac{dr}{\sqrt{1 - \frac{r_S}{r}}}$. Again, this diverges as $r \to r_S$. The distance between two r values, say, r_1 and r_2, is given by the integral $d[r_1, r_2, r_S] = \int_{r_1}^{r_2} dr \frac{dr}{\sqrt{1 - \frac{r_S}{r}}}$. This integral is elementary and the antiderivative is $\sqrt{r(r - r_S)} + r_S \ln[\sqrt{r} + \sqrt{r - r_S}]$ (Fig. 8.1).

The validity of the metric (8.28) ends at the horizon, but nothing special happens there. An observer right at the horizon of a large black hole would consider it a place like any other. A coordinate change named after Martin Kruskal and György Szekeres allows us to extend the validity (for a point mass) to all space-time (except for the central singularity). The transformation $(r, t) \to (u, v)$ is illustrated in Fig. 8.2. The

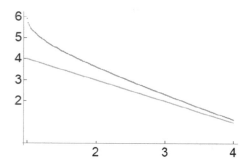

Fig. 8.1 The integral $d[r_1, r_2 = 5, r_S = 1]$ represents the coordinate distance between an observer at $r_2 = 5$ and a reckless companion at r_1, in units of r_S. Classically, when the companion approaches the horizon, the distance should approach $4r_S$. Instead, the relativistic trend is represented in the upper curve

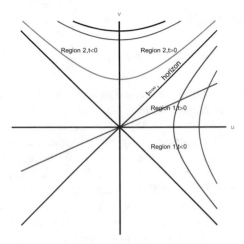

Fig. 8.2 The Kruskal diagram. Region 1 is split by the u axis into $t > 0$ above and $t < 0$ below. Region 1 is split by the v axis into $t > 0$ (right) and $t < 0$ (left). The straight border between the two regions corresponds to $r = r_S$ and $t = \infty$. In Region 1, the hyperbolas $v = \pm\sqrt{u^2 + 1 - \frac{2}{r_S}}$ are constant-r curves; the red one has $r = 1.1r_S$ and the blue one has $r = 1.2r_S$. The straight lines have constant t. In Region 2, the black hyperbola $v = \sqrt{1 + u^2}$ represents the singularity $r = 0$, the red one has $r = 0.5r_S$ and the pink one $r = 0.9r_S$

$u - v$ plane is divided into four regions by the lines $u = \pm v$. Region 1 represents the region outside the horizon and contains the positive u axis, while Region 2 represents the region inside the horizon and contains the positive v axis. The remaining half plane is not used by the above parametrization. The new coordinates u, v are defined in Region 1, that is, for $r > r_S$, by

$$u = \sqrt{\frac{r}{r_S} - 1}\, \exp\left(\frac{r}{2r_s}\right) \cosh\left(\frac{ct}{2r_s}\right) \tag{8.30}$$

$$v = \sqrt{\frac{r}{r_S} - 1}\, \exp\left(\frac{r}{2r_s}\right) \sinh\left(\frac{ct}{2r_s}\right), \tag{8.31}$$

and in Region 2, that is, for $r < r_S$ by

$$u = \sqrt{1 - \frac{r}{r_S}}\, \exp\left(\frac{r}{2r_s}\right) \sinh\left(\frac{ct}{2r_s}\right) \tag{8.32}$$

$$v = \sqrt{1 - \frac{r}{r_S}}\, \exp\left(\frac{r}{2r_s}\right) \cosh\left(\frac{ct}{2r_s}\right), \tag{8.33}$$

$$v^2 - u^2 = \left(1 - \frac{r}{r_S}\right) \exp\left(\frac{r}{r_S}\right). \tag{8.34}$$

With this transformation, the metric becomes:

$$ds^2 = \frac{4r_s^3}{r} e^{-\frac{r}{r_s}} (du^2 - dv^2) + r^2(d\theta^2 + \sin^2(\theta)d\phi^2), \tag{8.35}$$

where $r(u, v)$ is r expressed as a function of u and v. The angular dependence is $(r^2(d\theta^2 + \sin^2\theta d\phi^2)$, with $r = r(u, v)$.

One can verify that the metric (8.35) solves the Einstein equations everywhere outside the singularity. The diagram is quite effective in demonstrating the properties of the black hole. Suppose one wishes to send a signal from outside to the singularity. The starting space-time point must be chosen in Region 1. The propagation takes no proper time, so $ds = 0$ and $du = \pm dv$. Thus, the word line of the signal will be parallel to the $v = -u$ axis and will hit the singularity, necessarily after crossing the $t = \infty$ line. The physical meaning is that an observer in Region 1 will have to wait an infinite time to see the signal crossing the horizon. Similarly, if a material body is fired to the singularity, it will fall to it and be swallowed in a finite proper time; however, an outside observer will see it accelerating according to the Newtonian physics, but then decelerating, becoming dim and never reaching the horizon. The slowing down of clocks (or red-shift phenomenon) eventually wins.

8.5   Relativistic Delay of Signals in a Gravity Field

Shapiro, in 1964, suggested a test of General Relativity based on the following principle. A planet between Earth and the Sun or beyond the sun but visible near the solar disk can be used as a mirror that reflects a radio signal. The signal emitted from r_E and reflected at the planet in r_R will explore the metric (8.28) generated by the solar mass. The coordinate time length of the outward voyage can be obtained putting in (8.28) $d\tau = 0$ (since the signal travels at c) and $d\theta = d\phi = 0$. Integrating, and multiplying by 2 to include the return trip, one finds that $\Delta t = \frac{2}{c} \int_{r_E}^{r_R} \frac{dr}{1 - \frac{r_S}{r}}$.

This is the coordinate time interval. To convert it into the proper time, we use (8.6), do the integral and find

$$\delta\tau = \frac{2}{c}\sqrt{1 - \frac{r_R}{r_E}} \left[r_R - r_E + r_S \ln \left| \frac{r^E - r_S}{r^R - r_S} \right| \right]. \tag{8.36}$$

This is not the length of the trip divided by c. The gravity field of the sun causes a delay, which, for Mercury or Venus, is about $\frac{r_S}{c}[\ln(\frac{r_E}{r_R}) + \frac{r_E - r_R}{r_R}]$. Measurements performed on upper conjunctions of Venus (i.e. Venus close to the sun in the sky but beyond the sun) were already performed in 1970 and more measurements were done by the rockets Mariner 6 and Mariner 7, giving excellent agreement with the theory. One can be surprised by the apparent reduction of the speed of light from c to something like $c(1 - \frac{r_S}{r})$, where r is the distance from the sun. However this is an effect of curved space-time that can be rationalized in terms of an increase of the radial distances and a slowing down of the time in the neighbourhood of important masses. Any local measurement done by taking rods and clocks into the space close to the sun would give c for the speed of light, because rods and clocks would be affected themselves by the field.

8.6 Clocks in a Gravity Field, GPS and Gravitational Red Shift

Let us perform some thought experiments in the gravity field described by the Schwarzschild metric (8.28) of a massive body. The first experiment consists in the emission of a light signal from a point (r_E, θ_E, ϕ_E) at time t_E and its detection at the space-time point $(r_R, \theta_R, \phi_R, t_R)$. The light follows a null geodesic in space-time, and the time it takes, i.e. $t_R - t_E$, depends only on the two space points and can be obtained from the metric once the space trajectory of the light is specified; finding the trajectory is a problem of geometry that can be solved once and for all and does not concern us in this example. If the experiment is repeated after a delay Δt_E, the same delay affects the reception, so $\Delta t_R = \Delta t_E$. This entire story looks the same as that of the Newton dynamics, but it is different, because clocks measure the proper time τ, not t, and the two are related by Eq. (8.6). This implies that

$$\Delta \tau_R = \Delta \tau_E \sqrt{\frac{1 - \frac{r_S}{r_R}}{1 - \frac{r_S}{r_E}}}. \tag{8.37}$$

The actual repetition frequency is modified by the gravity field. In the next thought experiment, instead of measuring the repetition frequency we measure the color of the light and say that its frequency depends on the position in the field of the source. As before, the ratio of frequencies is

$$\frac{\nu_R}{\nu_E} = \sqrt{\frac{1 - \frac{r_S}{r_E}}{1 - \frac{r_S}{r_R}}}. \tag{8.38}$$

The weak field case can be obtained by expanding this result for small r_S; the same result is obtained setting $ds = cd\tau$ and $dr = 0$ in the approximate metric (8.27). The proper time τ of a clock is related to the time t measured by an identical clock outside the field by

$$d\tau = \sqrt{\left(1 + \frac{2V}{c^2}\right)} dt \sim \left(1 + \frac{V(r)}{c^2}\right) dt. \tag{8.39}$$

Since $U = 0$ at infinity and $U < 0$ in the field, a distant observer will conclude that the field makes the clocks run slow. Of course, the variable r that appears in these expressions is a coordinate radius, which is not simply the distance from the origin in the case of strong fields, near the horizon.

The gravitational redshift of spectral lines in the light coming from the sun, and particularly from dense stars like white dwarfs, was observed long ago.

For an atom A that emits at a proper frequency ν_A at distance r_A from a mass M, an observer B at distance r_B from M will see light at frequency ν_B

$$\frac{\nu_B - \nu_A}{\nu_A} \sim \frac{GM}{c^2}\left(\frac{1}{r_B} - \frac{1}{r_A}\right), \tag{8.40}$$

assuming that the fields are weak (the gravitational energy must be small compared to the rest energy). For an atom in the gravitational field of a star with radius R and Schwarzschild radius r_S, this implies a red shift of the frequency given by

$$\frac{\Delta\nu}{\nu} = \frac{r_0}{2R}. \tag{8.41}$$

For the Earth, $M = 6 \times 10^{24}$ kg, $r_S = 9 \times 10^{-3}$ m and $\frac{\Delta\nu}{\nu}$ is on the order of 10^{-9}, for the sun, $M = 2 \times 10^{30}$ kg, $R = 7 \times 10^8$ m, $r_S = 3$ km, and the shift is still one part in a million, but for a neutron star with one solar mass and a radius of order 10^4 m one expects $\frac{\Delta\nu}{\nu}$ is in the scale of 10^{-1}.

A measurement of the gravitational red shift in the laboratory was performed for the first time in 1965 by Pound and Snider using the Mössbauer effect, a laser and the Earth's gravity. Now this is routine. The clocks on the satellites used for the Geo-positioning System must be corrected to account for relativistic effects. They go slow (compared to clocks on Earth) by 7 ms/day because of the time-dilation effect of Special Relativity, but they go fast by 45 ms/day because of the gravitational effect. Such small times are important in practice, because light is so fast. Without the resulting correction of 38 ms/day the GPS of our cars would have errors on the order of kilometers.

8.7 Bending and Gravitational Lensing of Light

For the present purposes, it suffices to consider the propagation of a ray according to Eq. (8.17) in a weak field whose metric is given by (8.27) produced by a mass M centered at the origin. We shall keep the lowest order effects. Letting the ray come parallel to the x axis, with four-vector $k^\mu = (k^0(x), k^1(x), k^2(x), k^3(x))$ such that $k^1(-\infty) = k, k^2(-\infty) = k^3(-\infty) = 0$. Let the impact parameter be R. Assume that the ray is deflected by a small angle, which can be approximated by $\alpha = \frac{k^2(\infty)}{k^1}$. Since the covariant components of g are diag$(-(1+\frac{2V}{c^2}), (1-\frac{2V}{c^2}), (1-\frac{2V}{c^2}), (1-\frac{2V}{c^2}))$, the contravariant components are immediately found and we readily obtain the affine connection directly from Eq. (7.23). One finds $\Gamma_{11}^2 = \frac{\partial_2 V}{c^2}$. Therefore,

$$\frac{dk^2}{dx} = \frac{-2\omega}{c^3}\partial_2 V.$$ (8.42)

Integrating, one finds

$$\alpha = -\frac{4GM}{Rc^2} = \frac{r_S}{R},$$ (8.43)

where r_S is the Schwarzschild radius. The massless light is deflected like a particle of mass M that feels an acceleration $g = \frac{MG}{R^2}$ along a path length $2R$; indeed, setting $x \sim ct$, $y \sim \frac{gt^2}{2}$, one gets $y'(x) \sim \frac{GMx}{R^2c^2}$, and for $x = 2R$, $\tan(\alpha) \sim \alpha$ yields Eq. (8.43).

In the case of the sun, inserting $r_S = 3\,\text{km}$ and $R = 7 \times 10^5\,\text{km}$ in Eq. (8.43) one finds 8.57×10^{-6} which corresponds to 1.75 arc seconds. This was first verified approximately in an eclipse in 1919, and has been so repeatedly with increasing precision since then.

The empty space behaves like a non-dispersive dielectric medium because of the distortion due to gravity. This is very important in astrophysics. Galaxies are often seen to distort and amplify the images of background objects, often showing arcs, so-called Einstein rings or multiple images of the same object. In Sect. 9.5, we shall see that single photons may form multiple images corresponding to different routes from a Quasar around a Galaxy to Earth, and then interfere with themselves. This is a very large scale test of Quantum Mechanics!

8.8 Shift of the Perihelion

The Kepler problem leads to closed orbits for the reasons discussed in Sect. 2.6, but in the Solar System the perihelion of each planet shifts a little at each orbit. Already in the Newtonian approximation, the many-body problem is formidable because of the mutual interactions, and such systems tend to be unstable in the long run[7]; however, in our solar system the corrections that cumulate in one orbit of each planet are so small that one can simply add the effects of each binary interaction, and at the beginning of the XX century it was known that the sum of the corrections added to a total 5557 arc seconds per century, while the observed shift of the perihelion was 5600 arc seconds. This was indeed explained by Einstein as another small, additive correction of relativistic origin. One can model the situation as if Mercury was a test-particle moving in a geodesic of the Schwarzschild metric (8.28). Then, the nonzero elements of the metric are:

---

[7]The orbits of many body systems are chaotic over long timescales. The Solar System possesses a Lyapunov time, perhaps in the range of a hundred million years (the Lyapunov time is the characteristic timescale on which a dynamical system is chaotic) although we know that life has definitely existed for a longer time than that. However the relativistic shift has been observed through its cumulative effects on the scale of a century.

$$g_{00} = -\left(1 - \frac{r_S}{r}\right) = \frac{1}{g^{00}}, g_{11} = \left(1 - \frac{r_S}{r}\right) = \frac{1}{g^{11}},$$

$$g_{22} = r^2 = \frac{1}{g^{22}}, g_{33} = r^2 \sin^2(\theta) = \frac{1}{g^{33}}.$$

The motion can be taken to be confined to the plane $\theta = \frac{\pi}{2}$. The geodesic the planet follows can be derived from the Lagrangian (7.26), which, in this metric, is given by

$$\mathcal{L}_{Geo} = \frac{1}{2}\left[\frac{\dot{r}^2}{1 - \frac{r_S}{r}} + r^2\dot{\phi}^2 - \left(1 - \frac{r_S}{r}\right)\left(\frac{d(ct)}{d\tau}\right)^2\right]. \tag{8.44}$$

The proper time of the planet is $\int d\tau$ along the space-time geodesic, which is given by the action principle using the above Lagrangian; therefore, the *action integral* is proportional to the proper time. We define the proper time as given by:

$$\frac{\dot{r}^2}{1 - \frac{r_S}{r}} + r^2\dot{\phi}^2 - \left(1 - \frac{r_S}{r}\right)\left(\frac{d(ct)}{d\tau}\right)^2 = -c^2. \tag{8.45}$$

As in the classical case, we have two conservation laws arising from cyclic coordinates, namely, ϕ and t. The former yields $\frac{\partial L}{\partial \phi} = 0$, that is, the conservation of angular momentum

$$r^2\frac{d\phi}{d\tau} = L, \tag{8.46}$$

where L is a constant of integration, essentially the angular momentum of the planet divided by its mass; the latter conservation law is $\frac{\partial L}{\partial t} = 0$ and gives us

$$\left(1 - \frac{r_S}{r}\right)\dot{t} = Z, \tag{8.47}$$

where Z is another constant. We eliminate the time derivative by writing $\frac{dr}{dt} = \frac{dr}{d\phi}\dot{\phi}$. Putting (8.47) and (8.46) in (8.45), one arrives at:

$$c^2\left(1 - \frac{r_S}{r}\right)\left(\frac{\dot{t}}{\dot{\phi}}\right)^2 - \frac{\left(\frac{dr}{d\phi}\right)^2}{1 - \frac{r_S}{r}} - r^2 = \left(\frac{c}{\dot{\phi}}\right)^2.$$

Now we may set

$$\frac{\dot{t}}{\dot{\phi}} = \frac{dr}{d\phi}$$

and obtain the differential equation

$$\left(\frac{dr}{d\phi}\right)^2 = \frac{c^2 Z^2 r^4}{L^2} - r^2\left[1 + \left(\frac{cr}{L}\right)^2\right]\left(1 - \frac{r_S}{r}\right) = 0, \tag{8.48}$$

which determines the angular dependence of the radial coordinate. As long as we can identify r with the distance from the origin (which is correct for weak fields) this can be understood as the equation for the orbit. In order to make the equation slightly simpler, we set $u = \frac{1}{r}$, which implies $\frac{dr}{d\phi} = \frac{-1}{u^2}\frac{du}{d\phi}$., One finds, after multiplying both sides by u^4,

$$\frac{c^2(Z^2 - 1)}{L^2} + \frac{c^2 r_S}{L^2} u + r_S u^3 = u^2 + \left(\frac{du}{d\phi}\right)^2. \tag{8.49}$$

This could be integrated, but it is customary to differentiate with respect to ϕ and get:

$$\frac{d^2 u}{d\phi^2} + u = A + Bu^2, \tag{8.50}$$

where $A = \frac{GMm^2}{L^2} = \frac{c^2 r_S}{2h^2}$, $B = \frac{3GM}{c^2} = \frac{3}{2}r_S$ and $h = \frac{L}{m} = r^2\omega$ is the classical angular momentum of the planet divided by its rest mass; this is the generalization of Equation (2.41) of Classical Mechanics, that in he present notation becomes: $\frac{d^2 u}{d\phi^2} + u - \frac{GMm^2}{L^2} = 0$. Even in the full theory, the mass of the planet does not enter into it. Equation (8.50) can be studied numerically on a laptop, see Fig. 8.3. It turns out that for $B = 0$, we get the classical closed orbits which are circular if $A = 1$. The extra term B produces a rosetta-like orbit and the deviations from the Newtonian theory are large in the case of compact binary stars.

The great accuracy of the Newton theory makes it clear that B is very small, and so one can trivially solve the problem through perturbation theory, i.e., by approximating

Fig. 8.3 Orbits obtained by a polar plot of the inverse $r = \frac{1}{u}$ of numerical solutions of Equation (8.49). Top left: A = 1, B = 0; Top right: A = 3, B = 0; bottom left, several orbits with A = 3, B = 0.01; bottom right: several orbits with A = 3, B = 0.02

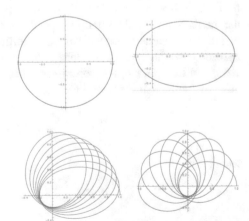

the factor of B by the Newtonian solution. One easily finds that the rate of precession of the perihelion is:

$$\frac{\Delta\phi}{T} = \frac{3\pi r_S}{Ta(1-e^2)},$$

(8.51)

where a is the semi-major axis of the ellipse and e its eccentricity. The two are bound by the relation $a(1-e^2) = \frac{L}{GMm^2}$, where L is the angular momentum. The calculated shifts are 42.9" per century for Mercury, 8,6" for Venus, 3.8" for the Earth, and 1.35" for Mars. They have been confirmed by the observations; the main residual uncertainties are due to small complications like the quadrupole moment of the sun.

8.9   Geodetic Effect

Consider a gyroscope on a satellite in a circular orbit around a mass, e.g. the Earth. An observer on the satellite is in free-fall, and therefore notices that the gravity field is locally removed; he establishes a frame of reference with the origin on the satellite and axes oriented in such a way that the fixed stars do not move. Finally, he sets a gyroscope in a fixed position with respect to his reference. According to the Newtonian mechanics, both the observer on the satellite and an observer on Earth agree that no force and no torque are acting on the gyroscope which should keep rotating around the Earth in the same way forever, keeping its axis forever in the same direction. In General Relativity, there are three distinct effects that cause tiny changes of the rotation axis at each rotation around the Earth. Two effects are related to the rotation of the Earth, and will be described briefly in the next section. This section is devoted to the geodetic effect (sometimes called the geodesic effect), which occurs because the satellite is rotating around a mass, independently of the rotation of the latter. The mass produces a Schwarzschild metric in the reference of the Earth; this metric does not mix time with the angles and there is no precession for gyroscopes on Earth. However, the instantaneous frame on the satellite is connected to the Earth frame by a Lorentz transformation, and this does mix space and time. Therefore, the metric tensor on the satellite depends on time; this produces a time-dependence of the components of the spin in the satellite reference. For symmetry reasons, a gyroscope which is set in the plane of the orbit must remain in the plane. The rate of change of the spin s of a gyroscope in orbit with radius r and velocity v due to this effect is found, using a weak field approximation, to be

$$\frac{ds}{dt} = \mathbf{\Omega}_G \wedge s,$$

(8.52)

where

$$\mathbf{\Omega}_G = \frac{3}{2}\frac{r_S}{r^3}\mathbf{r} \wedge \mathbf{v}.$$

(8.53)

If the gyroscope points away from Earth and turns anticlockwise, the spin is positive. Then, for each rotation, the axis gains an angle of order $\alpha = \frac{3}{2}\pi\frac{r_S}{r}$. The effect was verified experimentally in 2007 with an accuracy better than 0.5 per cent by the Gravity Probe B cooperation,[8] using a satellite in a polar orbit at 642 km from Earth; the shift was in the direction of motion and was as small as 0.0018 degrees per year.

8.10 Frame Dragging and Gravitomagnetic Field

Since all forms of energy, including kinetic energy, contribute to the gravity, one must expect that in the relativistic theory, the rotation of the sun must have some influence on the motion of the planets. This is true, but there is more. If the mass rotates, the metric (8.28) is superseded by the Kerr metric, discovered in 1963,

$$ds^2 = (r^2 + a^2\cos^2(\theta))\left[\frac{dr^2}{r^2 - 2Mr + a^2} + d\theta^2\right] + (r^2 + a^2)\sin^2(\theta)d\phi^2 - c^2dt^2$$
$$+ 2Mr\frac{(a\sin^2(\theta)d\phi - cdt)^2}{r^2 + a^2\cos^2(\theta)}, \tag{8.54}$$

where M is the mass, Ma is the z component of the angular momentum of the rotating body; this reduces to (8.28) for $a = 0$. This has also been used to describe a rotating black hole. The Kerr metric has the direction of the angular momentum as a preferred direction; the singularity is surrounded by an event horizon of radius $r = M + \sqrt{M^2 - a^2}$ and by an *ergosphere* delimited by a *limit of stationariness* within a surface of equation $r\theta = M + \sqrt{M^2 - a^2\cos^2\theta}$, ($\theta = 0$ in the direction of the angular momentum). The ergosphere is a region of extreme *frame dragging*, so it is time to introduce this phenomenon. Josef Lense and Hans Thirring already predicted a rotation of the orbit of a test particle around a rotating body already in 1918, and applied their findings to the solar system. The Lens-Thirring effect depending on the angular momentum J of the Sun consists of a precession of the nodal plane, which is the intersection of the orbital plane of a plane with the equatorial plane of the Sun; this is independent of the perihelion precession discussed above, which does not depend on the rotation of the Sun. It depends on the semimajor axis a of the orbit on its eccentricity e. The angular frequency of rotation is $\dfrac{2J}{a^3(1-e^2)^{\frac{3}{2}}}$ (Fig. 8.4).

The Lense-Thirring rotation on the orbit of a satellite around the Earth is extremely slow, but the effect has been verified by the LAGEOS satellite (Laser Geodynamics Satellite) in 2004, and with better accuracy by the Gravity Probe B cooperation more recently.

---

[8]See CWF Everitt et al., Class. Quantum Grav. 32 (2015) 224001.

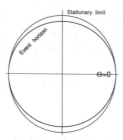

Fig. 8.4 The singularity in the Kerr solution is surrounded by an event horizon (black) and by a "limit of stationariness" (red); here $M = 1$ and $a = 0.6$; the angular momentum is in the horizontal direction

The most characteristic term in (8.54) is the one in $d\phi dt$. The reader might have noticed that a mixing of time and space in the metric occurs here and in the rotating platform problem. The Coriolis inertial force has a gravitational analogue, namely, gravito-magnetism. Actually, in the weak field case, $A = (h_{01}, h_{02}, h_{03})$, is like a three-vector in analogy with the electromagnetic vector potential; its curl H is the gravito-magnetic field H. This produces a *gravitational Lorentz force*

$$F = mv \wedge H, \tag{8.55}$$

which acts on the masses in a way that resembles the magnetic force acting on charges. The force (8.55) also produces a torque on gyroscopes. The effect is similar to the precession of a magnetic moment in a magnetic field. If we put the origin at a massive body with angular momentum J, the angular velocity of precession of a gyroscope at x can be shown to be $-\frac{H}{2}$.

A falling gyroscope, which is not rotating initially, defines an instantaneous inertial system, like any test particle. The remarkable effect is that if the field is produced by a rotating body, the instantaneous inertial system is rotating, while a non-rotating system would not be inertial. Therefore one speaks about a *dragging of inertial frames*. The rotation of the Earth causes the rotation of the inertial system in the satellites. The effects on artificial satellites are very small and tends to be masked by other effects, like irregularities in the Earth's field, however, the predictions of General Relativity have been tested and confirmed by accounting accurately for all the disturbing effects. According to predictions by Pugh and Schiff, this frame dragging effect produces a drift rate for a gyroscope on a satellite at distance r from the centre of the Earth[9]

$$\frac{ds}{dt} = \Omega_{fd} \wedge s \tag{8.56}$$

---

[9]See C.W. Everitt et al., Classical and Quantum Gravity **32**, 224001 (2015).

with

$$\boldsymbol{\Omega}_{fd} = \frac{GI}{c^2 r^3} \left[\frac{3r}{r^3} (\boldsymbol{\omega}_e \cdot \boldsymbol{r}) - \boldsymbol{\omega}_e \right], \tag{8.57}$$

where I is the moment of inertia and $\boldsymbol{\omega}_e$ the angular velocity of our planet.

In the geometry chosen by Gravity Probe B, as mentioned above, the orbit is polar and the frame dragging effect 1.1×10^{-5} degrees per year is well-separated from the Geodetic effect, because the two shifts are at right angles, with the frame dragging producing a shift towards the west while the Geodetic effect acts in the North-South direction. In the ergosphere of a rotating black hole, the frame dragging is so strong that all masses are forced to rotate around the black hole; however, in principle, an object can enter it and leave. Roger Penrose suggested that masses could be thrown in and extracted with increased energy, obtained at the expenses of the black hole rotation energy.

8.11 Gravitational Waves

It is time to resume, from Sect. 8.3.1, the analogy and the differences with electro-magnetism in the production of radiation. From Eq. (8.26), the linearised equations for the correction to $\eta_{\mu\nu}$ in vacuo are just the familiar wave equations

$$\Box h_{\mu\nu} = 0; \tag{8.58}$$

for waves propagating along the z axis, the solutions are of the form

$$h_{\mu\nu} = h_{\mu\nu}(kz - \omega t + \phi_{\mu\nu}), \tag{8.59}$$

where $k = \frac{\omega}{c}$ and $\phi_{\mu\nu}$ are constant phases. Using these, one must look for the metric tensor. One can show that there are two polarizations. The metric associated with the plane waves propagating along z is of the form

$$ds^2 = (1 + h_{xx})dx^2 + 2h_{xy}dxdy + (1 - h_{xx})dy^2 + dz^2 - c^2dt^2; \tag{8.60}$$

there are solutions of the xx type (with $h_{xy} = 0$) and solutions of the xy type (with $h_{xx} = 0$). The first kind tends to stretch and compress objects along the x and y axes in phase, while the second does the same at 45 degrees.

Since there are no negative masses, there are no dipoles and there is no dipole radiation. The leading contribution is quadrupolar. From the field equations, one can derive the energy P radiated per unit time when masses are accelerated.

$$P = \frac{G}{5c^5} \sum_{\alpha,\beta=1}^{3} \left(\frac{d^3 Q_{\alpha\beta}}{dt^3} \right)^2, \tag{8.61}$$

where

$$Q_{\alpha\beta}(t) = \int \rho(x,t)(x_\alpha x_\beta - \delta_{\alpha\beta}\frac{r^2}{3})d^3x \qquad (8.62)$$

are the components of the quadrupole tensor. Here is the Fourier transform of the formula describing the power radiation of electric quadrupole electromagnetic waves.[10]

$$P = \frac{ck^6}{360}\sum_{\alpha,\beta}|Q_{\alpha\beta}|^2. \qquad (8.63)$$

Apart from the factor in front of the summation, the similarity is evident. For a periodic motion, Eq. (8.61) implies that $P \sim \frac{G\omega^6}{c^5}\sum_{\alpha,\beta=1}^{3}Q_{\alpha\beta}^2$. The binary pulsar PSRB1913+16 discovered in 1974 by Russell Alan Hulse and Joseph Hooton Taylor provided an indirect but compelling confirmation of gravity waves; for this reason, they received the Nobel prize in 1993. It is a system of two neutron stars, one of which is a pulsar. Pulsars emits a narrow cone of radiation while rotating very fast, and the radiation arrives on Earth as a series of regular pulses. In this case, the period is about 59 ms. The two neutron stars are close to each other and rotate around their centre of gravity in just about 7.75 h. The pair emits energy in the form of gravitational energy, and this causes a measurable decrease of the period (and eventually, the two stars will collide and coalesce, producing a strong blast). The observations were in agreement with the calculations based on General Relativity.

Finally, gravity waves were detected directly for the first time in September 2015 by researchers of Caltec, MIT and LIGO; the data were interpreted as being due to the merge of two black holes that occurred 1,3 billion years ago. The signal frequency increased from 35 to 250 Hz during the last orbits of both objects approaching merge. The amplitude of the oscillations was in the range of 10^{-15} m. It was possible to determine the masses (29 and 36 solar masses) and it was estimated that about 3 solar masses where emitted in the form of gravitational waves. Several similar events have been reported since then, even in coincidence with the Virgo gravity wave observatory in Cascina (Pisa, Italy). For this enormous achievement, the Nobel Prize for Physics for 2017 has been awarded to the U.S. scientists Kip Thorne, Barry Barish and Rainer Weiss and the international cooperations Ligo and Virgo have been mentioned in the motivation. The detection of gravity waves was hailed as the beginning of a new branch of Astronomy. Indeed, in 2017 the LIGO and Virgo collaborations detected gravitational waves originating from the coalescence of a binary neutron star system, which also originated a gamma-ray burst observed by the Fermi gamma-ray Space Telescope; the finding was confirmed by optical telescope observations of the same event.

---

[10] see e.g. John D. Jackson, "Classical Electrodynamics", Sect. 9.3.

8.12   The Standard Model of Cosmology

Cosmology is extremely mind-bending application of Theoretical Physics to the Universe, but the starting point that embodies the simplest consequences of General Relativity in this field is rather simple. The Newtonian theory cannot describe properly an infinite Universe and in the case of a finite one, it cannot treat all points on the same footing. The relativistic standard model is based precisely on these symmetry properties, namely, isotropy and homogeneity, which allow for a simple solution of the Einstein equations which would otherwise be hard to handle. The stress-energy tensor will be independent of position; moreover, we can choose a co-moving reference, that is, we may set the velocity of matter to zero in every point; then, the stress-energy tensor (6.30) has only one nonzero component, namely, $T_0^0 = -(\rho c^2 + \rho U)$, and this component does not depend on position. The component g_{00} of the metric tensor determines the proper time of a body at rest, so it cannot depend on position. We know that the components $g_{0k}, k = 1, 2, 3$ have to do with rotations, and the above principles strongly suggest that we can safely take $g_{0k} = 0$. In conclusion, we are led to a metric of the form

$$ds^2 = -dl^2 + c^2 dt^2$$

and the 3d space metric $g_{0k}, i, k = 1, 2, 3$ yielding dl^2 must be such that the curvature is the same throughout. We have already met such constant-curvature 3d spaces in Sect. 7.0.2 and we know that basically we have the choice between a closed Universe with space metric proportional to (7.36) and an open Universe (7.38). The closed Universe arises from the space metric (7.36) yielding the space-time metric:

$$ds^2 = c^2 d\tau^2 - a^2[d\chi^2 + \sin^2(\chi)(d\theta^2 + \sin^2(\theta)d\phi^2)]. \tag{8.64}$$

This becomes flat for $a \to \infty$ (recall the definition $\sin(\chi) = \frac{r}{a}$).

Similarly, the open Universe arising from Equation (7.38) has the space-time metric

$$ds^2 = c^2 d\tau^2 - a^2[d\chi^2 + \sinh^2(\chi)(d\theta^2 + \sin^2(\theta)d\phi^2)]. \tag{8.65}$$

Alexander Friedmann, Georges Lemaître, Howard P. Robertson and Arthur Geoffrey Walker formulated this model before 1930, without the cosmological term Λ which was not supported by the data available then.

The Einstein equations for these models can be solved exactly; this can be done directly using the definition of the curvature tensor or using some tricks that are presented in Landau-Lifschitz and other classics, but the argument is still rather long.

In the case of (8.64) it is expedient to replace the time with a new variable, by putting

$$a = c\frac{dt}{d\eta}; \qquad (8.66)$$

then,

$$ds^2 = a^2[-d\eta^2 + d\chi^2 + \sin^2(\chi)(d\theta^2 + \sin^2(\theta)d\phi^2)]. \qquad (8.67)$$

Now $g_{\mu\nu} = \mathrm{diag}(g_{00}, g_{11}, g_{22}, g_{33}) = \mathrm{diag}(-a^2, a^2, a^2\sin^2(\chi)a^2\sin^2(\chi)\sin^2(\theta))$
and $g^{\mu\nu} = \mathrm{diag}(g^{00}, g^{11}, g^{22}, g^{33})$ with $g^{\mu\mu} = \frac{1}{g_{\mu\mu}}$.

The result depends on the total mass M of the Universe, which is a conserved quantity. We need the constant $a_0 = \frac{2kM}{3\pi c^2}$, where $k = c^4 G_N$. Eventually, one finds the parametric law governing the size of the Universe a, namely,

$$a(\eta) = a_0(1 - \cos(\eta)),$$
$$t = \frac{a_0}{c}(\eta - \sin(\eta)). \qquad (8.68)$$

It turns out that the Universe reaches a maximum size, then collapses to a point and restarts expanding. It is a pulsating Universe with period $T = \frac{2\pi a_0}{c}$. In the case (8.65) the Universe is open and expands forever:

$$a(\eta) = a_0(\cosh(\eta) - 1),$$
$$t = -\frac{a_0}{c}(\eta - \sinh(\eta)). \qquad (8.69)$$

It would be exciting to discover that we are in a curved Universe, possibly with a nontrivial topology; the data, however, say that Nature does not like such complications, and the Universe is accurately flat. For a flat Universe,

$$ds^2 = -c^2\tau^2 + b^2(dx^2 + dy^2 + dz^2), \qquad (8.70)$$

and one can show that

$$\left(\frac{db}{dt}\right)^2 = \frac{8\pi G}{3c^2}\rho(U + c^2)b^2, \qquad (8.71)$$

where it is reasonable to assume that $\rho(U + c^2)b^3$ is constant; then, b grows as a power of t (Fig. 8.5).

Einstein started to apply his theory to the large-scale structure of the Universe in 1917, before the two major discoveries by Edwin Hubble: in 1925 he established the existence of Galaxies outside the Milky Way, leading to a much larger Universe, and in 1929, he discovered the expansion of the Universe.[11] The expansion causes a

---

[11]In some way Theory and Experiment were in competition since Georges Lemaître had predicted the redshift-distance relation in 1927.

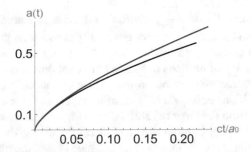

Fig. 8.5 $a(t)$ for the open Universe (red) and for the closed Universe (black) versus time. The red curve eventually increases to infinity linearly, while the black one reaches a maximum and then goes to zero symmetrically, leading to a pulsating Universe. Our Universe is still so young that the two curves (and the flat Universe) are hard to distinguish

redshift of the light of a galaxy at distance d; the redshift is like a Doppler shift due to a velocity v given by

$$v = Hd \tag{8.72}$$

where the Hubble constant H is on the order of 45–90 km per second per Megaparsec and 1 parsec is about 3.26 light years. Each Galaxy also has a peculiar velocity, therefore v is a sort of average value. There is some vague analogy between the galaxies in the expanding Universe and cue balls painted on the surface of an inflating balloon. However, the cue balls on the balloon should grow in size, while the sizes of atoms, the Earth, the Solar System and the Galaxy do not grow: the expansion creates space between bound systems, but not inside them. The discovery came unexpectedly, but it finally explained the old Olber's paradox. The astronomer Heinrich Wilhelm Olbers (1758–1840) remarked that if the Universe were infinite and static, there should be stars in every solid angle. Indeed the night sky is dark because the farthest stars are red-shifted by the Doppler effect.

In 1964, Penzias and Wilson made the fortuitous but very important discovery of the cosmic background radiation at 2.7 K, which confirmed the hot Big Bang, and at the time there was a very optimistic feeling that the concepts and the laws of Physics developed on this Planet were ruling the Universe and although largely incomplete the mosaic was under way. None could question that Cosmology was a branch of Physics. In the 60s, the exploration of the Solar System started. The relevant distances and masses had been estimated before the space era quite accurately by the available telescopes, the parallax and the knowledge of the Newton theory of gravity. The parallax of nearby stars is measurable, and since we know the size of the orbit of the Earth the distance can be measured. However, most stellar distances can be measured by more indirect means, and one must rely on the luminosity of peculiar stars (novae, Cefeid variable stars) that can be used as standard candles. This works for the near Galaxies, but the uncertainties grow when dealing with large intergalactic distances, on the order of many millions of light years. At still larger distances, one can rely on the *luminosity distance* $D_L = \sqrt{\frac{L}{2\pi l}}$, where L is the absolute luminosity and l is the power received from the source; another possibility is based on the angular size and the real size of the object. Both methods are affected by the possible evolution of the source in the course of millions or billions of years, and at any rate, the estimate must be based on some assumption on the metric of space-time. At large cosmic distances (many millions or several billions of light years), the relativistic effects are large, as is evident from the large red shifts due to the expansion of the Universe. In addition, it should be kept in mind that in General Relativity, the distance between two objects is meaningful only locally. The distance l loses the obvious meaning of everyday life unless the metric tensor is time-independent; this assumption cannot hold in an expanding Universe. In general one can define the element of distance dl and then compute $\int dl$ over a curved space-time, but the result depends on the path of integration. At any rate, the red shift makes distant objects dimmer and dimmer. In Special Relativity, nothing can go faster than light, but there is no such limitation to the expansion of the Universe. It is clear that the major part of the Universe recedes from us at super-luminal speeds; on the other hand, we can observe only the objects which recede with speed less than c. With such caveats, the age of the Universe is 13.8 billions of years, and taking the expansion into account, the edge of the observable Universe is estimated to be at 46 billion light years. The task of building a coherent scheme in which the observation could be rationalized appeared more feasible 40 years ago than it does now. The main issue then was to know whether the gravity-induced slow-down of the expansion was enough to stop it and produce a cosmic collapse or not; in this case, the expansion would continue forever. Much more powerful telescopes are operating now than decades ago, there are space telescopes producing breath-taking pictures, and new windows have been opened (infrared, radio, X-ray astronomy). As a result, the optimism about our ability to understand the Universe according to the laws established in the Laboratory has evaporated gradually. Vera Rubin studied the rotation curves of the nearby Galaxies. She discovered that the stars in the outskirts rotate faster than they should if they were attracted by the visible matter. There are also similar problems with Galaxy clusters and Fritz Zwicky had already proposed the existence of Dark Matter. There is no agreement about the nature of this stuff, which is actively sought for.

Even more astonishing was the discovery that the expansion actually seems to accelerate. The accelerated expansion was discovered in 1998, and was quite unexpected. Two independent projects, the Supernova Cosmology Project and the High-Z Supernova Search Team, both used distant type Ia supernovae as standard candles, hoping to measure a deceleration due to gravity. In summary, very high redshift supernovae appeared to have a larger luminosity distance than expected from the Standard Model. This suggests an accelerating Universe for the following reason. The light has traveled a greater distance in an accelerating Universe, than it would if the expansion rate were the same as in our neighbourhood. Thus, for the same redshift, a supernova is more distant in an accelerating Universe, and hence, dimmer, than in other non-accelerating Universe models. This conclusion restored Λ.

The cosmological constant does not change the symmetry of the model, thus the metric tensor remains diagonal. Thus, Λ can be fitted into the above Standard Model. One must increase the density by $\frac{\Lambda c^2}{8\pi G}$ and decrease the pressure by $\frac{\Lambda c^4}{8\pi G}$. So, Λ produces an extra energy, known as *dark energy*, associated with a negative pressure.[12]

In 1968, Zel'dovich proposed an explanation of the cosmological constant Λ in terms of vacuum energy. The idea of an energy content of vacuum comes from Quantum Mechanics and will be developed in Sect. 16.2. Zel'dovich argued that in Special Relativity, the energy-momentum tensor of the vacuum should be proportional to $\eta_{\mu\nu}$ which is a tensor that keeps the same form under Lorentz transformations. Indeed the vacuum should appear the same in all inertial systems. By the principle of General Covariance this implies that in General Relativity $T_{\mu\nu}$ must be proportional to $g_{\mu\nu}$ and produce a Λ term in the Einstein equations. In this way, the idea of a dark energy that becomes more and more dominant as the space grows could find an explanation. At the present time, Cosmologists suppose that the total mass - energy of the Universe contains 4.9% ordinary matter, 26.8% dark matter and 68.3% dark energy. It is embarrassing to admit that nobody is sure about the nature of dark energy and dark matter; but without them, the physical laws discovered in the laboratory and successful in the neighbourhood of the solar system appear to fail.

Even the isotropy of the Universe is somewhat paradoxical. This is the so-called horizon problem. How can we explain the fact that the cosmic microwave background follows Planck's law, and the temperature is the same (with minor fluctuations) in all directions? This implies that (1) there was equilibrium during the Big Bang, (2) there was equilibrium even between parts of the Universe which had not yet exchanged radiation. Moreover, equilibrium implies maximum entropy, while, according to Thermodynamics, the entropy of the Universe is increasing. There is also a flatness problem: why is the Universe (within a few percent) flat? These are questions about the initial conditions. In 1980 Alan Guth proposed the idea of an inflation field that should have worked at the beginning of the Universe. This became popular but has unwanted implications and has been questioned; the inflation field is an *ad hoc* assumption without any clear role in Physics. But at any rate, this topic is outside our scope.

In summary, we can observe only a tiny (probably negligible) part of the Universe, and the current understanding of the observations is rather poor. Maybe the subject is already running outside the realm of Physics, maybe it will lead to new Physics.

---

[12]However, Λ has a competitor, known as Quintessence. This is thought to be a hypothetical scalar field; it should provide a dynamic dark energy in the sense that it generally has a density and equation of state that varies through time and space.

Part II
Quantum Mechanics

Chapter 9
The Wave Function

This theory has many counter-intuitive features, but perhaps the
most amazing aspect is the absolute need for complex numbers,
those that their inventor Girolamo Cardano (1501–1576) called
'fictitious numbers'.

Among the fundamental constants of Physics, Planck's[1] constant h is an action
[h = energy $\times$ time = angular momentum]. It is a tiny action, $h \sim 6,62619610^{-34}$ J
s. Most often, one uses the notation $\hbar = \frac{h}{2\pi} = 1.05410^{-34}$ J s. All the phenomena in
which h is non-negligible are *quantum phenomena*, while sometimes phenomena in
which the characteristic actions are $\gg h$ may have *classical* aspects, which allow for
a simpler description; however all the microscopic mechanisms are fundamentally
quantum, and many phenomena such as magnetism, superfluidity and superconductivity are just macroscopic quantum phenomena; their quantum mechanical description is far from trivial. Historically, the breakdown of classical Physics (after many
triumphs) was urged by the clear inability of the latter to explain a few quantum
effects that where known by 1900. This breakdown was shocking, because well-
understood laws of Physics that had been validated by experiment in many ways
were seen to lead to wrong conclusions in a class of phenomena quite unexpectedly.

Without h, one cannot understand the stability of atoms. From electromagnetism,
we know that a charge e must irradiate a power

$$W = \frac{dE}{dt} = \frac{2}{3}\frac{e^2}{c^3}a^2, \tag{9.1}$$

---

[1] Max Planck (Kiel 1858-Gottinga 1947) German physicist, initiator of Quantum Mechanics, Nobel
for Physics in 1918.

© Springer International Publishing AG, part of Springer Nature 2018
M. Cini, *Elements of Classical and Quantum Physics*,
UNITEXT for Physics, https://doi.org/10.1007/978-3-319-71330-4_9

where E is the energy of the atom, and a is the acceleration; this law explains very well the working of an antenna. Now let us try to build an atomic model where the electron makes an orbit of radius r. Substituting $a = \frac{e^2}{mr^2}$, one gets $W = \frac{dE}{dt} = \frac{2}{3r^4}r_0^3\frac{e^2}{c^3}$, where $r_0 = \frac{e^2}{mc^2} \sim 2.8 \ 10^{-13}$ cm is the so-called classical radius of the electron. Now, the atom in a normal state does not radiate, yet the electron feels a force and is accelerated. Moreover, the electron should fall in the nucleus in a short time. The electrostatic energy of the atom is $E = -\frac{e^2}{2r}$, so

$$\frac{dE}{dt} = -\frac{r_0 mc^2}{2r^2}\frac{dr}{dt}.$$

Equating this to W, one finds that $r^2\dot{r} = -\frac{4}{3}cr_0^2$. We can integrate this with the initial condition $r(0) = a_0$ with the result that $r^3(t) = a_0^3 - 4r_0^2 ct$. The radius becomes zero at $t = 1.6 * 10^{-11}$ s. This is in very sharp contrast with the experiment! Classically, one understands the chemical bond even less. Atoms, molecules, nuclei, solids, and elementary particle interactions are all quantum effects. Later it became clear that the quantum theory was needed to understand macroscopic phenomena as well. Classically, the cohesive energy of metals cannot be explained.

It is easy to see that magnetism would be impossible. This was the result of the Bohr–van Leeuwen Theorem. The Lorentz force correctly describes the motion of charges. However, for a classical electron in a magnetic field, the canonical partition function reads:

$$Z = \int d^3x \int d^3p \exp\left[-\frac{(p_x - eHy/c)^2 + p_y^2 + p_z^2}{2mK_BT}\right];$$

this should be the starting point of description of magnetic effects in solids, but a trivial shift of the p_x variable yields

$$Z = V(2\pi mK_BT)^{3/2}.$$

The field has disappeared! Thus, classical physics predicts that magnetism does not exist, in sharp contrast with the experiment. Superconductivity and superfluidity are also macroscopic quantum effects. By contrast, the predictions of Quantum Theory are extremely accurate. In no other field has one gained a similar degree of knowledge; in some atomic phenomena, like the Lamb shift, 10 digits precision has been achieved.

This book is devoted to the non-relativistic version of the theory, although the relativistic one is well known. We shall be able to understand many important facts in a semi-quantitative fashion.

9.1 Corpuscles and Waves in Classical Physics

In classical physics, there are two types of basic objects, corpuscles and waves. The wave and particle concepts are well distinct. A corpuscle is a particle that obeys the canonical equations and follows a deterministic trajectory; if there is no force, it goes straight, with constant speed. The waves are described as solutions of a well-defined linear partial differential equation; they may be propagating in a elastic medium, or appear as continuous fields. In the presence of obstacles, they give rise to phenomena of interference and diffraction. Newton thought that light was made of corpuscles. This idea was natural in view of Geometrical Optics, based on light rays. This knowledge already enabled the building of powerful telescopes and microscopes that produced extraordinary progress. But at the half point of the nineteenths century, this type of geometrical optics was recognised as a limiting case of physical optics, that is, Maxwell's theory. It holds when all material objects are large compared to the wave length. One phenomenon that the corpuscular model of Newton could not explain is diffraction. Actually, it was known since 1665, when an Italian scientist Francesco Maria Grimaldi, a Jesuit priest, discovered that *'Light propagates or spreads not only in a straight line, by refraction, and by reflection, but also by a somewhat different fourth way: by diffraction.'*

A monochromatic plane wave of wave vector $\vec{k} = (k, 0, 0)$ parallel to the x axis impinges on a screen with a slit for $y \in (-a, a)$ and then exposes a photographic plate. To fix ideas, we take a wave polarized along the z axis with electric field

$$\mathbf{E} = (E_x, E_y, E_z) = \mathrm{Re}(0, 0, E e^{ikx}). \tag{9.2}$$

The frequency and polarization of the field do not change. However, the beam undergoes a *Fraunhofer diffraction*. This is borne out by the Maxwell equations and is evident for ka of order 1. On the photographic plate, no sharp picture of the slit appears, but rather the intensity is a function $I(\theta)$ of the angle of deflection:

$$dI \propto \frac{sin^2(ka\theta)}{(ka\theta)^2} d\theta.$$

This function is reminiscent of a representation (3.7) of Dirac's δ; the narrower the slit, the broader the diffraction pattern. Beyond the screen, the wave vector of the diffracted field makes a small angle θ with the x axis; actually, it has got a y component $|k \sin(\theta)| \sim k\theta$ (Fig. 9.1).

Classical Physics explains these facts very well. If the plane incident wave has constant amplitude E (independent of y), the part of the front cut off by the slit must depend on y as $\theta(a^2 - y^2)$.

Fig. 9.1 Graph of
$\sin^2(3x)/x^2$ for $x \in (-3\ 3)$

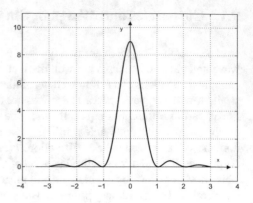

Fourier's theorem gives the wave packet representation

$$\theta(a^2 - y^2) = \frac{1}{\pi} \int_{-\infty}^{\infty} dq\, e^{-iqy} \frac{\sin(qa)}{q}.$$

Letting $E \rightarrow E\theta(a^2 - y^2)$ in Eq. (9.2) one obtains

$$E_z = E \int dq\, e^{i[kx-qy]} \frac{\sin(qa)}{\pi q}.$$

A point of the plate at an angle $\theta \ll 1$ receives a component

$$E_z \propto e^{i[kx-k\theta y]} \frac{\sin(k\theta a)}{\pi k\theta}$$

with a deflected wave vector, $\frac{q}{k} = \tan\theta \sim \theta$.

The intensity which is measured on the plate goes with the square of the field. Note carefully that

> E is an amplitude, the intensity is proportional to $|E|^2$,

Thus, it is clear that diffraction is a property of waves, and the amplitudes, not the intensities, are additive.

9.2 Dualism

If the intensity of the light is very weak, something happens that is not expected by Classical Physics. The pixels in the plate are exposed or not exposed in a yes or no fashion: there is no half measure. Higher and lower intensity of the radiation just changes the probability of each pixel of being exposed.

Max Planck introduced h in 1900 in order to fix the black body theory, affected by the *ultraviolet catastrophe*. Then, in 1905 Albert Einstein understood that the light is emitted and absorbed as quanta. This implies that it consists of quantized photons carrying each a quantized energy $E = h\nu = \hbar\omega$. This interpretation was forced by the *photoelectric effect*: experiments showed that the radiation picks electrons from a surface if its frequency is above a threshold, which depends on the material. The intensity of the electron current above the threshold is proportional to the intensity of the radiation. The relation between frequency and wavevector is $\omega = ck$, and the relation between energy and momentum is $E = cp$ for an electromagnetic wave. As a result of energy quantization, the photon momentum is also quantized and the momentum carried by a photon having wave vector k is $p = \hbar k$. We must accept the fact that a photon propagates as a wave, yet it is always absorbed as a corpuscle.

Then, in 1924, the French Duke Louis-Victor De Broglie[2] suggested that in analogy with the dualism between photons and waves, particles like an electron should also correspond to a wave function ψ. Most remarkably, he also argued that ψ must be an unobservabe, complex function

$$\psi_{\vec{p},E}(\vec{x},t) \approx \exp\left[\frac{i(\vec{p}\cdot\vec{x} - E(\vec{p})t)}{\hbar}\right], \qquad (9.3)$$

with $\vec{p} = \hbar\vec{k}$ as the electron momentum. The square modulus is proportional to the intensity of the electron beam and to the probability of detecting the electron. The introduction of a complex quantity that is not observable but contains the physical information was a wonderful, bold flash of insight. In 1927, Davisson–Germer at Bell Labs USA confirmed De Broglie's hypothesis. They fired electrons at a crystalline Ni target, and the resulting diffraction pattern was found to match the values predicted by the De Broglie formula. Since the late '50s, this is a technique known as Low Energy Electron Diffraction, used for the study of surfaces and their excitations. Here I use a more direct experiment to show the effect. This is feasible with the current technology but was impossible to do at the time (Fig. 9.2).

A beam of electrons of momentum **p** parallel to the x axis impinges on a screen and goes through a slit for $y \in (-a, a)$; the electrons meet a photographic plate (or better yet, a modern detector). The beam undergoes a Fraunhofer diffraction.. The intensity at a small angle θ from the x direction turns out to go like $dI = \frac{\sin^2(ka\theta)}{(ka\theta)^2}d\theta$. This is not expected classically, but agrees with the De Broglie hypothesis (and indeed, no diffraction is seen if $ka \gg 1$). Again, if the incident wave is è $\psi \propto e^{i\vec{k}\cdot\vec{r}}$, with initial wave vector $\vec{k} = (k, 0, 0)$, the transmitted wave is $\psi \propto \int dq e^{i[kx-qy]}\frac{\sin(qa)}{\pi q}$. For small angles, $|q| \approx k\theta$; evidently,

[2]De Broglie (1892–1987) won the Nobel Prize in 1929 for the very remarkable prediction about the existence and the form of a complex wave function for a free particle in a thesis work in 1924.

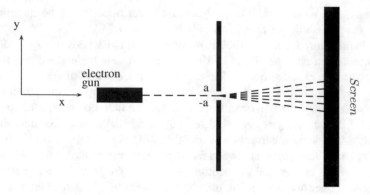

Fig. 9.2 Diffraction of an electron beam through a slit: the beam from gun C arrives at P in the detector plane (photographic plate, Geiger, etc.). The z axis is orthogonal to the plane of the figure

ψ is an amplitude, and the intensity goes like $|\psi|^2$.

The wave function is a probability amplitude. In a one-dimensional problem, we may be considering a detection of a particle at some location **x**; the detection experiment can be done many times using a beam of particles; otherwise, we may do a series of experiments involving just one particle, like an electron in an H atom, but repeating the experiment on many samples. In both instances, it makes sense if we ask for the probability that the detection occurs in a volume V. This is given by

$$P_V = \int_V d^3x |\Psi(\mathbf{x}, t)|^2,$$

where I have included a possible dependence on the time t because the system could evolve. In other words,

$$\rho(\mathbf{x}, t) = |\psi(\mathbf{x}, t)|^2 \tag{9.4}$$

is a probability density that must be normalized to 1 by

$$\int d\mathbf{x} \rho(x, t) \equiv 1,$$

integrating over all space, since the particle must be somewhere.

The quantum theory does not propose a *compromise* between the conflicting views of waves and particles. It does not attempt any pun. It makes extremely precise predictions based on a rigorous mathematical apparatus which has the precedence over any argument. Wave-like diffraction and particle-like detection take place in the same experiment (Fig. 9.3).

Fig. 9.3 The double slit experiment: the electron beam from gun C arrives to P on the photographic plate through two paths. The lengths are $l_1 = a + b$ and $l_2 = c + d$. An interference path is seen on the plate

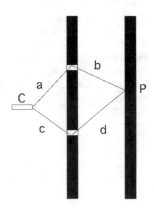

9.3 Which Way Did the Electron Pass Through?

The double slit experiment with electrons (Fig. 24.3) is an even more striking display of the wave-like properties of matter particles. The intensity is proportional to

$$1 + \cos[k(l_1 - l_2)],$$

where l_1, l_2 are the path lengths. When one of the slits is closed, the interference pattern between the two waves disappears, but one observes the diffraction pattern of the other slit. This is like the double slit experiment familiar in classical physical optics, but cannot be understood in terms of a beam of corpuscles. Moreover, the intensity of the electron beam can be made so low that in each moment, it is very unlikely that more than one electron is present. Then, the experiment must be continued for a long time, but when enough electrons reach the plate, the same results as before are obtained. Thus, the electron must somehow interfere with itself! Moreover one can never tell which slit a given electron went through; all attempts to modify the experiment and determine the electron trajectory destroy the interference. Currently,[3] double slit experiments are also done with atoms; often the double slit is not material, but is obtained by using laser beams.

9.3.1 Bohm-Aharonov Effect

In 1959, Bohm and Aharonov proposed a modification of the double slit experiment, as shown in Fig. 9.4.

The flux tube is a solenoid that produces a magnetic field inside a cylinder where the electrons cannot enter, but the field is screened, for instance, by coating the solenoid with a superconducting film. Therefore, the field outside vanishes and cannot

---

[3] See for instance A. Miffre et al., $arXiv/cond - mat$, 0506106.

Fig. 9.4 The double slit experiment modified by a flux tube (round dot). The interference at P depends on the flux collected within the path abdc, even if there is no magnetic field outside the flux tube

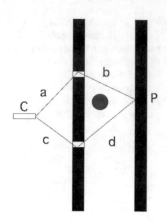

produce any Lorentz force; classically, no effect is expected. Nevertheless, there is a magnetic flux though a surface bounded by the path abdc. This induces a phase difference between the wave function along the path ab and the wave function along cd, which shifts the interference path on the screen (see Section 15.2).

9.3.2   Experiments by Deutsch on Photons

New enlightening experiments using photons have been performed by D. Deutsch et al.[4] The light from source S is split by the half silvered mirror SST (Fig. 9.5). Classically, detectors A and B should receive half of the intensity, as predicted by the Maxwell equations, and this happens if the intensity is such that the number of photons in the system at any given time is large. However one can decrease the intensity to one quantum per second, then no more than one photon is around. Then, the photon ends up either in A or in B, but with equal probability, so the classical prediction is verified only on average when many experiments have been conducted. But what happens each time?

It is natural to suppose that once in SST, the photon makes a random choice. Is that true? One can find an answer through the second experiment in Fig. 9.5, done with one photon at a time. The geometry is such that there is constructive interference in A and destructive interference in B. As a matter of fact, the lonely photon always goes towards A. We must conclude that 1) there is no random choice, and 2) the particle sommehow explores both paths and interferes with itself! If one of the paths is slightly modified, the detector in B starts counting photons. This is really striking. At any rate, the quantum behavior supersedes the classical description at low enough intensity.

---

[4]See for instance David Deutsch, Arthur Ekert and Rossella Lupacchini, math.HO/9911150.

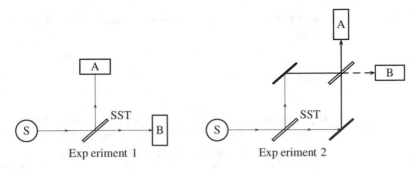

Fig. 9.5 experiment 1: by a semi-transparent mirror, the photon is sent to A or B with equal frequency. Experiment 2: By using a double semi-transparent mirror and two ordinary mirrors the photon is sent to A

The next question is: what is the photon wave function? The classical result at high intensity is correctly predicted by the Maxwell equations. In the Lorentz gauge, the vector potential $\vec{A}$ obeys the wave equation; from $\vec{A}$, one can derive the electric field $\vec{E} = \frac{\partial \vec{A}}{\partial t}$, and the light intensity is proportional to $\vec{E}^2$. There is a partial wave in both paths, and all the intensity goes to A if there is constructive interference in A and destructive interference in B.

Moreover, the wave function must not be an observable quantity. This suggests the correct answer: even at low intensity, the wave equation is useful and the square of $\vec{A}$ is related to the probability of detecting the photon. Maxwell's equations survive unharmed, but must be interpreted in a new way. This interpretation is also consistent with the first experiment of Fig. 9.5. A further implication is that each species of particle has its own wave function. But a further question arises naturally: how can we predict in the first experiment where the next photon will go? Here, the answer is: in no way! A similar experiment has been performed[5] with neutrons.

9.3.3 Plane Wave, Superposition Principle and Operators

A plane monochromatic electromagnetic wave was interpreted by Planck as a beam of photons, each with the same frequency, energy and momentum. The De Broglie analogy suggests that the incident electron *wave function* in a double slit experiment is the more accurate description of the propagation of what classically would be a train of electrons of equal speed moving like a series of gunshots. The analogy is admittedly coarse, since in the case of an electromagnetic wave, we need a four-potential as a wave function capable of containing all the information, while in the case of particles, we might expect one component, furthermore a complex one. Later, we shall find that even in the non-relativistic theory a single component is not

---

[5]M. Iannuzzi et al., Physical Review Letters (2006).

enough, because of spin. However, for simplicity, we start with the single-component
De Broglie wave of the form:

$$\psi_{\vec{p},E}(\vec{x},t) \approx \exp\left[\frac{i(\vec{p}\cdot\vec{x} - E(\vec{p})t)}{\hbar}\right], \tag{9.5}$$

with $\vec{p} = \hbar\vec{k}$, $E = \hbar\omega$. In the electromagnetic case, the light intensity goes with
the square of the field. By analogy, the probability of detecting an electron in $\vec{x}$, t
(that is, at $\vec{x}$ at time t) must be proportional to $|\psi_{\vec{p},E}(\vec{x},t)|^2$. Using (9.5), this is
constant, so actually we have no idea of where the electrons are. If the reader finds
that this information looks very incomplete, compared to what one would deem a
full classical description, I cannot deny that this is true.

By analogy with electromagnetism (in vacuo), one should very much hope that the
theory is linear, since nonlinear phenomena are much more complicated to describe.
People naturally tried with a linear theory and it worked: the superposition principle
is valid, and a linear combination of wave functions ia a possible wave function.

A wave function $\psi_{\vec{p},E}$ represents particles prepared in a state with well-defined
momentum, so it contains the information that the momentum is **p**. How can we
get the momentum **p** out of it *by means of a linear operation*? A logarithm, cannot
work, because of the linearity constraint. However we can introduce the *momentum
operator*

$$\vec{p}_{op} = (-i\hbar)\vec{\nabla}, \tag{9.6}$$

which gives us

$$\vec{p}_{op}\psi_{\vec{p},E}(x) = \vec{p}\,\psi_{\vec{p},E}(x). \tag{9.7}$$

Mathematically, this is a well-known equation, an eigenvalue equation: the proper
statement is that $\psi_{\vec{p},E}$ is an eigenfunction of the momentum operator and $\vec{p}$ is
the corresponding eigenvalue. This complex function represents a particle moving
in the direction of $\vec{p}$. We can get a real wave function by summing $\psi_{\vec{p},E}(\vec{x},t)$
with $\psi_{-\vec{p},E}(\vec{x},t)$. The opposite currents sum to zero. We shall see that a real wave
function cannot carry a current.

It is time to better specify what information can be extracted from ψ. Sometimes,
broady speaking, we say that ψ describes the particle. Actually, the average over
the wave function tells us about the result of a series of many measurement on the
sample containing many particles, rather than the properties of a single particle. One
can choose experimental conditions such that only one particle is present. Then, ψ
cannot predict the position of the particle. But if the single-particle experiment is
repeated many times, then ψ can be used to make statistical predictions. In a plane
wave state, the position of the particles is quite undetermined. A general state is a
superposition called a wave packet[6]:

---

[6]The wave function must always have a Fourier transform in the space of ordinary functions or in
the space of distributions.

$$\psi(\vec{x},t) = \frac{1}{(2\pi)^{3/2}} \int d\vec{p}\, g(\vec{p}) \exp\left[\frac{i(\vec{p}\cdot\vec{x} - E(\vec{p})t)}{\hbar}\right]. \qquad (9.8)$$

(The $\frac{1}{(2\pi)^{3/2}}$ factor will be explained below.) Now, the position of the packet must be computed as a statistical mean using the modified wave $\psi(\vec{x},t)$:

$$<\vec{x}> = \int d\vec{x}\,\vec{x}\,|\psi(\vec{x},t)|^2 = \int d\vec{x}\,\psi(\vec{x},t)^*\,\vec{x}\,\psi(\vec{x},t). \qquad (9.9)$$

Here, $g(\vec{p})$ is a probability amplitude (its square modulus is the probability) that a momentum measurement yields $\vec{p}$. In effects, $g(\vec{p})$ is the *wave function in the momentum representation*.

If $g(\vec{p})$ is peaked around $\vec{p}_0$, there is a large uncertainty about its location. At the limit, when the momentum is sharply defined, we are back to the De Broglie wave.

The information contained in $g(\mathbf{p}, t)$ is equivalent to the knowledge of $\psi(\mathbf{x})$, and we can treat $g(\mathbf{p}, t)$ as a wave function in the *momentum representation*.

Problem 24 Prove that

$$<\vec{p}> = \int d\vec{p}\,\vec{p}\,|g(\vec{p})|^2; \qquad (9.10)$$

yields the same result as

$$\int d\vec{x}\,\psi(\vec{x},t)^*(-i\hbar)\vec{\nabla}\psi(\vec{x},t) = <\vec{p}>. \qquad (9.11)$$

Solution 24

$$-i\hbar\nabla\psi(\vec{x},t) = \frac{1}{(2\pi)^{3/2}} \int d\vec{p}\, g(\vec{p})\vec{p} \exp\left[\frac{i(\vec{p}\cdot\vec{x} - E(\vec{p})t)}{\hbar}\right],$$

and from (9.8) we find that

$$\psi(\vec{x},t)^* = \frac{1}{(2\pi)^{3/2}} \int d\vec{p}'\, g^*(\vec{p}') \exp\left[\frac{-i(\vec{p}'\cdot\vec{x} - E(\vec{p}')t)}{\hbar}\right]. \qquad (9.12)$$

Set up the integral (9.11) and integrate over $d\vec{x}$, using

$$\int d\vec{x}\, e^{i[\vec{p}' - \vec{p}]\cdot\vec{x}} = (2\pi)^3 \delta(\vec{p} - \vec{p}').$$

The result is

$$\int d\vec{x}\,\psi(\vec{x},t)^*(-i\hbar)\vec{\nabla}\psi(\vec{x},t) = \int d\vec{p}\,d\vec{p}'\,\delta(\vec{p}-\vec{p}')g^*(\vec{p}')g(\vec{p})\vec{p},$$

which is equivalent to (9.10). We see that we needed the factor $\frac{1}{(2\pi)^{3/2}}$ introduced in (9.8).

This example shows how the average momentum in the coordinate representation $\psi(x)$ is obtained by the **momentum operator** introduced in (9.6),

$$\vec{p}_{op} = (-i\hbar)\vec{\nabla}.$$

The classical variables become quantum operators; the way the operators are written depends on the representation. We have already seen that when acting on a $\psi(x)$, the correspondence is $\vec{x} \to \vec{x}_{op} = x, \vec{p} \to \vec{p}_{op}$. In general, if A is a physical quantity (that is something that can be measured) we call it an *observable* . There are quantum observables (like the spin of electrons, the lepton number and the flavour of quarks) that do not correspond to classical variables. For each observable, there is an operator $\hat{A}$ such that

$$< A >= \int d\vec{x}\,\psi(\vec{x},t)^*\,\hat{A}\,\psi(\vec{x},t)$$

is the average value of many measurements. The De Broglie wave (9, 5) is an example of a general rule: the eigenfunction of $\hat{A}$ with eigenvalue a is denoted by ψ_a; if the physical magnitude corresponding to $\hat{A}$ is measured, the result is granted to be the eigenvalue a. It also gives a hint as how to find the operator of kinetic energy $\hat{T} = \frac{p^2}{2m}$. Similarly, the angular momentum operator $\hat{L} = \mathbf{r} \wedge \mathbf{p}$, which depend on coordinate and momenta is obtained letting $\mathbf{p}$ become $(-i\hbar)\vec{\nabla}$. However, in general caution is necessary in transforming classical observables into operators, since quantum operators may fail to commute, as we shall see next.

Commutator

Using a test function $\phi(x)$ one can observe that the commutator $[p,x]_-\phi(x) \equiv (px - xp)\phi(x) = -i\hbar\phi(x)$, does not vanish. The operators $\hat{x}, \hat{p}$ do not commute:

$$\boxed{[p,x]_- = px - xp = -i\hbar.}$$

This *fundamental commutation rule* replaces the classical fundamental Poisson bracket. We shall see that this analogy is profound.

Example 5

$$(x - ip)(x + ip) = x^2 + p^2 + \hbar.$$

9.4 The Copenhagen Interpretation

Quantum Mechanics is a physical theory, and its purpose is the description of Nature and the prediction of the outcome of experiments. The theory is so successful that everyone believes it must be true. It allows us to predict the results of all experiments done so far in particle, nuclear, atomic, molecular and condensed matter Physics so well that no failure of Quantum Mechanics has ever been reported. The only exception is that nobody knows how to reconcile it with General Relativity.

The fundamental laws of Nature are so different from the picture of reality it that we form in the everyday life that we feel confused. The great Richard Feynman declared that *nobody understands Quantum Mechanics*. It is not just the random character of the results of some measurements; that is a fact, which requires a statistical treatment. The trouble is that everyone would like to infer from the experimental results a cartoon of what really happens at the microscopic level, and this requires a totally new way of thinking about particles and interactions. In Physics, the accuracy of measurements and the quantitative predictions of the theories are of fundamental importance, but qualitative arguments are equally essential. This is what we mean by interpretation.

The use of probability theory had appeared in Theoretical Physics when Gibbs invented Statistical Mechanics. Here, the meaning is quite different, and has nothing to do with ensemble averages, while the Temperature is not involved.[7] In Classical Physics we need statistics when we have incomplete knowledge of a system. Knowledge of the internal state of a slot machine, enables us to predict the right moment for inserting the coin; since the internal state is secret, one can only speak about probability. Is it possible to go beyond the bizarre rules of the game and fill the gap between what appears from the measurements and the underlying reality?

In the *Copenhagen interpretation*, proposed by Niels Bohr and Werner Heisenberg in the '20s, the reply to the last question is a radical no. The wave function ψ contains a statistical description, which, however is a full description of reality. The particle does not possess such properties as a classical trajectory. The only values that can be given to observables are those of operators for which ψ happens to be an eigenfunction. The micro world is so unusual that a particle can have many non-classical properties, including spin, isospin, parity, strangeness, color charge; on the other hand, particles lack familiar properties like a sharply defined trajectory. In the case of the slot machine, the missing information is hidden, but the precise position of the electron prior to measurement is fuzzy.

Suppose somebody wants to measure the electron-proton distance in the normal state of the H atom, by sending fast charged particles and measuring the change of their momentum, which depends on the field they come across. The measurement on

---

[7]Indeed there is a link between Quantum and Statistical Mechanics, which are unified by now e.g. in the Keldysh theory. This exploits the analogy between the Boltzmann factor and the exponential $\exp \frac{-iEt}{\hbar}$ which appears e.g. in the De Broglie wave and in stationary states (see below). However conceptually these remain quite different exponentials.

a single H atom can give any value. Preparing many atoms in the same conditions and repeating the measurement many times, one gets a well-defined distribution of electron-proton separations. The result can be interpreted as a $|\psi(\vec{r})|^2$ and allows us to calculate the statistical average $\langle r \rangle = \int |\psi(\vec{r})|^2 d^3r$. This is not the same as repeating the measurement many times on the same atom, because the measurement itself perturbs the system, while classically, this kind of complication is assumed to be negligible. More precisely, the measurement of an observable $\hat{O}$ yielding the result o leaves the system in an eigenstate ψ_o of $\hat{O}$ with exactly that eigenvalue.[8] In other words, the wave function has collapsed to ψ_o. The *collapse of the wave function* is one of the most striking predictions of Quantum Mechanics and leads, among others, to paradoxes to be discussed in Chap. 27.

There are also quantities that are sharply defined under some circumstances. If a sample of H atoms in the ground state is used to measure the binding energy, the result is about 13.59 eV and is always the same within the experimental error.

Using Eq. (9.4), one obtains the *mean value* of any function $f(x)$ defined by

$$\langle f(x) \rangle = \int_{-\infty}^{\infty} dx \rho(x,t) f(x).$$

This is commonly referred to as the *expectation value*, even if everybody understands that this is a misnomer. For instance, a probability distribution symmetric around x_0 could vanish right there, so x_0 would be unattainable but still the expectation value.

The average position at time t is

$$\langle x(t) \rangle = \int_{-\infty}^{\infty} dx \rho(x,t) x,$$

which implies

$$\langle \{x - \langle x \rangle\} \rangle = \int_{-\infty}^{\infty} dx \rho(x,t) \{x - \langle x \rangle\} = 0.$$

As in every statistical treatment, the width of the distribution is given by the standard deviation σ defined by

$$\sigma^2 = \langle \{x - \langle x \rangle\}^2 \rangle = \int_{-\infty}^{\infty} dx \rho(x,t) \{x - \langle x \rangle\}^2.$$

One finds that

$$\sigma^2 = \langle x^2 \rangle - \langle x \rangle^2,$$

where

---

[8] If the measurement is repeated afterwards the system is found in a state $\psi_o(t)$ that is evolved from ψ_o, and if $\hat{O}$ commutes with the Hamiltonian, the eigenvalue is still o. This statement may be somewhat obscure to the reader but will be clear after studying the evolution in the next chapter.

$$\langle x^2 \rangle = \int_{-\infty}^{\infty} dx \rho(x,t) x^2.$$

The wave function is complex and can be taken to be real only in special cases. This is far from obvious since the measurements always yield real results. However we have seen that the De Broglie wave is complex, and is the eigenfunction of momentum, which is itself a complex operator. We shall see that the phase is crucial for angular momentum eigenstates, for the existence of a current, for enabling a change of gauge and a change of reference in a Galileo transformation; later on, we shall discover more subtle ways the phase is necessary (topological phase). On the other hand, the wave function, like the electromagnetic potentials, is not observable.

In simple systems like a particle in a quantum well, one measurement of energy can tell us which one is the state. In general, the task of labelling a system with a quantum state can be demanding. Complex systems with several degrees of freedom may require a series of measurements. How many quantities and which quantities are needed? This complicated issue is studied by *Quantum Tomography*.

9.4.1 The Bohm Formulation of Quantum Theory

Moreover, David Bohm, in 1952, has proposed an alternative interpretation of Quantum Mechanics in which a particle obeys Classical Mechanics but is acted upon by a quantum potential. The wave function is written in the form

$$\psi = R \exp\left(\frac{iS}{\hbar}\right), \tag{9.13}$$

with $R \geq 0$ and real action S; for one particle, the action is given by the classical Hamilton–Jacobi equation with an additional non-local quantum potential

$$U = -\sum_k \frac{\hbar^2}{2m} \frac{\nabla^2 R}{R}. \tag{9.14}$$

This nonlinear framework determines particle trajectories. The theory, which has also been formulated in the many-particle case, is considered as an alternative interpretation of Quantum Theory. It gives the same physical results. It is non-local and highly non-linear, and this is the price to pay to keep a strict analogy with Classical Mechanics.

9.5   Quantum Eraser Experiments and Delayed Choice Experiments

In double-slit experiments, the interference pattern vanishes if, by the use of light or by other means, one can in principle detect which slit the particle passes, but whenever this information is erased the interference pattern returns. There are other ways to study the interplay of particle-like and wave-like properties. Let us go back to Sect. 9.3.2 and to Fig. 9.5. In Experiment 1, a photon arriving from SST to A comes from the vertical path, and a photon hitting B comes from the horizontal path; in both cases, it behaves as a particle, and there is no interference. Experiment 2 shows interference: by changing the optical paths, one modulates the intensity of the light in both detectors. John Wheeler noticed some interesting implications of this experiment. In the latter case, the wave properties are evident, which implies that the photon goes both ways. In some sense, the photon behaves as a particle or as a wave depending on the presence or absence of the second beam splitter. One can say that by introducing the second beam splitter, the path information has been *erased*. Erasing produces interference phenomena at detection screens *A* and *B* positioned just beyond each exit port, where the count rate depends on the optical paths. However, it takes time to go from *S* to *A* or *B*. What happens if the second beam splitter is inserted into the left panel (or removed from the right panel) while the photon is in flight?

Experiment says that when the second beam splitter is present, the photon behaves as a wave even if it is inserted during the flight; if it is removed during the flight the photon behaves as a particle even if it must initially have started as a wave. Some people think that it is possible to change the *decision* of the photon in a retroactive way. This is actually an upsetting conclusion. A cosmic version of the experiment is even more impressing. Multiple images of a quasar are produced in the field of a Galaxy by gravitational lensing (Sect. 8.7). One can orient a telescope in such a way that only one image is visible; in this case, only the photons that according to one interpretation *decided* to travel as particles can contribute (left part of Fig. 9.6. However one can also produce an interference pattern out of two images (right part of Fig. 9.6) and in this case the photons must have *decided* to travel as a wave. Is it possible that the experimental arrangement on Earth reverses a *decision* made millions of years ago?

The situation is clarified by an enlightening paper published online by David Ellerman, entitled: "A Common Fallacy in Quantum Mechanics: Why Delayed Choice Experiments do NOT Imply Retrocausality". Essentially, the argument is that the photons do not have to *decide* their mode of propagation (particle-like or wave-like). Rather, let us go back to Fig. 9.5: the wave functions are a quantum superposition of propagation along the vertical path, propagation along the horizontal path and propagation along both paths; it is the choice of the method of detection that makes the wave function collapse in a particular way. In this way, the upsetting perspective of a violation of causality and/or retrocausality is removed.

Fig. 9.6 Left: only one of the images of the quasar enters the telescope. Right: collecting both left and right images of the quasar and superposing them, we erase the path information and produce an interference pattern

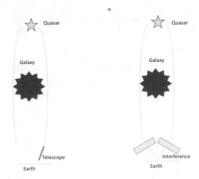

9.5.1 Is the Interpretation a Closed Issue?

There are still subtle problems, not chitchat but conceptual problems, that could lead to observable consequences, and thus concern Science. Einstein thought that the theory is not complete and refused the use of statistical means (his motto *God does not play dice* is famous). The developments of his ideas, the coexistence with relativity and some aspects of the problem will be discussed later.

Among the obscure problems, the border between classical and quantum physics stands out, and the issue is involved. Superfluidity is a macroscopic quantum phenomenon whereby liquid He creeps up the walls of a cup. Superconductivity is a macroscopic quantum phenomenon whereby trains are levitated. Magnetism is a purely quantum effect due to electron correlation. So, it is not just a matter of size. The Schrödinger's cat is a well-known riddle. An atom can be prepared in a mixed state, or superposition $\psi = c_1\psi_1 + c_2\psi_2$, where c_1, c_2 are complex numbers and ψ_1, ψ_2 orthogonal eigenfunctions of some observable $\hat{A}$ corresponding to eigenvalues a_1, a_2. A measurement of A collapses ψ, and the system must fall in one of the eigenvalues. Suppose the cat is closed for an hour in a box with a small quantity of a radioactive element, tailored in such a way that there is a 50 per cent chance that a nucleus decays in one hour. An infernal apparatus in the case of decay poisons the poor cat. Is this a way of preparing a fifty-fifty superposition of a live cat and dead cat, until the box is opened and the cat is observed? (Of course in order to assign a significance to the wave function one should prepare a large number of cats and infernal machines.) It is clear that Quantum Mechanics does not describe all the reality, but just says what one can tell on the basis of a specific experiment; there is no paradox in the use of probability, and if the cat were replaced by a two-state quantum system, there would be no paradox. One problem is that we feel that the cat should be treated by classical probability. Where is the border between classical and quantum? The second problem is that the observer has a special role in this story, yet the observer is a part of reality, which remains in the shadow. Could the cat be an observer? In 2012, the Nobel Prize was awarded to Serge Haroche and

David J. Wineland for experimental studies in the field of Quantum Optics in which systems consisting of atoms and photons were kept 50 ms ia a state of quantum superposition. In a sense, such systems are Schrödinger cats. After 50 ms, due to the interaction with the environment *decoherence* occurs, which means, a classical description prevails. Such studies could pave the way to a quantum computer.

Chapter 10
The Eigenvalue Equation and the Evolution Operator

It is a strange theory that predicts particles that interfere with themselves and propagate through forbidden regions. But that is just what happens in reality.

The operators that represent variables of classical dynamics are built by analogy with classical analogues.[1] Sometimes, there are complications due to the fact that some operators fail to commute; such cases will be noted below. Therefore, for the simple case of one particle in a potential,

$$\hat{H} = -\frac{\hbar^2}{2m}\nabla^2 + V(\hat{x}) \tag{10.1}$$

is the Hamiltonian operator. By the same criterion, one can write down the Hamiltonian operator for systems of particles. The first term $H = \frac{p^2}{2m}$ is the kinetic energy operators, which by itself has the eigenvalue equation

$$\hat{H}\psi_{\vec{p},E} = H\psi_{\vec{p},E}; \tag{10.2}$$

the eigenfunctions are just the De Broglie waves, which have the known property

$$\hat{\vec{p}}\,\psi_{\vec{p},E} = \vec{p}\,\psi_{\vec{p},E}. \tag{10.3}$$

---

[1] Since the classical description, as we know, can be changed by canonical transformation, this statement implies that in Quantum Mechanics, we must enjoy the same freedom. We shall see later how this arises.

© Springer International Publishing AG, part of Springer Nature 2018
M. Cini, *Elements of Classical and Quantum Physics*,
UNITEXT for Physics, https://doi.org/10.1007/978-3-319-71330-4_10

The eigenfunctions of (10.1) depend on the potential V, and we shall see several examples. Inspired by the De Broglie wave, where the energy is at the exponent, Schrödinger decided that the energy operator $\hat{E}$ must be

$$\hat{E} = i\hbar \frac{\partial}{\partial t}. \tag{10.4}$$

Indeed, $\hat{E}$, acting on ψ_E, yields

$$\hat{E}\psi_{\vec{p},E} = E\psi_{\vec{p},E}. \tag{10.5}$$

In time-independent problems, the energy must be conserved and classically $H = E$. Schrödinger proposed that

$$\boxed{\hat{H}\psi = \hat{E}\psi.} \tag{10.6}$$

Actually, the Schrödinger equation (10.6) is one of the most important equations of history, and describes the evolution of single-particle and many-particle systems. It remains valid also for problems which are not stationary; actually it dictates the evolution of all states. If H is time-independent, we may seek wave function with sharp energy that are eigenstates of both H and $\hat{E}$. The above sentence is not a derivation, since a general theory cannot be deduced from a special case, but a heuristic argument.

For a particle in a (possibly time-dependent) potential $V(x, t)$ in one space dimension, one must solve this linear partial differential equation:

$$\left[\frac{\hat{p}^2}{2m} + V(\hat{x}, t) \right] \psi(x, t) = i\hbar \frac{\partial}{\partial t} \psi(x, t), \tag{10.7}$$

where $\hat{p} = -i\hbar \frac{d}{dx}$; the extension to 3 dimensions is immediate, with $\hat{p} \to -i\vec{\nabla}$.

It is common to speak about waves and wave mechanics, but the Schrödinger equation (often referred to as the S.E.) is no wave equation; actually it becomes a diffusion equation after the transformation $t \to i\tau$.

The De Broglie plane wave (9.3) solves the Schrödinger equation for $V = 0$. Note that $\psi_{-\vec{p},E}$ has the same energy as $\psi_{\vec{p},E}$, so any linear combination of the two still has energy E.

It proves to be an excellent idea to introduce the *time evolution operator* $\hat{U}(t, t')$ such that

$$\psi(x, t) = \hat{U}(t, t')\psi(x, t'); \tag{10.8}$$

the evolution operator then obeys

$$\left[\frac{\hat{p}^2}{2m} + V(\hat{x}, t) \right] U(t, t') = i\hbar \frac{\partial}{\partial t} U(t, t'). \tag{10.9}$$

Perhaps, this way to rephrase the problem in terms of operator does not immediately appear to be a clever idea; it proves very useful in the development of the theory, since $\hat{U}$ has several interesting properties that have been exploited with much ingenuity. To start with, the formal solution of Equation (10.9)

$$U(t, t') = 1 - \frac{i}{\hbar} \int_0^t H(t_1) U(t_1, t) \tag{10.10}$$

lends itself to an iterative expansion. The so-called Group property

$$U(t, t_a) U(t_a, t') = U(t, t') \tag{10.11}$$

which is obviously true for any t_a, is the starting point of the Path Integral formalism, as I show below (Sect. 15.2). For time independent problems, one can write

$$U(t, t') = \exp\left[-i \frac{H(t - t')}{\hbar}\right]. \tag{10.12}$$

When $\psi(\hat{x}, t = 0)$ is known, (10.7) determines its evolution in the future and in the past. Linearity implies the superposition principle. Equation (10.6) is second-order in the space derivatives and first-order in time, so it breaks the principle of Relativity. However, when relativistic effects are important, it lends itself to generalizations (Klein–Gordon equation, Dirac equation) that work extremely well. Already in the present form, it explains a lot of phenomena. The equation can be used for electrons, provided that spin does not enter into play.

Potential Wall

In one space dimension, one readily solves the problem with a potential wall $V(x) = \infty$, $x < 0$, $V(x) = 0$, $x > 0$, with the condition $\psi_E(x, t) = 0$, $x < 0$;

$$\psi_E(x, t) \approx \sin(kx) \exp\left[\frac{-i E(p)t}{\hbar}\right], \quad p = \hbar k. \tag{10.13}$$

This wave function represents an electron beam fired against a potential wall. The particles bounce, and there is a wave[2] with momentum p and a reflected wave with momentum $-p$. By the interference between incident and reflected waves, the wave function vanishes for kx multiple of π. The particles pass through such points, but cannot be detected there! The particles propagate as waves, and no corpuscles then exist; when the particles are detected somewhere the wave does not exist any more.

---

[2]By the Euler formula, $\sin(\alpha) = \frac{e^{i\alpha} - e^{-i\alpha}}{2i}$.

10.1   Stationary State Equation and Its Resolvent

Any state $\psi(x, t)$ with a well defined energy E is eigenstate of the energy operator $\hat{E}$

$$\hat{E}\psi(x, t) = E\psi(x, t),$$

so its time-dependence is through the phase factor $e^{-i\frac{Et}{\hbar}}$. The probability density of such a state is time-independent. Therefore the energy eigenstates are called stationary states.

Substituting

$$\psi(x, t) = \psi(x)e^{-i\frac{Et}{\hbar}}, \tag{10.14}$$

into (10.7), we separate the variables. The time-independent equation for $\psi(x)$ reads as:

$$\hat{H}\psi(x) = E\psi(x). \tag{10.15}$$

For a particle which is known to be at x at time 0, the amplitude to find it at y at time t is $\langle y|\hat{U}(t)|x\rangle$, where $\hat{U}(t)$ is the evolution operator defined in the last section. The Fourier transform $\langle y|\frac{1}{\omega-\hat{H}}|x\rangle$ can be phrased as the amplitude to go from x to y at frequency ω, and the resolvent

$$\hat{G}(\omega) = \frac{1}{\omega - \hat{H}}, \tag{10.16}$$

which is closely related to Green's functions, will be useful later.

10.2   Continuity Equation

The Continuity Equation is well-known from Electromagnetism, where it states the conservation of charge. The charge density $\rho(x, t)$ and the current density $\vec{J}(x, t) = \rho\vec{v}$ (where $\vec{v}$ is the velocity) obey

$$\frac{\partial\rho}{\partial t} + \nabla\vec{J} = 0. \tag{10.17}$$

Integrating over volume V and using the Gauss Theorem,

$$\frac{d}{dt}\int_V \rho d^3x + \int_V \nabla\vec{J} d^3x = \frac{dQ}{dt} + \int_S \vec{J} d\vec{S} = 0. \tag{10.18}$$

The change total charge Q in a given volume depends on the flux of current across the boundary. In the single-particle Schrödinger theory, the relevant probability density is

$$\rho(x, t) = |\psi(x, t)|^2,$$

and we must find the current density $J(x, t)$ associated with it. For charged particles, these expressions multiplied by the electron charge e naturally represent the charge and current densities associated with the wave function; again there is the need to average over many particles or many instances before allowing this correspondence.

In one dimension, the task of obtaining the current density is simplest; multiply by $\psi^*(x)$ the S.E.

$$-\frac{\hbar^2}{2m}\frac{d^2\psi}{dx^2} + V(x, t)\psi = i\hbar\frac{\partial\psi}{\partial t},$$

then multiply by ψ the conjugated equation

$$-\frac{\hbar^2}{2m}\frac{d^2\psi^*}{dx^2} + V(x, t)\psi^* = -i\hbar\frac{\partial\psi^*}{\partial t}$$

and subtract: the result is

$$-\frac{\hbar}{2m}\left[\psi^*\frac{d^2\psi}{dx^2} - \psi\frac{d^2\psi^*}{dx^2}\right] = i\left[\psi^*\frac{\partial\psi}{\partial t} - \psi\frac{\partial\psi^*}{\partial t}\right], \tag{10.19}$$

that is,

$$-\frac{\hbar}{2mi}\frac{d}{dx}\left[\psi^*\frac{d\psi}{dx} - \psi\frac{d\psi^*}{dx}\right] = \frac{\partial|\psi|^2}{\partial t}.$$

Therefore, the probability current density is

$$J = \frac{\hbar}{2mi}\left[\psi^*\frac{d\psi}{dx} - \psi\frac{d\psi^*}{dx}\right]. \tag{10.20}$$

This result looks less unfamiliar if one reasons that $J = \frac{1}{2m}(\psi^* p\psi + c.c.)$. Since the probability of finding the particle becomes 1 if one looks for it everywhere, the wave function must be *normalized* to unity:

$$\int_{-\infty}^{\infty} dx|\psi(x, t)|^2 = 1. \tag{10.21}$$

This makes sense provided that the normalization is forever; we shall check that this is indeed the case provided that the potentials are real; in other words, the evolution is unitary.

The above is easily extended to three dimensions.

$$J = \frac{\hbar}{2mi}\left[\psi^*\nabla\psi - \psi\nabla\psi^*\right]. \tag{10.22}$$

If ψ is real, the current vanishes. Complex wave functions are needed to describe transport phenomena.

10.3   Schrödinger and Heisenberg Formulations of Quantum Mechanics

The *matrix mechanics* was formulated in 1925 by Werner Heisenberg.[3] In the same year the quantum theory of the Hydrogen atom was proposed by Wolfgang Pauli[4] in a beautiful but complicated applications of Group theory, assuming quantum commutation relations. In 1926 Erwin Schrödinger[5] wrote the fundamental paper "Quantisierung als ein Eigenwert Problem" introducing the S.E.

Initially, Schrödinger found the Heisenberg *matrix mechanics* unconvincing, while Heisenberg thought that Schrödinger's theory was rubbish. It did not take a long time before it became clear that the different formulations gave the same results because they where mathematically equivalent. Actually, each of them clarified the others. Both are equivalent to Feynman's Path Integral formulation that will be presented in Sect. 15.2 below.

---

[3] Werner Heisenberg (Würzburg, 1901-München 1976), received Nobel prize in 1932.

[4] Wolfgang Pauli (Vienna 1900-Zurich 1958) received the Nobel prize in 1945.

[5] Erwin Schrödinger (Vienna 1887-Vienna 1961) succeeded to Max Planck as professor of Physics in Berlin in 1927; he won the Nobel Prize in 1933.

Chapter 11
Particle in One Dimension

Simple cases trigger Intuition.

11.1 Deep Potential Well

One dimensional problems illustrate in simple fashion many aspects of the theory, and for this reason, every beginner in Quantum Mechanics is traditionally exposed to them. Some interesting one-particle problems in higher dimensions can be solved by separation of variables, thereby reducing them to 1d subproblems, and we shall encounter several examples in other chapters. However, in recent times there have been important technological developments in nano-devices and low-dimensional objects (like nano-wires and nano-tubes) that have made these problems much more directly relevant. In classical physics, a wire is one-dimensional in the limited sense that its length is many times its diameter. However, in Quantum Mechanics, the low dimensionality of thin films and 1d objects has a much stronger meaning. The next problem (deep potential well) shows that the confinement implies quantized levels divided by gaps that become wider the narrower the well. This implies that at low enough temperatures, the electrons behave as if the transverse dimensions were effectively frozen out. In this way, low d problems are no longer academic.

Consider a Schrödinger particle trapped between $x = 0$ and $x = a$ by the potential

$$V(x) = \begin{cases} \infty \text{ for } x < 0, \\ 0 \quad \text{for } 0 < x < a, \\ \infty \text{ for } x > a. \end{cases}$$

© Springer International Publishing AG, part of Springer Nature 2018
M. Cini, *Elements of Classical and Quantum Physics*,
UNITEXT for Physics, https://doi.org/10.1007/978-3-319-71330-4_11

We must solve

$$\begin{cases} -\frac{\hbar^2}{2m}\frac{d^2}{dx^2}\psi_n(x) = E_n\psi_n(x), \\ \psi_n(0) = \psi_n(a) = 0. \end{cases}$$

The result is

$$\psi_n(x) = C\sin(\chi_n x), \quad \chi_n = n\frac{\pi}{a}, \tag{11.1}$$

where $n = 1, 2, \ldots$ is an integer quantum number (positive, since negative numbers give the same solutions). The energy eigenvalues are:

$$E_n = \frac{\pi^2\hbar^2}{2ma^2}n^2. \tag{11.2}$$

Remarks:

1. Classically, the motion in this potential is too stupid to have the status of a problem, but the quantum version is very interesting. The energy spectrum is discrete. *Natura facit saltus!*
2. ψ_n has n-1 nodes (that is, $\psi_n(x) = 0$ has $n - 1$ roots.) As N grows, λ gets shorter while E grows.
3. E_1 is the ground state, the others are excited states. Classically, the particle could be somewhere with zero velocity and zero energy. The minimum quantized energy is $E_1 > 0$; there is a minimum amount called *zero point energy*. This is in line with the *uncertainty principle* $\Delta x\,\Delta p \sim \hbar$, where Δx is the spread in position and Δp is the spread in momentum. Since the particle is known to be between $x = 0$ and $x = a$, that is $\Delta x \sim a$, the momentum has uncertainty $\Delta p \sim \frac{\hbar}{a}$; therefore, $\Delta\frac{p^2}{2m} \sim E_1$.
4. $\psi_n(x)$ is real and there is no current. One can imagine that the particle bounces and creates two opposite currents that make up zero.

The normalization constant C is found by the condition

$$\int_0^a |\psi_n(x)|^2 dx = C^2\int_0^a dx\,\sin^2(\chi_n x) = 1.$$

Since

$$\int dx\,\sin^2(x) = \frac{x}{2} - \frac{\sin(2x)}{4},$$

one finds that the normalized eigenfunctions are

$$\psi_n(x) = \sqrt{\frac{2}{a}}\sin\left[n\pi\frac{x}{a}\right]. \tag{11.3}$$

Several remarks are in order:

1. $\psi_n(x)$ is the sum of opposite momentum plane waves. The mean momentum is:

$$\langle p \rangle = \langle \psi_n(x) | p | \psi_n(x) \rangle = \int_0^a \psi_n(x)(-i\hbar)\frac{d}{dx}\psi_n(x)dx.$$

This is imaginary. Are we in trouble? No, since $\langle p \rangle = 0$. Indeed, $\frac{1}{2}\int_0^a dx \frac{d}{dx} \psi_n(x)^2 = 0$.

2. $\psi_n(x)$ is eigenfunction of p^2 for x inside the well. The same argument applies to any bound state.

3. $\{\psi_n(x)\}$ is just the complete orthonormal set in the interval $(0, a)$ familiar from the theory of Fourier series.

11.2 Free Particle and Continuum States

It is time to worry about orthonormalization and the physical meaning of the De Broglie waves, which represent the particle in a perfectly monochromatic, collimated beam; in one dimension,

$$\psi_k(x, t) = \frac{exp[i(kx - \omega(k)t)]}{\sqrt{2\pi}}.$$

11.2.1 Normalization

We encountered the factor $\frac{1}{\sqrt{2\pi}}$ in (9.8), where it comes with the third power because in that case the problem is 3d (three-dimensional). It is there to ensure the correct mean value $\langle p \rangle$ according to (9.11). The normalization method is different from the condition

$$1 = \int |\psi(x)|^2 dx,$$

which we use for bound states. Indeed, $\psi_k(x, t)$ is not $\in L^2$ (square integrable functions) over the x axis; we normalize it over the δ function:

$$\int_{-\infty}^{\infty} dx \psi_k(x, t)^* \psi_{k'}(x, t) = \delta(k - k'). \tag{11.4}$$

However, this is not the only method in use. One can always pretend that the electron gun and the entire experimental setup is contained in a box of side L; in one dimension, the plane wave function is

$$\psi_k(x,t) = \frac{exp[i(kx - \omega(k)t)]}{\sqrt{L}}.$$

We choose the discrete k values such that the wave functions are orthogonal over the Kronecker δ :

$$\delta_{m,n} = \begin{cases} 1, & m = n, \\ 0, & m \neq n. \end{cases}$$

This is granted by periodic boundary conditions

$$e^{ikL} = 1; \tag{11.5}$$

these are verified for a discrete set of k such that

$$k_n = \frac{2\pi n}{L}, \quad n \text{ integer},$$

and the orthogonality is ensured by the theory of Fourier series. If L is large, the box has no influence on the experiment.

Dirac's δ yields a formal method for going to the infinite box limit. In 3d, the volume is $V = L^3$, and

$$\psi_k(x,t) = \frac{exp[i(kx - \omega(k)t)]}{\sqrt{V}}.$$

11.2.2 Phase Velocity

The dispersion law is the law $E(p)$, where E is the kinetic energy and p the momentum; it depends on the particle and also on the theory. For a corpuscle in classical mechanics, $E(p) = \frac{p^2}{2m}$, but in the relativistic theory, $E(p) = \sqrt{p^2c^2 + m^2c^4}$; for a photon, $\omega = ck$. More exotic, anisotropic dispersion laws describe the propagation of excitations (band electrons, dressed electrons, vibrations called phonons, mixed photon-vibration modes called polaritons) in condensed matter.

Although Schrödinger's is not the wave equation, nevertheless, the De Broglie plane wave is still of the form $f(x - v_p t)$, where v_p is called the phase velocity. For a photon it is the actual velocity c, but with the particle dispersion law $\hbar\omega(k) = \frac{p^2}{2m}$, the phase velocity $v_p = \hbar\frac{k}{2m}$ is half the physical speed $v = \frac{p}{m}$. In the plane wave the position is not defined, so the best one can do to measure the speed is to measure p and divide by m. The velocity operator is, indeed, $v = \frac{p}{m}$.

11.2.3 Group Velocity of a Wave Packet

Consider a free particle ($V = 0$). If we prepare a wave packet at time $t = 0$

$$\psi_0(x) \equiv \psi(x, 0) = \int dk \phi(k) e^{ikx}, \tag{11.6}$$

it will evolve at time t to become

$$\psi(x, t) = \int_{-\infty}^{\infty} dk \phi(k) e^{ikx} e^{-i\omega(k)t};$$

to see that, we could plug it into the S.E., but it is enough to observe that each k component in (11.6) is evolving as a De Broglie wave.

Suppose $\phi(k)$ is strongly peaked around $k = k_0$. This is a practically feasible approximation of a plane wave state. Only k values close to k_0 matter; in this neighborhood,

$$\omega(k) \sim \omega(k_0) + v(k - k_0), \quad v = \frac{d\omega(k)}{dk}\big|_{k_0}. \tag{11.7}$$

Therefore, to a good approximation,

$$\psi(x, t) = \int_{-\infty}^{\infty} dk \phi(k) e^{ikx} e^{-i[\omega(k_0)+v(k-k_0)...]t},$$

that is,

$$\psi(x, t) \sim \exp[i(-\omega(k_0)t + k_0 vt)] \int dk \phi(k) \exp[i(k[x - vt])]$$
$$= \exp[i(-\omega(k_0)t + k_0 vt)]\psi_0(x - vt).$$

The probability density $|\psi|^2$ propagates with the group velocity v, which is therefore the closest approximation to the classical behavior.

Continuing the expansion in (11.7), however, one discovers that the wave packet gradually broadens in real space.

11.3 Solutions of the Stationary State Equation in 1d

It follows from the S. E. that if the potential is continuous with all the derivatives (that is, $V(x) \in C^\infty$) the wave function $\psi(x)$ is also $\in C^\infty$. Depending on the form of the potential, one may find localized states, or running waves on the left or on the right. Localized states occur when $\psi(x) \to 0$ for $x \to \pm\infty$ and correspond to discrete energy eigenvalues for bound states. Many potentials of physical interest

Fig. 11.1 Discretised
potential

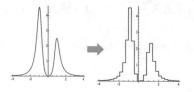

are scattering problems[1] in which for $x \to \pm\infty$ V tends towards constants and
consequently ψ becomes a de Broglie wave. Then, all energies are possible, i.e. the
spectrum of eigenvalues is continuous. In general, there is no closed-form analytical
solutions. Thus, we can approximate $V(x)$ by a piecewise constant, histogram-like
function (see Fig. 11.1). This introduces artificial steps, but makes the solution easy. In
any interval in which $V(x)$ is taken to be constant, $\psi(x)$ is a linear combination of two
exponentials. In classically allowed regions where the energy E exceeds $V(x)$, these
are $e^{\pm ikx}$, where $\hbar^2 k^2 = 2m(E - V)$. Classically, the particle cannot be found where
$E < V(x)$, but in those classically forbidden regions, the wave function continues;
however, the exponentials become real, namely, $e^{\pm \lambda x}$ where $\hbar^2 \lambda^2 = 2m(V - E)$.
The need for a normalizable function forbids growing exponentials for $x \to \infty$ and
decreasing exponentials for $x \to -\infty$.

At a step discontinuity of the potential at $x = x_0$, the S.E. implies a step-like contribution to $\psi''(x)$ proportional to $\theta(x - x_0)$, but $\psi'(x)$ and $\psi(x)$ remain continuous.
In the case of the hard wall and at the ends of the deep well $\psi'(x)$ is discontinuous,
but there V makes an infinite jump.

11.4 Potential Step

A particle in the potential

$$V(x) = \begin{cases} 0, & x < 0 \\ V > 0, & x > 0 \end{cases} \tag{11.8}$$

is free to go to infinity (both classically and in Quantum Mechanics); therefore, the
spectrum of $\hat{H}$ is continuous. Suppose an electron gun fires a beam of particles from
the left ($x = -\infty$) to a detector to the far right. We are looking for the stationary
states, with a flux of particles reaching the barrier; classical bullets would bounce
elastically if their kinetic energy E_k is less than the threshold V, otherwise they
should go to the right with kinetic energy $E_k - V$.

---

[1]There is indeed a variety of interesting problems with V, which is periodic (electrons in perfect
crystals), aperiodic (e.g. electrons in glasses): here, I consider only the simplest examples.

11.4.1 Energy Below Threshold

If $E_k = \frac{\hbar^2 k^2}{2m} < V$, the solution is

$$\psi(x) = \begin{cases} Ae^{ikx} + Be^{-ikx} \text{ con } k > 0 \text{ per } x < 0, \\ Ce^{-\lambda x}, \text{ per } x > 0, \end{cases} \tag{11.9}$$

with $\hbar^2 k^2 = 2mE$, $\hbar^2 \lambda^2 = 2m(V - E)$. The positive exponential would make the wave function blow up at infinity and was discarded.

Computing the current with the term in A, one finds that $J = |A|^2 \frac{p}{m} > 0$, where $p = \hbar k$; this is just the current from the gun, that is set by the experimenter; therefore we may take $A = 1$. The wave functions and its derivative are continuous:

$$\begin{cases} 1 + B = C, \\ ik(1 - B) = -\lambda C \implies 1 - B = \frac{i\lambda}{k} C. \end{cases} \tag{11.10}$$

The sum of these equations yields $C = \frac{2k}{k+i\lambda}$; hence, $B = C - 1 = \frac{-i\lambda+k}{i\lambda+k} = -\frac{\lambda+ik}{\lambda-ik}$. Since $|B|^2 = 1$, B is a phase factor, and we may set $B \equiv e^{i\beta}$, with real β, while $C = 1 + e^{i\beta}$. Therefore, $C = e^{i\frac{\beta}{2}}(e^{-i\frac{\beta}{2}} + e^{i\frac{\beta}{2}})$ and the whole wave function becomes real if we multiply it by $e^{-i\frac{\beta}{2}}$. A constant phase factor is irrelevant, but it makes evident that this, like any real wave function, carries zero current. All the particles bounce. This is also evident from the fact that $\psi \to 0$ for $x \to \inf$: since $\frac{\partial \rho}{\partial t} = 0 \Rightarrow \frac{dJ}{dx} = 0$, $J = 0$ everywhere.

As in the infinite wall case, there are places where the particle cannot be found because of interference. On the other hand, there is a chance of detecting the particle in the classically forbidden region $x > 0$. The classical prohibition is absolute, since a corpuscle with $E = T + V$ and $V > E$ must have a negative kinetic energy, which is impossible. Even in Quantum Mechanics, the kinetic energy is a *positive definite operator* $\hat{T}$, that is, an operator such that all the eigenvalues are positive; any measurement of a positive definite operator must return a positive value. The paradox can be understood as follows. The classical concept that the particle at a given point has some kinetic energy is not tenable in Quantum Mechanics, since $\hat{T}$ does not commute with the position x. We shall see in the next chapter that two observables exist simultaneously when the corresponding operators commute. The kinetic energy matrix taken over a complete set has positive eigenvalues, but the x values in the barrier do not constitute a complete set, so the idea of a *negative kinetic energy under the barrier* is not sound. On the other hand, the particle propagates as a wave ψ with some energy E, and the corpuscle exists only after the detection. In the same way, in the double slit experiment, the wave goes through both slits, and the corpuscle exists only where the interference pattern is recorded. These remarks remove the paradox, but the quantum behavior remains qualitatively different!

11.4.2 Energy Above Threshold

If $E > V$,

$$\psi(x) = \begin{cases} e^{ikx} + Be^{-ikx}, & x < 0 \\ Ce^{i\chi x}, & x > 0, \end{cases}$$

with $\hbar^2 k^2 = 2mE, \quad \hbar^2 \chi^2 = 2m(E - V.)$

From the continuity conditions for ψ and $\frac{d\psi}{dx}$ in $x = 0$

$$\begin{cases} 1 + B = C, \\ ik(1 - B) = i\chi C \Longrightarrow 1 - B = \frac{\chi}{k}C, \end{cases}$$

one finds that

$$\begin{cases} C = \frac{2k}{k+\chi}, \\ B = \frac{k-\chi}{k+\chi}. \end{cases}$$

The incident current is $A^2 k = k$; the reflected current is $|B|^2 k$; by definition, the ratio is the reflection coefficient

$$R = |B|^2 = \left| \frac{\chi - k}{\chi + k} \right|^2.$$

Even if the barrier is lower than the energy eigenvalue E, a reflected current arises; this would not happen classically. The transmitted current is $|c|^2 \chi$; by definition, the transmission coefficient is the ratio with the incident current and is given by:

$$T = |C|^2 \frac{\chi}{k} = 4\chi k \left| \frac{1}{\chi + k} \right|^2.$$

Therefore, $R + T = 1$.

11.4.2.1 Descending Step $V < 0$

Classically, for a negative step ($V < 0$), all the particles should fall in with increased energy.

Quantum mechanically, the above solution still holds and there is a reflected current, too. This is observed experimentally when a beam of electrons is diffracted through the surface atoms of a solid.

11.4.3 Potential Well of Finite Depth

Consider the symmetric potential

$$V(x) = \begin{cases} 0, & x < -a \\ -W < 0, & -a < x < a \\ 0, & x > a. \end{cases}$$

An unsymmetric one would also be solvable, but would not add much to the interest. The symmetric case is special, but also instructive, because it offers an elementary example of the use of symmetry, which is a very important and far-reaching subject. An even potential $(V(-x) = V(x))$ allows for even solutions $(\psi(-x) = \psi(x))$ but also odd solutions $(\psi(-x) = -\psi(x))$.

Bound States $E < 0$

The bound solutions have negative energies, and there is a finite number of them; they come with two different symmetries:

even :   odd :

$$\psi(x) = \begin{cases} Be^{kx}, & x < -a, \\ D\cos(lx), & -a < x < a, \\ Be^{-kx}, & x > a, \end{cases} \qquad \psi(x) = \begin{cases} -Be^{kx}, & x < -a, \\ D\sin(lx), & -a < x < a, \\ Be^{-kx}, & x > a. \end{cases}$$

$$\tag{11.11}$$

where

$$k = \sqrt{\frac{-2mE}{\hbar^2}}, \qquad l = \sqrt{\frac{2m(E+W)}{\hbar^2}}; \tag{11.12}$$

Imposing the continuity conditions for ψ and $\frac{d\psi}{dx}$ in $x = a$, we are sure they are satisfied in $x = -a$ as well. Therefore (Fig. 11.2),

Fig. 11.2 Potential well of depth W

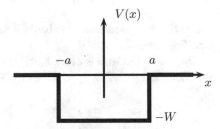

even :   odd :

$$\begin{cases} D\cos(la) = Be^{-ka}, \\ Dl\sin(la) = kBe^{-ka}; \end{cases} \qquad \begin{cases} D\sin(la) = Be^{-ka}, \\ Dl\cos(la) = -kBe^{-ka}. \end{cases} \qquad (11.13)$$

The eigenvalue conditions are:

$$\begin{array}{cc} \text{even :} & \text{odd :} \\ l\tan(la) = k & l\cot(la) = -k. \end{array} \qquad (11.14)$$

Let us put

$$z = la = \sqrt{\frac{2m(E+W)}{\hbar^2}}\,a$$

in order to simplify the argument of transcendental functions. Equation (11.12) allows us to obtain k in terms of l. Using $l^2 = \frac{2mW}{\hbar^2} - k^2$,

$$k^2 a^2 = \frac{2mW}{\hbar^2}a^2 - l^2 a^2 = z_0^2 - z^2 > 0$$

with

$$z_0^2 = \frac{2mW}{\hbar^2}a^2;$$

so,

$$\frac{k}{l} = \frac{ka}{la} = \frac{\sqrt{z_0^2 - z^2}}{z}.$$

Equation (11.14) becomes

$$\begin{array}{cc} \text{even :} & \text{odd :} \\ \tan(z) = \frac{\sqrt{z_0^2-z^2}}{z}, & \cot(z) = -\frac{\sqrt{z_0^2-z^2}}{z}. \end{array} \qquad (11.15)$$

These equations are readily solved numerically or graphically, and there are simple limiting cases.

If $z_0 \gg 1$, the well is deep, and if, in addition, we look at $z \ll z_0$, we obtain the lowest levels. As a first approximation, for $\zeta_0 \to \infty$,

$$\tan(z) \to \infty \; per \; z_n \to (2n+1)\frac{\pi}{2}, n \in \mathcal{N},$$
$$\cot(z) \to \infty \; per \; z_n \to (2n)\frac{\pi}{2}, n \in \mathcal{N},$$

where $\mathcal{N}$ is the set of natural numbers $1, 2, 3, \dots.$. The condition

$$z^2 = \frac{2m(E + W)a^2}{\hbar^2} = (2n + 1)\frac{\pi^2}{4}$$

yields

$$E = -W + \frac{\hbar^2\pi^2}{2m(2a)^2}(2n + 1)^2.$$

Taking into account that the width is $2a$, these are the odd n solutions of the deep well, which are spatially even; $\cot(z) \to \infty$ yields the even n solutions. When n grows this approximation deteriorates.

Since $0 < z < z_0$, if $z_0 \ll 1$, z must also be $\ll 1$. For spatially even states, we replace $z\tan(z) = \sqrt{z_0^2 - z^2}$ with $\tan(z) \approx z$, that is, $z^2 = \sqrt{z_0^2 - z^2} \implies z^4 \approx z_0^2 - z^2 \implies z^2 \approx z_0^2 - z_0^4$. There is always at least one solution. In the spatially odd case we consider $z\cot(z) = -\sqrt{z_0^2 - z^2}$; there is no solution with $\cot(z) \approx \frac{1}{z}$. A very weak potential in one dimension always has a bound state, which is even.

The normalization requires, as usual, $\int_{-\infty}^{\infty} |\psi_n(x)|^2 = 1$.

11.4.4 Continuum States

For $E > 0$, all energies are allowed, and each eigenvalue is twice degenerate, that is, there are two orthogonal states with the same energy. This reflects the experimenter's freedom to choose between shooting with a gun from left or right. Let us choose the first arrangement. Classically, a marble thrown from $-\infty$ would proceed beyond the well with the same speed; because of quantum effects, part of the wave fired from $-\infty$ is reflected. Letting $\hbar k = \sqrt{2mE}$,

$$\psi(x) = \begin{cases} Ae^{ikx} + Be^{-ikx}, & x < -a, \\ C\sin(lx) + D\cos(lx), & -a < x < a, \\ Fe^{ikx}, & x > a, \end{cases}$$

and the incident amplitude is A, since e^{ikx} has the current $\frac{k}{m} > 0$ which goes to the right; B is the reflected amplitude and F the transmitted one. A is not really unknown, since the incident flux is chosen at will by the experimenter. The conditions in in $-a$ yield:

$$\begin{cases} Ae^{-ika} + Be^{ika} = -C\sin(la) + D\cos(la), \\ ik\left[Ae^{-ika} - Be^{ika}\right] = l\left[C\cos(la) + D\sin(la)\right], \end{cases} \tag{11.16}$$

and those in a give

$$\begin{cases} C\sin(la) + D\cos(la) = Fe^{ika}, \\ l\,[C\cos(la) - D\sin(la)] = ikFe^{ika}. \end{cases}$$

According to the Cramer[2] rule, since the determinant of coefficients is $-l$,

$$\begin{cases} C = Fe^{ika}\left[\sin(la) + i\frac{k}{l}\cos(la)\right], \\ D = Fe^{ika}\left[\cos(la) - i\frac{k}{l}\sin(la)\right]. \end{cases} \tag{11.17}$$

Solving (11.16), one finds

$$\begin{cases} A = \frac{e^{ika}}{2k}\left[-(Ck + iDl)\sin(la) + (Dk - iCl)\cos(la)\right], \\ B = \frac{e^{-ika}}{2k}\left[(-Ck + iDl)\sin(la) + (Dk + iCl)\cos(la)\right]; \end{cases} \tag{11.18}$$

and substituting (11.17),

$$\begin{cases} A = \frac{-ie^{2ika}}{2kl}F\left[(k^2 + l^2)\sin(2la) + 2ikl\cos(2la)\right], \\ B = \frac{-iF}{2kl}(k^2 - l^2)\sin(2la). \end{cases} \tag{11.19}$$

Finally,

$$F = \frac{e^{-2ika}}{\cos(2la) - i\frac{\sin(2la)}{2kl}(k^2 + l^2)}. \tag{11.20}$$

The reflection coefficient is $R = |\frac{B}{A}|^2$ and the transmission one is $T = |\frac{F}{A}|^2$; of course, $R + T = 1$.

11.5 The δ Potential

The limit of a deep narrow potential

$$V(x) = -a\delta(x), \quad a > 0 \tag{11.21}$$

is also interesting and lends itself to a simple calculation of the discrete spectrum. Very short range interactions also arise from the Weak Interaction. The S.E. for $x \neq 0$ reads as

$$\frac{d^2\psi}{dx^2} = k^2\psi(x), \quad k^2 = -\frac{2mE}{\hbar^2} > 0.$$

---

[2]Gabriel Cramer (Geneve 1704–Bagnols sur Céze (France) 1752).

The wave function is $\psi(x) = A[e^{kx}\theta(-x) + e^{-kx}\theta(x)]$. When doing the first derivative, the θ functions originate δ functions, which, however, cancel. When doing the second derivative, we get the δ function which is required by the Schrödinger equation. We find that the δ potential has one bound state, with

$$E = -\frac{ma^2}{\hbar^2}.$$

The normalization constant A is easily obtained and and

$$\psi(x) = \frac{\sqrt{ma}}{\hbar} \exp\left[-\frac{ma|x|}{\hbar^2}\right].$$

11.6 Potential Barrier and Tunneling

Consider a beam of particles fired from a gun, say, on the left, towards the potential bump

$$V(x) = \begin{cases} 0 & x < 0, \\ V > 0 & 0 < x < S, \\ 0 & x > S. \end{cases}$$

with the detector on the right. When the beam energy $E > V$, that is, the kinetic energy is higher than the barrier, the problem is easily worked out, but the results are qualitatively similar to the $S \to \infty$ case that we have already considered above, with some reflected and some transmitted current. The famous tunnel effect arises when the energy is below threshold.

With $E < V$, the general integral is a combination of real exponentials $e^{\pm\lambda x}$ in the barrier, while outside, the exponentials are complex $e^{\pm ikx}$, $k > 0$. The $e^{\pm ikx}$ waves carry current $\pm k$. Since the gun is on the left, no particle comes from $x \to \inf$ and the wave e^{-ikx} is absent for $x > S$. Using the rules of Sect. 11.3, the solution is found to be:

$$\psi(x) = \begin{cases} \alpha e^{ikx} + \beta e^{-ikx} & x < 0, \\ \gamma e^{-\lambda x} + \delta e^{\lambda x} & 0 < x < S, \\ \epsilon e^{ikx} & x > S, \end{cases}$$

with $\hbar^2 k^2 = 2mE$, $\hbar^2\lambda^2 = 2m(V - E)$. We may fix the incoming current by setting $\alpha = 1$. The continuity relations give us (Fig. 11.3):

$$\begin{cases} 1 + \beta = \gamma + \delta, \\ ik(1 - \beta) = -\lambda\gamma + \lambda\delta, \\ \gamma e^{-\lambda S} + \delta e^{\lambda S} = \epsilon e^{ikS}, \\ -\lambda\gamma e^{-\lambda S} + \lambda\delta e^{\lambda S} = ik\epsilon e^{ikS}. \end{cases} \tag{11.22}$$

Fig. 11.3 Potential barrier

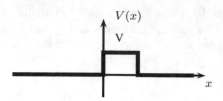

The last two give us γ and δ: letting

$$\eta = \frac{ik}{\lambda},$$

one finds that

$$\begin{cases} \gamma = \frac{\epsilon e^{ikS}}{2}(1 - \eta)e^{\lambda S}, \\ \delta = \frac{\epsilon e^{ikS}}{2}(1 + \eta)e^{-\lambda S}. \end{cases}$$

The first two (11.22) give us

$$2 = \gamma(1 - \frac{1}{\eta}) + \delta(1 + \frac{1}{\eta}).$$

Hence,

$$2 = \frac{\epsilon e^{ikS}}{2}\{(1 - \eta)e^{\lambda S}(1 - \frac{1}{\eta}) + (1 + \eta)e^{-\lambda S}(1 + \frac{1}{\eta})\}$$
$$= \frac{\epsilon e^{ikS}}{2}\{2\cosh(\lambda S) - (\eta + \frac{1}{\eta})\sinh(\lambda S), \}$$

and finally,

$$\epsilon e^{ikS} = \frac{1}{\cosh(\lambda S) - \frac{1}{2}(\eta + \frac{1}{\eta})\sinh(\lambda S)}.$$

The classically impenetrable barrier is overcome with probability $|\epsilon e^{ikS}|^2 \sim e^{-2\lambda S}$. This is the famous **tunnel effect**. In Sect. 11.4.1 I have already discussed the paradox of a particle that is detected in a classically forbidden region; the tunnel effect is an even more striking piece of evidence, since one can measure the energy E of the particle beyond the barrier even at a long distance from it, find that $E < V$ and reach the conclusion that the particle has crossed a forbidden region. This can be understood only if the particle propagates as a wave ψ with some energy E and the corpuscle exists as such only after the detection in some position.

There are countless examples and applications of tunnel phenomena in Science and Technology. Here is a short list:

1. Nuclear decay. A rough idea of α and β decays is provided by a potential well with a transparent wall (Fig. 11.4). One can start by evaluating the bound states ϕ_n that the well would have if the wall were infinitely thick. When the actual

Fig. 11.4 Potential well
with a penetrable wall

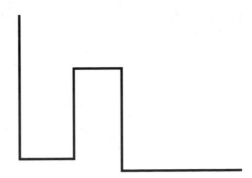

potential is restored, ϕ_n is no longer an eigenstate, but rather a wave packet called
a *resonance*, with a width ΔE in energy. Sooner or later the particle will escape
to infinity. The interested reader can find more about resonances in Chap. 24. The
life time Δt of the resonance can vary from a fraction of a second to several billion
years.

2. Hydrogen bond. In liquid H_2O, a proton can tunnel between two Oxygen atoms
   like a particle in a double well potential. This gives a significant contribution
   to the chemistry of water and other liquids with OH groups. Normally isotope
   effects are negligible, but beware! Heavy water D_2O with Deuterium replacing
   Hydogen is poisonous!

3. Josephson effect. In superconductors, electrons form bound pairs called Cooper
   pairs (see Sect. 25.6.1). This happens at low temperatures when the Coulomb
   repulsion is exceeded by an effective attractive interaction due to vibrations, or
   to special electron correlation effects. If two superconductors separated by a thin
   insulating barrier are placed in a circuit, a d.c. potential difference V produces an
   alternating current

$$I = I_0 \sin(\frac{2eV}{\hbar}t),$$

while under some condition, an a.c. $V(t)$ produces a d.c. current (Shapiro effect).
Such interesting phenomena can be understood in terms of Cooper pairs tunneling the barrier as such. Their effective wave function $\Psi(x, t)$ is called the
superconducting order parameter.

There are also many applications. For example:

1. SQUID (Superconducting Quantum Interference Devices) are extremely sensitive
   magnetometers. They can measure 10^{-14} T, while for instance the magnetic fields
   produced by the heart are of order 10^{-10} T and those produced by the brain can
   also be measured. The SQUID is based on the Josephson effect.

2. In nanocircuits, quantum effects are important. The metal-insulator-metal junc-
   tion devices, and the MOSFETs (metal oxide semiconductor field effect transistor), which are currently widely used in electronics, use an electric field to

Fig. 11.5 SET (Single electron tunneling) transistor. The grain is a metallic island of nanoscopic size (10^{-6} cm or less)

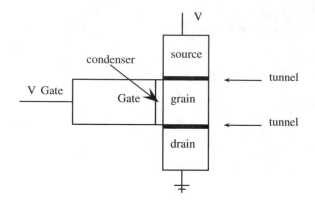

modulate the tunnel effect through a thin insulating barrier to change the electron population and conductivity.

3. The single electron transistor Fig. 11.5 is also based on the tunnel effect. The gate potential modulates the barrier and therefore also the current produced by the potential V.

4. The tunnel effect microscope Fig. 11.6 allows us to scan and image a surface with atomic resolution with a metal tip. It was invented in 1981 by IBM scientists Gerd Binnig e Heinrich Rohrer, who won the Nobel for Physics in 1986. While the tip scans the surface, its distance from it, that is of the order of atomic sizes, is recorded. The precise positioning of the tip is achieved by the piezoelectric effect. This effect arises in some solids like SiO_2 with a low-symmetry ionic crystalline structure that produces an electrical potential if deformed. Very sensitive transducers are built on this principle. A potential difference between the tip and the surface drives a tunnel current: the vacuum acts as barrier. When the tip is positive, the current comes from filled electronic states, when it is negative electrons are injected into empty states of the sample. This current depends exponentially on the distance from the surface. In this way, one gains information about morphology and electronic structure. A gallery of STM images can be found in the site

http://www.almaden.ibm.com/vis/stm/gallery.html.

Fig. 11.6 Block diagram of the tunnel microscope. In the enlarged detail, it is seen that the tip ends in a single atom from which the current passes to the surface

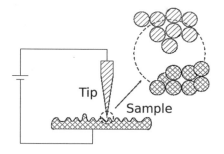

Chapter 12
The Postulates of Quantum Mechanics: Postulate 1

The axiomatic formulation of Quantum Mechanics is generally presented as a set of four postulates, introduced by John von Neumann. The physical meaning of each of them requires a nontrivial, careful enquiry.

Here is first postulate.

The state of any system is represented by a complex wave function $\Psi_a(x, t)$, where: x stands for the set of all the coordinates, t denotes time, and a is a (possibly empty) set of constants of the motion (the so called quantum numbers). If a is the set of all the observables that are compatible, the quantum state is uniquely determined.

12.1 The Wave Function

The system referred to above could be a particle, an atom, or even a macroscopic superconductor (then, x stands for a very large set of coordinates), so the statement is quite strong and general. In any case, all the information that is available from all possible experiments is in $\Psi_a(x, t)$. The wave function must be taken to be *normalized*. In the case of a single degree of freedom $\Psi_a(x, t)$, the normalization condition reads as $\int |\Psi(x, t)|^2 dx = 1$, while in general, one must integrate the square modulus over all the variables. The function is complex, therefore $\Psi(x, t) = |\Psi(x, t)|e^{i\phi(x, t)}$, where $\phi(x, t)$ is the phase. One can change $\phi(x, t)$ by a constant phase factor (for instance, multiplying $\Psi(x, t)$ by i) but the physical state remains the same; nevertheless, the phase difference between two wave functions does matter a lot, since the

© Springer International Publishing AG, part of Springer Nature 2018
M. Cini, *Elements of Classical and Quantum Physics*,
UNITEXT for Physics, https://doi.org/10.1007/978-3-319-71330-4_12

wave functions do interfere. We shall see that the phase can be changed in several ways (e.g. rotations, Galileo transformations, gauge changes).

The set of possible wave functions of a given system is a vector space; all linear combinations of wave functions representing alternative states of the same system $\Phi = \alpha \Psi + \beta \chi$ with complex coefficients α, β, once normalized, are other possible wave functions. The scalar product

$$(\Psi|\Phi) \equiv \langle \Psi|\Phi \rangle = \int \Psi(x,t)^* \Phi(x,t) dx \qquad (12.1)$$

is a complex number. In this equation I present two alternative notations in use. The above notation $\langle \Psi|\Phi \rangle$ is the bra-ket notation introduced by Dirac, with $\Psi(x)$ as a bra and $\Phi(x)$ as a ket. When $\langle \Psi|\Phi \rangle = 0$, the functions are said orthogonal. The maximum number d of orthogonal wave functions is the dimension of the vector space; for many systems $d = \infty$. The normalization condition is $\langle \Psi|\Psi \rangle = 1$. Most often, it is preferable to work with an orthonormal basis set of functions. Let us see how to achieve such a set.

12.2   Gram–Schmidt Orthogonalization

The Gram–Schmidt Orthogonalization is the most obvious. Let $(v_1, \cdots, v_n)$ be a set of normalized vectors, taken in any order. Using any orthonormal basis, $(v_i|v_j) = \sum_\alpha v_{i,\alpha}^* v_{j,\alpha}$ denotes the scalar product. We choose $w_1 = v_1$ as the first vector of the orthogonalized set and remove the component along v_1 from the remaining vectors by $v_i \to v_i - (v_i|v_1)v_1$. We normalize all the new vectors, then set $w_2 = v_2$ as the second vector of the orthogonalized set; next we proceed to orthogonalize the vectors $(v_3 \cdots v_n)$ to w_2, and so on. The procedure fails to give the basis if the initial set does not span a n-dimensional space. The result of this procedure is a basis of orthogonal vectors, but if we change the order if the v_i we get a different basis.

12.3   Löwdin Symmetric Orthogonalization

A smart orthogonalization method, proposed by Löwdin in 1950, is often preferred for its aesthetic and also practical advantages. It is called symmetric orthogonalization, since it starts from a set of normalized vectors and treats all of them on equal footing. In the case of just two vectors v_1 and v_2, it consists in setting $w_1 = \alpha v_1 + \beta v_2$, $w_2 = \beta v_1 + \alpha v_2$ and seeking complex numbers α and β such that w_1 and w_2 are orthonormal. Löwdin proposed an elegant, systematic method to do that for n vectors. To start with, one computes the overlap matrix S with entries of the scalar products

$$S_{km} = (v_k|v_m), \qquad (12.2)$$

with the normalized non-orthogonal vectors $(v_1, \cdots, v_n)$. Then, one looks for a set of coefficients L_{kj} of a linear transformation

$$w_i = \sum_{k=1}^{n} L_{kj} v_j \qquad (12.3)$$

such that

$$(w_i|w_j) = \sum_{k,m} L_{ik}^* L_{jp} S_{pk} = \delta_{ij}. \qquad (12.4)$$

There is an infinity of sets of orthogonal vectors w_i. We can exploit this arbitrariness to put conditions on L. Now we note that if we assume that L is a Hermitian matrix, $L_{ki}^* = L_{ik}$, then we can rewrite this simply as $LSL = 1$, where 1 stands for the unit matrix. So, let us look for a Hermitean L. Next, we note that if L and S were numbers, the result would be $L = S^{-\frac{1}{2}}$. We must find a meaning of $L = S^{-\frac{1}{2}}$ in the case of Hermitean matrices. This is not hard to do. Since S is Hermitean, it can be diagonalized: $S_d = U^\dagger S U = \mathrm{diag}(\lambda_1, \lambda_2, \cdots, \lambda_n)$, where the columns of U are the eigenvectors of S and λ_i the ith eigenvalue. Then, we can denote by the symbol $S_d^{-\frac{1}{2}}$ the diagonal matrix $S_d^{-\frac{1}{2}} = \mathrm{diag}(\lambda_1^{-\frac{1}{2}}, \lambda_2^{-\frac{1}{2}}, \cdots, \lambda_n^{-\frac{1}{2}})$. Hence, $S^{-\frac{1}{2}} = U S^{-\frac{1}{2}} U^\dagger$ does the job. Here is the desired transformation:

$$w_j = \sum_{k=1}^{n} \left(S^{-\frac{1}{2}} \right)_{kj} v_k. \qquad (12.5)$$

Besides being convenient for numerical work, this method has been shown to minimize $\sum_i^n |w(i) - v(i)|^2$; in this way, the orthogonal basis is as close as possible to the original one.

12.4 Schwarz Inequality

Let $\vec{a}, \vec{b}$ denote vectors in real vector space. The norms are denoted by $a = \sqrt{\vec{a} \cdot \vec{a}}$ and $b = \sqrt{\vec{b} \cdot \vec{b}}$. Then, $\vec{a} \cdot \vec{b} = ab\cos(\theta)$; since the square of the cosine of any real angle is ≤ 1, one finds the *Schwarz inequality*

$$|\vec{a} \cdot \vec{b}|^2 \leq (\vec{a} \cdot \vec{a})(\vec{b} \cdot \vec{b}).$$

This is elementary, but the interesting thing is that it holds for any real or complex vector space independently of the dimension d. Even for state vectors (that is, wave functions),

$$|\langle \alpha|\beta \rangle|^2 \leq \langle \alpha|\alpha \rangle \, \langle \beta|\beta \rangle = 1.$$

12.5   Hilbert Spaces

Every state of the system is represented by a normalized function $\Psi_a(x, t)$; the function is a normalized vector $|a\rangle$ in a possibly infinite space of functions. The existence of a norm requires the notion of the length of a vector, or equivalently, the notion of the distance between two points. So, the space is a normed space. The expansion of a vector over an infinite basis is a series expansion and requires a limit to exist; physically, we must require that each converging series represents a possible state. Therefore, we must ask that the limit of every convergent series belongs to the space. However, this is not granted in an infinite-dimensional normed vector space. For instance, the Taylor series for $\sin(x)$ is a polynomial series, but $\sin(x)$ is not a polynomial, so the space of polynomials fails to satisfy this requirement. When every convergent series belongs to the space, we speak of a complete space. A complete normed space is called a Banach space. However there are many ways to assign a norm in a Banach space. For example, the space of continuous functions $f(x)$ defined in a closed interval (for instance, $x \in [0, 1]$) is a Banach space with the norm $||f|| = sup|f(x)|$. The space of square integrable functions where the wave functions belong is called L^2. The completeness of L^2 is shown by the Fourier theorem. It is a special Banach space in which the norm is defined in terms of the scalar product; in one dimension, it is $||f(x)|| = \langle f|f \rangle = \int_{-\infty}^{\infty} |f(x)|^2$; this definition extends to many dimensions in the obvious way. Such a Banach space is called a Hilbert space.

Suppose a system S_1 has a Hilbert space H_1 of dimension d_1 with basis $a_1, a_2, \cdots$ a_{d1}, and a second system S_2 has a Hilbert space H_2 of dimension d_2 with basis $b_1, b_2, \cdots b_{d2}$. Now, the two systems might interact. To handle the compound system $S_1 \bigcup S_2$ we need the tensor product $H_1 \otimes H_2$ with basis $a_1 \otimes b_1, a_1 \otimes b_2, \cdots$ $a_{d1} \otimes b_{d2}$.

12.6   Quantum Numbers

Depending on the way a system is prepared, the wave function Ψ may be an eigenfunction of certain operators. Then, it is very helpful to label Ψ as $\Psi_a(x)$, where a stands for the set of eigenvalues of those operators. This might be an empty set, otherwise $a = \{A, B, C, \ldots\}$ and its entries are called quantum numbers. We shall see that the labels must correspond to compatible observables, and their operators must commute. In the case in which Ψ is an eigenfunction of a time-independent Hamiltonian $\hat{H}$, the set $a = \{A, B, C, \ldots\}$ comprises the energy and constants of the motion that must coexist; they are referred to as compatible observables; they must commute both with $\hat{H}$ and with each other. For a given $\hat{H}$, there are, at times, alternative choices of the set of operators, and the label a is a subset of one of those. For the simple case of a free Schrödinger particles, as we shall see, the state can be labelled by energy and 3 momentum components, or by energy, the square of angular momentum and one of its components, but the square of angular momentum is not compatible with the components of the momentum, and the 3 components of the angular momentum are not compatible.

If a contains all the operators of a set of compatible operators, the quantum state is uniquely identified; this implies that two different quantum states must differ by at least one quantum number.

Chapter 13
Postulate 2

Any observable Q corresponds to an operator $\hat{Q}$ acting on the wave functions. These operators are linear, hermitean (that is, $\hat{Q}^\dagger = \hat{Q}$), and must possess a complete set of eigenvectors. The expectation value of Q in the state $|\Psi\rangle$ is $\langle\Psi|\hat{Q}|\Psi\rangle \equiv \langle\Psi|\hat{Q}\Psi\rangle$.

13.1 Mathematical Context: Operators, Matrices, and Function Spaces

$D \equiv \frac{d}{dx}$, and also $\hat{x}$ (which multiplies by x), are examples of linear operators $\hat{O}$: $\hat{O}(\Phi + \Psi) = \hat{O}\Phi + \hat{O}\Psi$. Coordinates and momenta of Classical Mechanics wear a hat and become quantum operators, and we shall meet more. The scalar product of $\hat{Q}|\Psi\rangle$ with $|\Phi\rangle$, which is $\langle\Phi|\hat{Q}|\Psi\rangle$, may be regarded as the element M_{ij} of some matrix M with i and j replaced by the indices Φ and Ψ, which are functions $\in L^2$; it is called a matrix element of $\hat{Q}$; the only real difference with a conventional matrix M_{ij} is that the indices are often continuous and the matrix most often has an infinity of rows and columns. Werner Heisenberg initially formulated the theory in terms of matrices.

13.2 Why Hermitean Operators?

Let α, β denote vectors of the Hilbert space. The Hermitean adjoint $\hat{A}^\dagger$ of $\hat{A}$ is defined by

$$\langle\hat{A}^\dagger\alpha|\beta\rangle = \langle\alpha|\hat{A}\beta\rangle \quad \forall\alpha, \beta. \tag{13.1}$$

© Springer International Publishing AG, part of Springer Nature 2018
M. Cini, *Elements of Classical and Quantum Physics*,
UNITEXT for Physics, https://doi.org/10.1007/978-3-319-71330-4_13

In the 1d case, the notation is simplest: then, we write

$$\int dx (\hat{A}^\dagger \alpha)^* \beta = \int dx \alpha^* \hat{A}\beta,$$

and taking the complex conjugate of both sides,

$$\left(\int dx \alpha^* \hat{A}\beta \right)^* = \int dx \beta^* \hat{A}^\dagger \alpha;$$

this suggests the general rule for taking the complex conjugate of any matrix element:

$$\langle \alpha | \hat{A} | \beta \rangle^* = \langle \beta | \hat{A}^\dagger | \alpha \rangle \quad \forall \alpha, \beta. \tag{13.2}$$

Self-adjoint or Hermitean operators $\hat{Q}$ have a special importance. A Hermitean operator is such that $\hat{Q}^\dagger = \hat{Q}$; hence all Hermitean operators have the property that $\langle \Phi | \hat{Q}\Psi \rangle = \langle \hat{Q}\Phi | \Psi \rangle$. If we set in Eq. (13.2) $\alpha = \beta$, we conclude that all the expectation values of Hermitean operators must be real. Since measures quantities must be real, the reader can have the impression that observables correspond to Hermitean operators. We shall see that this is true, and that the eigenvalues are also real. This is not the case for $D = \frac{d}{dx}$; in fact,

$$\langle f | Dg \rangle = -\langle Df | g \rangle.$$

and D is *anti-Hermitean*, while the momentum $\hat{p} = -iD$ is Hermitean[1] over L^2.

13.3   Orthogonal Spaces

Different eigenvalues of any Hermitean operator belong to orthogonal eigenvectors. Indeed, starting with

$$\begin{cases} \hat{A}|m\rangle = a_m |m\rangle, \\ \hat{A}|n\rangle = a_n |n\rangle; \end{cases}$$

and taking scalar products, we get

$$\begin{cases} \langle n|\hat{A}|m\rangle = a_m \langle n|m\rangle, \\ \langle m|\hat{A}|n\rangle = a_n \langle m|n\rangle. \end{cases}$$

---

[1] $\hat{p}$ fails to be Hermitean on the space $P(N)$ of polynomials of degree N, but this does not matter because all polynomials are outside L^2.

The complex conjugate of the second is $\langle n|\hat{A}^\dagger|m\rangle = a_n^*\langle n|m\rangle$, and since $A = A^\dagger$, it follows that $\langle n|\hat{A}|m\rangle = a_n\langle n|m\rangle$. Subtracting from the first equation, we find that $0 = (a_n - a_m)\langle n|m\rangle$. So, if $(t_n - t_m) \neq 0$, it follows that $\langle n|m\rangle = 0$.

13.4 Completeness

The eigenvectors of an observable operator $\hat{T}$ represent states such that $\hat{T}$ has a sharp value. The set of eigenvectors must be complete; this point will be expanded when dealing with the third postulate. The completeness grants that every $|\Psi\rangle$ can be expanded:

$$\Psi\rangle = \sum_m |m\rangle\langle m|\Psi\rangle. \tag{13.3}$$

Here, $\sum_m$ is a sum where the spectrum of $\hat{T}$ is discrete and an integral where it is continuous. In general, $|\Psi\rangle$ and $|m\rangle$ depend on many variables, but in one-dimensional problems, this expansion is just the familiar Fourier theory:

$$f(x) = \int_{-\infty}^{\infty} dq \frac{e^{-iqx}}{2\pi} \tilde{f}(q).$$

This is the continuous expansion in the complete set of plane waves over the real axis, which would be replaced by a Fourier series in the case of a periodic system. The plane waves have sharp energy and momentum. Now we can see that if $\hat{T}$ is an observable, it is *physically necessary* that the set $\{m\}$ of eigenvectors be complete. The need for completeness arises because the knowledge of $f(q)$ must be equivalent to that of $f(x)$, since this is a canonical transformation already in Classical Mechanics.

Formally, we may rewrite (13.3) in the form

$$\mathbf{1} = \sum_m |m\rangle\langle m|, \tag{13.4}$$

where $\mathbf{1}$ is the identity operator. Therefore, if $A_{ij} = \langle \Phi_i|\hat{A}\Phi_j\rangle \equiv \langle \Phi_i|\hat{A}|\Phi_j\rangle$ is the matrix element of $\hat{A}$ between two elements of a basis set, and B_{ij} is the matrix element of $\hat{B}$, the product $\hat{A}\hat{B}$ turns out to be $\langle \Phi_i|\hat{A}\hat{B}|\Phi_j\rangle = \langle \Phi_i|\hat{A} \sum_k |k\rangle\langle k| \hat{B}|\Phi_j\rangle = \sum_k A_{ik}B_{kj}$ and *the matrices of the operators must by multiplied row by column.* Mathematically, the matrices are a *representation* of the operators. The operator is known if the matrix and the basis set are specified.

13.5 Commutators: Angle and Angular Momentum

Recall that, classically, the Poisson bracket of canonically conjugated variables is 1. In Quantum theory, couples of canonically conjugated operators have commutators $-i\hbar$. For instance, $[\hat{p}, \hat{x}] = \hat{p}\hat{x} - \hat{x}, \hat{p} = -i\hbar$, where $\hbar = \frac{h}{2\pi}$. Now we meet another remarkable pair: angular momentum and angle.

We can start from the example of the *plane rigid rotator*, which can rotate around the z axis, with moment of inertia[2] I. In terms of the angle ϕ, the Lagrangian is

$$L(\phi, \dot{\phi}) = \frac{1}{2}I\dot{\phi}^2.$$

The momentum $\hat{L}_z = I\dot{\phi}$ is canonically conjugated to the angle ϕ. The Hamiltonian operator is like the classical function

$$\hat{H} = \frac{L_z^2}{2I}, \tag{13.5}$$

where $\hat{L}_z$ is the z component, $\hat{L}_z = xp_y - yp_x$.

In plane polar coordinates,

$$x = \rho \cos\phi, \ y = \rho \sin\phi.$$

The components of the momentum $-i\hbar\nabla$ can be obtained from the *chain rule*

$$\frac{\partial}{\partial x} = \frac{\partial\rho}{\partial x}\frac{\partial}{\partial\rho} + \frac{\partial\phi}{\partial x}\frac{\partial}{\partial\phi}, \quad \frac{\partial}{\partial y} = \frac{\partial\rho}{\partial y}\frac{\partial}{\partial\rho} + \frac{\partial\phi}{\partial y}\frac{\partial}{\partial\phi}.$$

So,

$$\rho = \sqrt{x^2 + y^2}, \ \phi = \arctan\frac{y}{x}.$$

Differentiating the radius,

$$\frac{\partial\rho}{\partial x} = \frac{x}{\rho} = \cos\phi, \ \frac{\partial\rho}{\partial y} = \frac{y}{\rho} = \sin\phi,$$

and since

$$\frac{d\arctan u}{du} = \frac{1}{1 + u^2},$$

---

[2]For a body with density $\rho(\vec{x})$, the moment of inertia relative to the z axis is $I = \int d^3x\rho(\vec{x})(x^2 + y^2)$.

we obtain $\frac{\partial\phi}{\partial x} = \frac{1}{1+(\frac{y}{x})^2}\frac{d}{dx}\left(\frac{y}{x}\right) = -\frac{y}{x^2+y^2}$, $\frac{\partial\phi}{\partial y} = \frac{1}{1+(\frac{y}{x})^2}\frac{d}{dy}\left(\frac{y}{x}\right) = \frac{x}{x^2+y^2}$; in polar coordinates,

$$\frac{\partial\phi}{\partial x} = -\frac{sin\phi}{\rho}, \quad \frac{\partial\phi}{\partial y} = \frac{cos\phi}{\rho};$$

so,

$$\frac{\partial}{\partial x} = cos\phi\frac{\partial}{\partial\rho} - \frac{sin\phi}{\rho}\frac{\partial}{\partial\phi}, \quad \frac{\partial}{\partial y} = sin\phi\frac{\partial}{\partial\rho} + \frac{cos\phi}{\rho}\frac{\partial}{\partial\phi}.$$

Now, $\hat{L}_z = xp_y - yp_x$ gives us

$$\hat{L}_z = -i\hbar\left\{\rho\cos(\phi)\left[sin\phi\frac{\partial}{\partial\rho} + \frac{\cos\phi}{\rho}\frac{\partial}{\partial\phi}\right]\right.$$

$$\left. -\rho\sin(\phi)\left[\cos\phi\frac{\partial}{\partial\rho} - \frac{\sin\phi}{\rho}\frac{\partial}{\partial\phi}\right]\right\}.$$

A spectacular simplification yields the final result:

$$\hat{L}_z = -i\hbar\frac{\partial}{\partial\phi}.$$

So, $-i\hbar\frac{\partial}{\partial\phi}$ has the fundamental commutation rule with ϕ. The eigenvalue equation

$$\hat{L}_z\psi_m(\phi) = m\hbar\psi_m(\phi)$$

yields the eigenfunctions

$$\psi_m(\phi) = \frac{e^{im\phi}}{\sqrt{2\pi}}.$$

$\hbar$ is the basic angular momentum, and the condition $\psi_{2\pi}(\phi) = \psi_m(0)$ requires that the *azimuthal quantum number* m be the integer, $m = 0, \pm1, \pm2, \cdots$ The result says that the angular momentum of any system must be a multiple.[3] Classically, it is obvious that by tilting the rotation axis a little bit, the component of L must change with continuity. Instead the measured values are discrete, while the probabilities of measuring those values vary with continuity, as we shall see. The energy eigenvalues of the rotor are $E_m = \frac{1}{2I}m^2\hbar^2$.

---

[3]This is the orbital angular momentum; the spin has another nature and will be discussed later.

13.6 Commutators: Properties and Tricks

From the definition, it is obvious that

$$[\hat{A}, \hat{B}]_- = -[\hat{B}, \hat{A}]_-;$$

$$[\hat{A}, \hat{B} + \hat{C}] = [\hat{A}, \hat{B}] + [\hat{A}, \hat{C}];$$

$$[\hat{A}\hat{B}, \hat{C}]_- = \hat{A}[\hat{B}, \hat{C}]_- + [\hat{A}, \hat{C}]_-\hat{B}; \qquad (13.6)$$

$$[\hat{A}, \hat{B}\hat{C}]_- = \hat{B}[\hat{A}, \hat{C}]_- + [\hat{A}, \hat{B}]_-\hat{C}.$$

The analogy with the Poisson brackets is also evident. Simple commutators are readily obtained. The result $[x^n, \hat{p}] = i\hbar n x^{n-1}$ can be obtained by acting over a $f(x)$ or else by starting with $[x^2, p]_- = 2i\hbar x$; otherwise, one can prove the result by induction, starting with $A = x$, $B = x^2$, $C = p$, $[x^3, p]_- = x[x^2, p] + [x, p]x^2 = 3i\hbar x^2$, and so on. From the definition of the angular momentum $\vec{L} = \vec{r} \wedge \vec{p}$

$$L_x = y p_z - z p_y, \quad L_y = z p_x - x p_z, \quad L_z = x p_y - y p_x,$$

and from the fundamental commutator $[p_z, z] = -i\hbar$, one finds

$$[L_x, L_y] = y p_x [p_z, z] + [z, p_z] p_y x = i\hbar L_z,$$

ad cyclic results, that can be summarized as follows:

$$\vec{L} \wedge \vec{L} = i\hbar \vec{L}. \qquad (13.7)$$

Similarly, one finds:

$$[\vec{L}^2, L_x] = [L_y^2, L_x] + [L_z^2, L_x],$$

$$[L_y^2, L_x] = L_y[L_y, L_x] + [L_y, L_x]L_y = -i\hbar(L_y L_z + L_z L_y) = -i\hbar\left[L_y, L_z\right]_+,$$

$$[L_z^2, L_x] = L_z[L_z, L_x] + [L_z, L_x]L_z = i\hbar\left[L_y, L_z\right]_+.$$

Thus, it turns out that

$$[\vec{L}^2, L_x] = 0. \qquad (13.8)$$

In other words, the square $\vec{L}^2$ of the angular momentum and any one of the components exist simultaneously.

Problem 25 For given operators $\hat{A}$, $\hat{B}$, $\hat{C}$, $\hat{D}$, $\hat{E}$, $\hat{F}$, compute

$$[\hat{A}\hat{B}\hat{C}, \hat{D}\hat{E}\hat{F}]_-.$$

Solution 25 One finds

$$\begin{aligned}
[\hat{A}\hat{B}\hat{C}, \hat{D}\hat{E}\hat{F}]_- &= \hat{A}\hat{B}\{[\hat{C}, \hat{D}]_-\hat{E}\hat{F} + \hat{D}[\hat{C}, \hat{E}]_-\hat{F} + \hat{D}\hat{E}[\hat{C}, \hat{F}]_-\} \\
&+\hat{A}\{[\hat{B}, \hat{D}]_-\hat{E}\hat{F} + \hat{D}[\hat{B}, \hat{E}]_-\hat{F} + \hat{D}\hat{E}[\hat{B}, \hat{F}]_-\}\hat{C} \\
&+\{[\hat{A}, \hat{D}]_-\hat{E}\hat{F} + \hat{D}[\hat{A}, \hat{E}]_-\hat{F} + \hat{D}\hat{E}[\hat{A}, \hat{F}]_-\}\hat{B}\hat{C}.
\end{aligned}$$

13.7 Angular Momentum in 3 Dimensions

13.7.1 The Algebra

From Eq. (13.7), one can find everything else about angular momentum. We can have one diagonal component and choose L_z; so there is a subspace of functions $|\lambda, m\rangle$, where λ is the eigenvalue of L^2 and $L_z|\lambda, m\rangle = m|\lambda, m\rangle$. It is obvious that the subspace must be finite and there is some positive integer l such that $m \leq l$.

Introduce the *shift operators* $L_\pm$, such that

$$L_+ = L_x + iL_y, \quad L_- = L_x - iL_y.$$

Since

$$[L_z, L_\pm] = \pm\hbar L_\pm,$$

one finds,

$$L_z L_\pm|\lambda, m\rangle = (m \pm 1)\hbar L_\pm|\lambda, m\rangle,$$

therefore

$$L_\pm|\lambda, m\rangle = C_\pm|\lambda, m \pm 1\rangle. \tag{13.9}$$

Since λ is the maximum eigenvalue, we run into a contradiction, unless

$$L_+|\lambda, l\rangle = 0.$$

The identity

$$L_- L_+ + L_z^2 + \hbar L_z = L^2$$

applied to $|\lambda, l\rangle$ yields $\lambda = l(l + 1)$. However, the usage is to denote the eigenstates by $|l, m\rangle$ instead of writing $|l(l + 1), m\rangle$.

Scalar multiplication of (13.9) by itself gives us

$$C_\pm^2 = [l(l+1) - m(m \pm 1)]\hbar^2,$$

and we obtain the useful result

$$L_\pm |l, m\rangle = \hbar\sqrt{l(l+1) - m(m \pm 1)}|l, m \pm 1\rangle. \tag{13.10}$$

13.7.2 Angular Momentum Matrices

We can now find the matrices of $\vec{L}$ on the $|l, m\rangle$ basis. They are:

$$\langle l_1 m_1 | L^2 | l_2 m_2 \rangle = \hbar^2 l_1(l_1+1)\delta_{l_1,l_2}\delta_{m_1,m_2} \tag{13.11}$$

$$\langle l_1 m_1 | L_z | l_2 m_2 \rangle = \hbar m_1 \delta_{l_1,l_2}\delta_{m_1,m_2} \tag{13.12}$$

$$\langle l_1 m_1 | L_\pm | l_2 m_2 \rangle = \hbar\sqrt{l_2(l_2+1) - m_2(m_2 \pm 1)}\delta_{l_1,l_2}\delta_{m_1,m_2\pm1}. \tag{13.13}$$

Example 6 Matrices of di $l = 1$ in $\hbar$ units.

$L_z = \text{diag}(1, 0, -1);\ L_+|1, -1\rangle = \sqrt{2}|1, 0\rangle;\ L_+|1, 0\rangle = \sqrt{2}|1, 1\rangle;\ L_+|1, 1\rangle = 0.$
We assign the basis vectors with:

$$|1, 1\rangle \rightarrow \begin{pmatrix} 1 \\ 0 \\ 0 \end{pmatrix}, |1, 0\rangle \rightarrow \begin{pmatrix} 0 \\ 1 \\ 0 \end{pmatrix}, |1, -1\rangle \rightarrow \begin{pmatrix} 0 \\ 0 \\ 1 \end{pmatrix},$$

and find (denoting the matrices with the names of the operators):

$$L_+ = \begin{pmatrix} 0 & \sqrt{2} & 0 \\ 0 & 0 & \sqrt{2} \\ 0 & 0 & 0 \end{pmatrix}; \tag{13.14}$$

it follows that

$$L_- = \begin{pmatrix} 0 & 0 & 0 \\ \sqrt{2} & 0 & 0 \\ 0 & \sqrt{2} & 0 \end{pmatrix}, \tag{13.15}$$

and so

$$L_x = \frac{1}{\sqrt{2}} \begin{pmatrix} 0 & 1 & 0 \\ 1 & 0 & 1 \\ 0 & 1 & 0 \end{pmatrix}, \tag{13.16}$$

$$L_y = \frac{1}{\sqrt{2}} \begin{pmatrix} 0 & -i & 0 \\ i & 0 & -i \\ 0 & i & 0 \end{pmatrix}. \qquad (13.17)$$

In general, in terms of matrices of size $2l + 1$, one can represent the angular momentum over the $|l, m\rangle$ basis. The matrices of $\vec{L}$ obey the same commutation rules and the same eigenvalues as the operators; they form a *representation* of angular momentum.

13.8   Traslations, Rotations, Boosts and Unitary Operators

Let us see how to change the reference frame in Quantum Mechanics. In one dimension, a shift by a is done by an operator $\hat{T}_a$ such that for any analytic function f,

$$T_a f(x) = f(x + a).$$

The function is translated to the left if $a > 0$.

The matrix which rotates a point around the z axis by $\Delta\phi$ is

$$M_{\Delta\phi} = \begin{pmatrix} \cos(\Delta\phi) & \sin(\Delta\phi) & 0 \\ -\sin(\Delta\phi) & \cos(\Delta\phi) & 0 \\ 0 & 0 & 1 \end{pmatrix}.$$

Note that $\vec{x}' = M_{\Delta\phi}\vec{x}$ is rotated clockwise if $\Delta\phi > 0$; the rotation of a function by $\Delta\phi$ is defined as $f(x, y, z) \rightarrow R_{\Delta\phi}f(x, y, z)$, where

$$R_{\Delta\phi}f(x, y, z) = f(x\cos(\Delta\phi) + y\sin(\Delta\phi), y\cos(\Delta\phi) - x\sin(\Delta\phi), z). \quad (13.18)$$

Note that $R_{\Delta\phi}$ is rotated counterclockwise. Finally, the transformation $f(x) \rightarrow f(x + vt)$ takes a scalar function to a new reference system moving with speed $-v$ with respect to the original one.

Are we done? No! We have written these changes of the reference system as operations on x. We must be able to represent them as quantum operators $\hat{A}$ acting on f itself. These are unitary that is,

$$\hat{A}^\dagger = \hat{A}^{-1}. \qquad (13.19)$$

This property ensures that a normalized wave function in sent to a normalized wave function, since $\langle A\psi | A\psi \rangle = \langle \psi | A^\dagger A\psi \rangle = \langle \psi | \psi \rangle$.

Translations

Since

$$f(x + a) = \sum_{n=0}^{\infty} \frac{a^n}{n!} \left(\frac{d}{dx} \right)^n f(x),$$

we may write

$$\hat{T}_a = \sum_{n=0}^{\infty} \frac{a^n}{n!} \left(\frac{d}{dx} \right)^n = e^{a \frac{d}{dx}} = e^{ia \frac{p}{\hbar}}. \tag{13.20}$$

This is a *hyperdifferential operator* (it does derivatives of any order up to ∞).

The momentum is the *generator* of the infinitesimal translations. It commutes with H and is conserved if the system is invariant under translation. Besides, since p is Hermitean, $\hat{T}_a^{\dagger} = \hat{T}_{-a}$; this verifies the unitarity.

Rotations

$\hat{R}$ can be written in terms of $\vec{L}$. To see how, we consider $\Delta \phi$ as the effect of a succession of $n \to \infty$ infinitesimal rotations around z, $\delta \phi = \frac{\Delta \phi}{n}$. An infinitesimal rotation with $\cos(\Delta \phi) \sim 1$, $\sin(\Delta \phi) \sim \Delta \phi$, produces a change that is related to $\vec{L}_z$, since

$$f(x, y, z) \underset{R_{\delta \phi}}{\overrightarrow{}} f(x', y', z) \sim f(x + y\delta\phi, y - x\delta\phi, z)$$

$$\sim f(x, y, z) + \delta\phi \left\{ y \frac{\partial f}{\partial x} - x \frac{\partial f}{\partial y} \right\} = f(x, y, z) + \delta\phi(-i)L_z f(x, y, z)$$

and

$$f(x, y, z) \underset{R_{\delta \phi}}{\overrightarrow{}} \left[1 - i\delta\phi L_z \right] f(x, y, z).$$

To sum up, the operator of the infinitesimal rotation is

$$R_{\delta \phi} = \left[1 - i\delta\phi L_z \right]$$

and L is the generator. For a finite rotation, we need $R_{\Delta \phi} = \overset{n}{\underset{\delta \phi}{\Pi}} R_{\delta \phi}$. By the known identity

$$e^{-a} = \underset{n \to \infty}{Lim} \left(1 - \frac{a}{n} \right)^n,$$

one obtains the operator that does a finite rotation $\Delta \phi$ around z in the form

$$R_{\phi} = e^{-i \frac{\Delta \phi L_z}{\hbar}}.$$

More generally,

$$R_{\vec{\phi}} = e^{-i\vec{\Delta\phi}\cdot\vec{L}}. \tag{13.21}$$

Besides, since L is Hermitean, $R_{\vec{\phi}}^{\dagger} = R_{-\vec{\phi}}$; this verifies the unitarity.

The set of all the operators (13.21) constitutes a mathematical structure called the O(3) Group. This means that, multiplying the operators we get the operator that corresponds to doing the rotations one after another, with each operator having an inverse. The symbol O(3) means that the rotations are in 3 dimensions and can be represented by orthogonal matrices. Any rotationally invariant system has the property that $[H, \vec{L}]_{-} = 0$. The angular momentum is conserved, as in Classical Mechanics.

Non-relativistic Boosts

The unitary operator $\hat{U}(v) = \exp(imv.x)$ adds the momentum mv to each momentum component of a wave function $\psi(mx)$; this is equivalent to a boost to a new reference system moving with speed $-v$ with respect to the original one.

13.9 General Uncertainty Principle and Compatible Observables

If we wish to measure an observable $\hat{A}$ in a system described by the wave function ψ, the outcome of the experiment can be summarized by a mean value $\langle\hat{A}\rangle$ and by a standard deviation $\sigma_A = \sqrt{\langle[A - \langle A\rangle]^2\rangle}$. This is also familiar in Classical Physics. The novelty of Quantum Mechanics is that the existence of a wave function creates an interdependence among the measurements of different observables. The precision with which two observables $\hat{A}$ and $\hat{B}$ can be measured simultaneously is limited by the following inequality[4]:

$$\sigma_A^2 \sigma_B^2 \geq \left(\frac{\langle\psi|[\hat{A}, \hat{B}]_{-}|\psi\rangle}{2i}\right)^2 \quad \forall\psi, \tag{13.22}$$

where $\sigma_A^2 = \langle[A - \langle A\rangle]^2\rangle$. This implies that $\hat{A}$ and $\hat{B}$ are compatible only if they commute.[5] Heisenberg proposed this as a fundamental principle of Physics, while Einstein disagreed strongly, and thought that a complete theory should disclose the missing information. The result shows that in Quantum Mechanics, two observables are compatible (= measurable simultaneously and exactly) only if their operators commute.

---

[4]If one picks $\psi =$ eigenstate of $\hat{A}$, say, then $\sigma_A = 0$; in this limiting case σ_B blows up.
[5]If they fail to commute, it can still happen that the r.h.s. of (13.22) vanishes for some ψ.

To prove this, one can assume[6] that $\langle \hat{A} \rangle = 0$, $\langle \hat{B} \rangle = 0$; Now, we may write $\sigma_A^2 = \langle \psi | \hat{A}^2 | \psi \rangle = \langle \hat{A}^\dagger \psi | \hat{A} \psi \rangle = |\langle f | f \rangle$, where $|f\rangle = \hat{A}\psi\rangle$, and similarly, $\sigma_B^2 = |\langle g | g \rangle$, where $|g\rangle = \hat{B}\psi\rangle$.

According to the Schwarz inequality,

$$\sigma_A^2 \sigma_B^2 = \langle \psi | \hat{A}^2 | \psi \rangle \langle \psi | \hat{B}^2 | \psi \rangle = \langle f | f \rangle \langle g | g \rangle \geq |\langle f | g \rangle|^2 = |\langle \psi | \hat{A}\hat{B} | \psi \rangle|^2,$$

since $\langle f | g \rangle = \langle AB \rangle$. To reach the final form, we need another trivial inequality. For every complex number z,

$$|z|^2 = (Re(z))^2 + (Im(z))^2 \geq (Im(z))^2 = \left(\frac{z - z^*}{2i} \right)^2.$$

Letting $z = \langle f | g \rangle$, one obtains

$$\sigma_A^2 \sigma_B^2 \geq \left(\frac{\langle f | g \rangle - \langle g | f \rangle}{2i} \right)^2.$$

Now, since $\langle f | g \rangle = \langle AB \rangle$, and $\langle g | f \rangle = \langle BA \rangle$, (13.22) results. The energy-time uncertainty deserves a separate discussion, (see Sect. 15.1.1).

Minimal uncertainty

There are no limits to ignorance: one can easily conceive states with large uncertainty both in x and in p. Then, $\sigma_A \sigma_B$ is much larger than the theoretical limit. It is interesting to find out the condition on ψ that grants a minimum value of $\sigma_A \sigma_B$; it is assumed that ψ is not an eigenstate of either observable. This condition emerges if we reverse-engineer the proof in the last paragraph. The main step there was the use of Schwartz inequality on the vectors $|f\rangle = (\hat{A} - <A>)|\psi\rangle$, $|g\rangle = (\hat{B} - <B>)|\psi\rangle$; the inequality becomes an equality for parallel vectors, therefore we take $|f\rangle$ proportional to $|g\rangle$. Moreover, we inserted a $\geq$ sign when we neglected the real part of $z^2 = \langle f | g \rangle^2$, but this becomes an $=$ sign if $\langle f | g \rangle$ is imaginary. Thus, picking

$$|g\rangle = i\lambda |f\rangle, \quad \text{with real } \lambda,$$

we obtain

$$\sigma_A^2 \sigma_B^2 = \left(\frac{\langle \psi | [\hat{A}, \hat{B}]_- | \psi \rangle}{2i} \right)^2.$$

This depends on the parameter λ because there is a family of states ranging between sharp B with ill-defined A to the opposite case.

---

[6]Indeed, $\langle \hat{A} - \langle \hat{A} \rangle \rangle = 0$ and $\sigma_{\hat{A} - \langle \hat{A} \rangle}^2 = \sigma_{\hat{A}}^2$; moreover, the commutator of $\hat{A}$ and $\hat{B}$ is not changed if we subtract the mean values from the operators.

In the special case $A = \hat{p}$, $B = \hat{x}$, the condition for minimum uncertainty reads

$$(\hat{p} - \langle p \rangle)\Psi = i\lambda(x - \langle x \rangle)\Psi; \qquad (13.23)$$

to find out the shape of this wave function we may shift the origin and write

$$(\hat{p} - \langle p \rangle)\Psi = i\lambda x \Psi.$$

One can see by inspection that there is a family of solutions, $\Psi = \exp[2\pi\frac{i\langle p \rangle x}{h}]\Psi_0$
where

$$-i\hbar\frac{d}{dx}\Psi_0(x) = i\lambda x \Psi_0.$$

Equation (13.23) is solved by the gaussian shifted packet (to be normalized)

$$\Psi(x) = \exp\left[-\lambda\frac{(x - \langle x \rangle)^2}{2\hbar} + \frac{i\langle p \rangle x}{\hbar}\right].$$

The ground state of the Harmonic oscillator is of this form, as we shall see.

Suppose the two operators $\hat{A}$ and $\hat{B}$ have a complete common set ψ_{ab} of eigenfunctions. Then,

$$\hat{A}\psi_{ab} = a\psi_{ab}, \quad \hat{B}\psi_{ab} = b\psi_{ab} \quad \forall \psi_{ab} \implies \hat{A}\hat{B}\psi = ab\psi = \hat{B}\hat{A}\psi.$$

Any relation that holds for the full complete set is an operator relation. So,

$$\implies [\hat{A}, \hat{B}]_- = 0.$$

The matrices taken on the set ψ_{ab} are diagonal, hence the matrices also commute.

We know that a unitary operator U is such that

$$UU^\dagger = 1 = U^\dagger U. \qquad (13.24)$$

Now consider the two hermitean matrices A and B, which represent the operators on the basis ψ_{ab}. We can transform to any other basis by a unitary transformation U by writing $A \to \tilde{A} = UAU^\dagger$. Since $UAU^\dagger UBU^\dagger = UBU^\dagger UAU^\dagger \Leftrightarrow [\tilde{A}, \tilde{B}] = 0$, the two matrices commute over any basis. Then, the commuting observables are compatible, while non-commuting ones are not.

13.9.1 Canonical Transformations in Quantum Mechanics

An immediate consequence of the above comments is that a unitary transformation U preserves the commutation rules; this is like the classical canonical transformations that preserve the Poisson brackets. U gives an equivalent description of the same physics. It can be a mere change of reference, but in several cases, it is a sort of *deus ex machina* which leads to the solution of otherwise intractable problems.

Chapter 14
Postulate 3

Any measurement of an observable $\hat{Q}$ must yield an eigenvalue λ of $\hat{Q}$ (solution to the eigenvalue equation $\hat{Q}|\Psi_\lambda\rangle = \lambda|\Psi_\lambda\rangle$). The measurement prepares the system in a state in which $\hat{Q} = \lambda$. If the system is in a state $|\Phi\rangle$, the probability of a particular λ is $P(\lambda) = \langle\Phi|\Psi_\lambda\rangle|^2$.

Note that the system might have a large number of degrees of freedom, yet one can make a measurement involving one of them, like one component of angular momentum, which has an eigenvalue equation depending on a single angle ϕ. If $|\Phi\rangle$ is an eigenstate belonging to a discrete eigenvalue λ, $P(\lambda) = 1$; in such cases, Quantum Mechanics gives certainties. Otherwise, after the measurement Φ *collapses* in an eigenstate of the operator, and if the measurement is repeated immediately (i.e., before the system evolves) the result is again λ. Quantum Mechanics gives no information about the way in which the interaction with the classical measurement apparatus produces the collapse, which is considered as a sort of instantaneous evolution which is outside the scope of the Scrödinger equation. Many people dislike the fact that the observer is not a part of the story. As an alternative to the collapse, Hug Everitt proposed a many-worlds interpretation, in which all the possible outcomes of the measurement take place in some Universe and the measurement takes us to one of these. This idea has influential supporters and opposers. But, while we must be ready to accept unobservable mathematical tools like vector potentials and wave functions, the description of the reality should not depend on unobservable parts of the reality itself. Science is not compatible with occultism. Thus, it appears to me that the remedy is worse than the disease.

© Springer International Publishing AG, part of Springer Nature 2018
M. Cini, *Elements of Classical and Quantum Physics*,
UNITEXT for Physics, https://doi.org/10.1007/978-3-319-71330-4_14

14.1 Reasons Why the Set Must be Complete

Since the measurement must necessarily return some value, this postulate requires

$$\sum_\lambda |\langle \Phi | \Psi_\lambda \rangle|^2 = \sum_\lambda P(\lambda) = 1;$$

this demands the expansion

$$|\Phi\rangle = \sum_\lambda |\Psi_\lambda\rangle \langle \Psi_\lambda | \Phi \rangle.$$

Therefore, we must assume the completeness

$$\sum_\lambda |\Psi_\lambda\rangle \langle \Psi_\lambda| = 1.$$

The formalism works smoothly in most cases, but sometimes one must use it with a proper choice of the function space and physical vision. For example, x is an observable. We must distinguish the variable x, the operator $\hat{x}$, and the eigenvalue that is the result of a measurement; for clarity, let us call x' the eigenvalue. Letting $\varphi^{(x')}(x)$ denote the eigenfunction, the eigenvalue equation reads as

$$\hat{x}\varphi^{(x')}(x) = x'\varphi^{(x')}(x).$$

Now, $\varphi^{(x')}(x) = \delta(x - x')$ does not belong to L^2 but to the space of the distributions. This also provides the required complete set, since

$$\forall \psi, \ \langle \varphi^{(x)} | \psi \rangle = \int dx' \varphi^{(x)}(x') \psi(x') = \psi(x).$$

But $\delta(x - x')$ is not a normal wave function, and, for instance, we cannot take the square modulus. The difficulty stems from the fact that x is a continuous variable, and it is impossible to make a measurement giving the position of a particle quite exactly; that would require infinite energy.

With this proviso, we must be ready to work with discrete and continuous observables. In the discrete case, the eigenvalue equation reads as $\hat{Q}|e_n\rangle = \lambda_n |e_n\rangle$, with $n = 1, 2, 3, \ldots$, and $\langle e_m | e_n \rangle = \delta_{mn}$; besides, $|\Psi\rangle = \sum_n c_n |e_n\rangle$ with $c_n = \langle e_n | \Psi \rangle$. Then, $P(n) = |c_n|^2$ is the probability of the eigenvalue λ_n. One notable example is provided by angular momentum. Any direction can be taken as the z axis, and $\hat{L}_z$ is the orbital angular momentum operator.[1] Any measurement must return $\frac{hm}{2\pi}$, where the azimuthal quantum number m is an integer.

---

[1]The spin angular momentum will be introduced in Chap. 18.

In the continuous case, the eigenvalue equation reads as $\hat{Q}|e_k\rangle = \lambda_k|e_k\rangle$, with $-\infty < k < \infty$, but $\langle e_h|e_k\rangle = \delta(h-k)$; besides, $|\Psi\rangle = \int dk c_k|e_k\rangle$, where $c_k = \langle e_k|\Psi\rangle$. So, $dP = |c_k|^2 dk$ is the probability of finding the eigenvalue within dk from λ_k.

We have just discussed the operator $\hat{x}$; as another example, we may take the momentum $\hat{p}$, having eigenfunctions $e_p(x) = \frac{e^{ikx}}{\sqrt{2\pi}}$, where $p = \hbar k$. The p component of $|\Psi\rangle$ is $c_p = \langle e_p|\Psi\rangle = \int_{-\infty}^{\infty} e_p(x)\Psi(x)dx = \Psi(p,t)$; this is the wave function in momentum space.

Chapter 15
Postulate 4

The time evolution of ψ is governed by the Schrödinger equation

$$i\hbar\frac{\partial\psi}{\partial t} = \hat{H}(t)\psi(t), \tag{15.1}$$

where $\hat{H}(t)$ is the Hamiltonian.

15.1 Time Derivative of an Observable and F = ma

I anticipate here that this statement (like all the postulates) holds for single particles and many particle systems as well (although in the latter case we have to specify in later chapters how the equation works in detail.)

The Schrödinger equation is first-order in time; therefore, if ψ is given at one time, it can be evolved in the future or in the past at all times. A second initial condition that is necessary in the classical case is not needed. It is also second-order in space variables, which implies that the theory is not acceptable from the standpoint of Relativity. Proper extensions (Klein–Gordon equation for spinless particles, Dirac equation for spin $\frac{1}{2}$ Fermions, Maxwell equations for Photons, the Proca equation for massive spin-1 particles, and so on,) are known since a long time; they are quite important, but do not spoil the relevance of the non-relativistic approximation to many interesting phenomena.

Note that even the matrix elements of time-dependent operators generally depend on the time because of the evolution of the wave function. The derivative $\frac{d\hat{Q}}{dt}$ of an operator $\hat{Q}$ is by definition an operator that satisfies:

© Springer International Publishing AG, part of Springer Nature 2018
M. Cini, *Elements of Classical and Quantum Physics*,
UNITEXT for Physics, https://doi.org/10.1007/978-3-319-71330-4_15

$$\left\langle \Psi | \frac{d\hat{Q}}{dt} | \Psi \right\rangle \equiv \frac{d}{dt} \left\langle \Psi | \hat{Q} | \Psi \right\rangle = \left\langle \Psi | \frac{\partial}{\partial t} \hat{Q} | \Psi \right\rangle + \left\langle \frac{\partial}{\partial t} \Psi | \hat{Q} | \Psi \right\rangle + \left\langle \Psi | \hat{Q} | \frac{\partial}{\partial t} \Psi \right\rangle.$$

Using the Schrödinger equation one finds $\hbar | \dot{\psi} \rangle = -i H | \psi \rangle \implies \hbar \langle \dot{\psi} | = i \langle \psi | H$, and so,

$$\frac{d\hat{Q}}{dt} = \frac{\partial \hat{Q}}{\partial t} + \frac{i}{\hbar} [\hat{H}, \hat{Q}]. \tag{15.2}$$

We know that in Classical Physics, the evolution of a function F of the canonical variables is given by a similar equation, namely, $\dot{F} = \frac{\partial F}{\partial t} + \{F, H\}$. The commutators replace the Poisson brackets and, if $\frac{\partial \hat{Q}}{\partial t} = 0$, the operators that commute with the Hamiltonian are conserved. For example, consider a particle in 1d in a potential; $\hat{H} = \frac{\hat{p}^2}{2m} + V(x)$. The velocity is $\frac{d\hat{x}}{dt} = \frac{i}{\hbar} [\hat{H}, \hat{x}] = \frac{p}{m}$, as expected. So, we find

$$m\ddot{x} = \hat{F}, \tag{15.3}$$

where $\hat{F} = -\frac{dV}{dx}$ is the force. The Newtonian equation of motion is still valid! Albeit, with a different interpretation, of course.

15.1.1 Energy-Time Uncertainty

The uncertainty principle of Sect. 13.3 does not involve the time, since there is no Hermitean operator for t. However, consider the uncertainty principle,

$$(\Delta A \ \Delta B)^2 \equiv \sigma_A^2 \sigma_B^2 \geq \left(\frac{\langle \psi | [\hat{A}, \hat{B}] | \psi \rangle}{2i} \right)^2,$$

for $A = \hat{H}$, the Hamiltonian, $B = \hat{Q}$, where $\hat{Q}$ is a time-independent observable. The commutator with H appears in (15.2), so one finds that

$$\Delta E \Delta Q \geq \frac{1}{2} \hbar | < \Psi | \frac{dQ}{dt} | \Psi > |. \tag{15.4}$$

The derivative of Q is due to the evolution of the wave packet. If for $|\Psi >$, we choose an eigenstate of H, there is no evolution and the derivative vanishes. For a system with a continuous spectrum, this never happens; experimentally, $|\Psi >$ is always a wave packet. Suppose that someone tries to devise an experiment in which the wave packet is prepared and then Q is measured so fast that the evolution barely begins. Then,

$$\Delta Q \approx < \frac{dQ}{dt} > \Delta t.$$

Putting this into (15.4), one finds

$$\Delta E \, \Delta t > \frac{1}{2}\hbar. \tag{15.5}$$

One cannot have a small Δt and simultaneously an arbitrarily small ΔE; being fast requires mixing different energies. Conversely, a well-defined energy requires an experiment of sufficient length. This is the energy-time uncertainty principle. Mathematically, this principle reflects known properties of the Fourier transform. A finite piece of a sinusoid mixes all frequencies.

Only long-lived atomic states have sharp energies; therefore, atomic clocks use so-called *forbidden* transitions, which actually take place but take long time. Several years ago, an error of one second in one million years was claimed. In 2015 a metrology laboratory at Riken in Japan has built a clock which misses a second in 15 billion years, which is longer than the age of the Universe.

Problem 26 A particle is in the ground state of an oscillator with Hamiltonian $H[1] = \frac{p^2}{2m} + \frac{1}{2}m\omega_0^2 x^2$. Calculate the probability $P(\eta)$ that it is in the ground state of a harmonic oscillator with Hamiltonian $H[\eta] = \frac{p^2}{2m} + \frac{1}{2}m(\eta\omega_0)^2 x^2$.

Solution 26 Using

$$\psi_0(x) = \frac{1}{\sqrt{x_1\sqrt{\pi}}}e^{-\frac{x^2}{2x_1^2}},$$

with $x_\eta^2 = \frac{\hbar}{m\eta\omega} = \frac{x_1}{\eta}$, letting $\frac{1}{x_a^2} = \frac{1}{2}(\frac{1}{x_1^2} + \frac{1}{x_\eta^2})$, one finds

$$\langle\psi_\eta|\psi_1\rangle = \frac{\sqrt{\pi}x_a}{\sqrt{\pi x_1 x_\eta}} = \sqrt{2}\frac{\eta^{\frac{1}{4}}}{\sqrt{1+\eta}}$$

and therefore

$$P(\eta) = 2\frac{\sqrt{\eta}}{1+\eta} = \frac{2}{\sqrt{\eta} + \frac{1}{\sqrt{\eta}}}.$$

15.1.2 Adiabatic and Quasi-adiabatic Evolution

The adiabatic theorem (Kato 1949) deals with the solution of the S.E. with a time-dependent Hamiltonian $H(t)$ in the case of a discrete and non-degenerate spectrum. The theorem says that if the system is prepared in the nth eigenstate and the evolution of H is sufficiently slow, then the system will remain forever in the nth eigenstate. The Kato theorem has many important and far reaching consequences. The reader could be surprised by this statement, since the requirement that the spectrum be discrete and non-degenerate looks very restrictive. In fact, this limitation can be circumvented by Gell-Mann and Low, and the many body Green's function theory of the electron liquid has been formulated as if the Coulomb interaction were turned on adiabatically, starting from a non-interacting system in the far past. Other reasons why adiabatic treatments are important will be apparent in Chap. 23. Since the theorem is so plausible I omit the proof, which may be found elsewhere[1].

It is important to investigate the relation between the solution of the time-dependent S.E. and the complete set of solutions of the **stationary state** equation with the *instantaneous Hamiltonian* $H(t)$, namely,

$$H(t)\phi_n(t) = \epsilon_n(t)\phi_n(t). \tag{15.6}$$

Let us consider the evolution between time 0 and time t. Substituting the formal expansion

$$\psi(t) = \sum_n \alpha_n(t)\phi_n(t) \exp\left[\frac{-i}{\hbar} \int_0^t \epsilon_n(t')dt'\right], \tag{15.7}$$

we readily find the condition

$$\sum_n (\dot{\alpha}_n\phi_n + \alpha_n\dot{\phi}_n) \exp\left[\frac{-i}{\hbar} \int_0^t \epsilon_n(t')dt'\right] = 0. \tag{15.8}$$

To find the coefficients, we take the scalar product with the instantaneous $\phi_k(t)$. We obtain

$$\dot{\alpha}_k = -\sum_n \alpha_n \langle\phi_k|\dot{\phi}_n\rangle \exp\left[\frac{-i}{\hbar} \int_0^t (\epsilon_n(t') - \epsilon_k(t'))dt'\right]. \tag{15.9}$$

We need $\langle\phi_k|\dot{\phi}_n\rangle$; to this end, we differentiate (15.6) and scalar multiply by $\phi_k(t)$, for $k \neq n$. The result is

$$\langle\phi_k|\dot{\phi}_n\rangle = \frac{\langle\phi_k|\frac{\partial H}{\partial t}|\phi_n\rangle}{\epsilon_n - \epsilon_k}. \tag{15.10}$$

This result allows us to estimate the amplitude to jump to different states when the adiabatic condition is weakly violated. We see that if $\langle\phi_k|\frac{\partial H}{\partial t}|\phi_n\rangle$ is small compared to the energy separation $\epsilon_n - \epsilon_k$, the system has little chance to be found in k if it is prepared originally in n. Now it is verified that if the spectrum is discrete and non-degenerate, a finite gap must be overcome before the system can make any transition; therefore, any evolution can be done so slowly that the system remains in n all the time.

---

[1] The original paper by Kato is somewhat involved, but for a simple proof see D.J. Griffiths, "Introduction to Quantum Mechanics", Prentice Hall.

15.1.3   Sudden Approximation

Suppose one knows the (one-body or many-body) normalized wave function $\psi(t_0)$ of some system and the Hamiltonian $H(t_0)$ at some time t_0; if H stays constant for $t > t_0$, we can write down immediately the evolution of $\psi(t_0)$ in terms of the eigenfunctions ϕ_n of H at time t_0, setting $\hbar = 1$,

$$\psi(t) = \sum_n \phi_n e^{-i\epsilon_n(t-t_0)} \langle \phi_n | \psi \rangle, \qquad (15.11)$$

where the sum can imply an integration in the case of a continuus spectrum. Typically, this result is useful if $\psi(t_0)$ is the wave function resulting from the evolution of the system for times before t_0 with a different Hamiltonian H'. This means that we are considering a model in which the Hamiltonian changes abruptly at time t_0, inevitably leaving the system in a mixture of eigenstates of the new Hamiltonian. While it is clear that Hamiltonians cannot really change instantly, there are many situations in which such a scheme is very useful. For instance, when a many-electron atom in its ground state suffers a nuclear decay that leads to a change of the atomic number, the change is fast enough to be considered as practically instantaneous when one wishes to calculate the probability that the atom gets ionized in the process; indeed, the characteristic electronic times are long compared to those of the much more energetic nuclear transitions. In addition, the sudden approximation offers a practical scheme for the solution of arbitrary time-dependent problems by am iteration of the above idea. Instead of solving the problem with $H(t)$ for $t \in (t_i, t_f)$, one divides the time interval, introducing intermediate times such that $t_i < t_1 < t_2 < \cdots < t_f$, approximating H(t) with a constant Hamiltonian in each sub-interval. If each sub-interval is short compared to the characteristic times of the system and of the change of the Hamiltonian, this is a convenient, practical way to compute the result.

Problem 27   A particle of mass m in 1d is in the ground state of the potential well:

$$V(x) = \begin{cases} \infty & \text{per } x < -\frac{a}{2}, \\ 0 & \text{per } 0 < x < a, \\ \infty & \text{per } x > \frac{a}{2}. \end{cases}$$

At time t = 0, the potential becomes $V(x) = 0$ everywhere. Write the wave function $\psi(x, t)$ at all times $t > 0$.

Solution 27   Since $\psi_0(x) = \sqrt{\frac{2}{a}} \sin(\frac{\pi(x-\frac{a}{2})}{a})$ and the plane wave is

$$\phi_k(x, t) = \frac{e^{i(kx-\omega(k)t)}}{\sqrt{2\pi}},$$

con

$$\hbar\omega = \frac{(\hbar k)^2}{2m},$$

the expansion $\psi_0(x) = \int dk \phi_k(x) \langle \phi_k | \psi_0 \rangle$ requires that

$$\langle \phi_k(x, 0) | \psi_0(x) \rangle = \alpha(k).$$

Using

$$\alpha(k) = \int_{-\frac{a}{2}}^{\frac{a}{2}} dx \frac{e^{-ikx}}{\sqrt{\pi a}} \sin\left(\frac{\pi(x - \frac{a}{2})}{a}\right) = 2\sqrt{\pi a} \frac{\cos(\frac{ak}{2})}{a^2 k^2 - \pi^2},$$

one obtains

$$\psi(x, t) = \int_{-\infty}^{\infty} \alpha(k) \frac{e^{i(kx - \omega(k)t)}}{\sqrt{2\pi}} dk = \sqrt{\frac{a}{2}} \int_{-\infty}^{\infty} \frac{\cos(\frac{ak}{2})}{a^2 k^2 - \pi^2} e^{i(kx - \omega(k)t)} dk.$$

15.1.4 Galileo Transformation

In Chap. 6, we performed thought experiments using the station and the train frames to introduce the Lorentz transformation. Here, we can do the same in the non-relativistic case. If $\psi_S(x, t)$ is the wave function of a particle in an inertial reference (the station), what is the wave function $\psi_T(x', t)$ in a reference (the train) moving with speed v? The Galileo transformation $x = x' + vt, t = t'$ applies. For a free particle, $p = p' + mv$, which implies that $\epsilon_p = \epsilon_{p'} + v.p' + \frac{1}{2}mv^2$.

This leads to

$$\psi = \psi' \exp\left[\frac{imv}{\hbar}x'\right] \exp\left[\frac{imv^2 t}{2}\right]. \tag{15.12}$$

Thus, the transformation of the wave function requires a phase factor $\exp[\frac{imv^2 t}{2}]$; since this is p-independent, the same law applies to any wave packet. In the presence of a potential, the transformation still applies in the present form if the potential is scalar, that is, independent of the reference. Otherwise, one should calculate the wave packet with the potential in the system of the station and then transform the result to the train. We have found one more use for phase factors.

15.1.5 Gauge Invariance

Quantum Mechanics must be formulated in terms of complex wave functions, and one could reasonably expect that much of the information about the state of a particle must reside in the phase. However, strikingly, for a charged particle we can change the phase as we please through a transformation

$$\psi(\boldsymbol{r}, t) \to \psi'(\boldsymbol{r}, t) = \psi(\boldsymbol{r}, t) \exp\left(\frac{ie\chi(\boldsymbol{r}, t)}{\hbar c}\right), \tag{15.13}$$

where χ is an arbitrary function of space-time. Such a transformation entails $p\psi \rightarrow (p - \frac{e}{c}A)\psi'$ and $i\hbar \frac{\partial}{\partial t}\psi \rightarrow (\frac{\partial}{\partial t} + \frac{e}{c}\frac{\partial \chi}{\partial t})\psi'$.

The Schrödinger equation for a charge in a field is

$$H\psi = \left(\frac{(p - \frac{eA}{c})^2}{2m} + eV\right)\psi = i\hbar\frac{\partial}{\partial t}\psi, \qquad (15.14)$$

so all that happens is a gauge transformation to new potentials $A' = A + \text{grad}\chi(r, t)$, $V' = V - \frac{1}{c}\frac{\partial \chi}{\partial t}$. In the classical equations of motion, only the fields appear and the gauge invariance of the theory is obvious; instead, the wave function is affected by the change; however, the matrix elements of the coordinates calculated with ψ are identical to those calculated with ψ'; and those of the mechanical momentum $p - \frac{e}{c}A$ calculated with ψ are identical with those of the mechanical momentum $p - \frac{e}{c}(A + \nabla\chi)$ calculated with ψ'. Consequently, the Physics is unaffected by the change.

15.1.6   No Cloning Theorem

The concept of probability in quantum theory refers to a distribution of results of measurements on many samples. The need for many identical samples is evident since the act of measurement collapses the wave function. This could be avoided if we could clone a given sample, with wave function $|\phi\rangle$ i.e. if we could produce many copies in the same quantum state without knowing it. The copies would conceivably be obtained by starting with systems having wave functions $|\alpha_i\rangle$, $i = 1, \ldots N$, which depend on the same variables as ϕ.

A quantum cloning machine is defined to operate in analogy with a photocopy machine, except that the contents of the original page is assumed unknown. Initially one has an unknown quantum state $|\phi\rangle$. For the sake of argument I assume here that it is a spin, but it could be anything. Besides, one has n spins $|\alpha_i\rangle$, $i = 1, \ldots n$ which have the same role as the white sheets. Here, E is the state of the machine and the environment. So, the initial state of the system at time $t = 0$ is

$$|s_\phi\rangle = |E\rangle|\phi\rangle|\alpha_1\rangle \cdots |\alpha_n\rangle.$$

The machine Hamiltonian has an evolution operator U such that at the end we must have

$$U(t)|s_\phi\rangle = U(t)|E\rangle|\phi\rangle|\alpha_1\rangle \cdots |\alpha_n\rangle = |E_\phi\rangle|\phi\rangle|\phi\rangle \cdots |\phi\rangle,$$

where E_ϕ is the final state of the machine and the environment. In 1982 Wootters and Zurek[2] showed that it is impossible to make such a machine. The essential reason

---

[2] Wootters, W.K. and Zurek, W.H.: A Single Quantum Cannot be Cloned. Nature 299 (1982), pp. 802–803.

is that one starts with a state α such that $\langle\phi|\alpha\rangle < 1$ and we want to evolve it until $\langle\phi(t)|\alpha(t)\rangle = 1$; however $\langle\phi(t)|\alpha(t)\rangle = \langle\phi(0)|U^\dagger(t)U(t)\alpha(0)\rangle$ where U is the evolution operator; but the evolution is unitary and so $\langle\phi(t)|\alpha(t)\rangle = \langle\phi(0)|\alpha(0)\rangle$. This simple argument is incomplete because it does not take into account the evolution of the machine and of the environment. Therefore the proof is slightly longer. Suppose that besides $|\phi\rangle$ we also clone a second state $|\psi\rangle$. The initial state would be

$$|E\rangle|\psi\rangle|\alpha_1\rangle \cdots |\alpha_n\rangle$$

and the final state

$$U(t)|s_\psi\rangle = U(t)|E\rangle|\psi\rangle|\alpha_1\rangle \ldots |\alpha_n\rangle = |E_\psi\rangle|\phi\rangle|\phi\rangle \cdots |\phi\rangle.$$

The overlap would be $\langle s_\phi|s_{psi}\rangle = \phi|\psi\rangle$ at $t = 0$; finally it should become $\langle s_\phi|U^\dagger U|s_\psi\rangle = \langle E_\phi|E_\psi\rangle\langle\phi|\psi\rangle^n$. Since U is unitary, the overlap is the same as the initial one. Hence we get:

$$\langle\phi|\psi\rangle = \langle E_\phi|E_\psi\rangle\langle\phi|\psi\rangle^n \rightarrow 1 = \langle E_\phi|E_\psi\rangle\langle\phi|\psi\rangle^{n-1}$$

for any n, and this cannot be true. Therefore, U does not exist.

15.2 Feynman Path Integral Formulation of Quantum Mechanics

At the end of the 40s, Richard Feynman[3] proposed the *path integral* formulation of Quantum Mechanics, based on a remark by Dirac. It is equivalent to the Heisenberg and Schrödinger formulations, but reveals in a clearer way the connection to Classical Mechanics. It is important in field theory, but for simplicity, I discuss its application to one-particle problems.

We introduced the quantum evolution operator $U(0, t)$ from time 0 to time t in Eq. (10.8); now, consider the amplitude $A(t) = \langle a|U(0, t)|b\rangle$ for the particle to go from state a at time $t = 0$ to a state b at time t. By using the Group property (10.11) of the evolution operator, one can slice the time interval $(0, t)$ finely to $N \gg 1$ sub-intervals of small duration ϵ, and the limit $N \rightarrow \infty$, $\epsilon \rightarrow 0$ is understood at the end of the procedure. For small ϵ, one can safely assume that H is constant within each slice even if $H = H(t)$. Consequently, by the reasoning that led us to Eq. (10.12), $U(n\epsilon, (n + 1)\epsilon) = \exp(-iH\epsilon)$, where H is understood to be taken at the time slice.

The result is a terrific multiple integral:

$$A(t) =$$

---

[3]Richard Feynman (New York City 1918–1988) won the Nobel in 1965 for the invention of his diagrams and his contributions to Quantum Electrodynamics.

$$\int dq_1 \ldots \int dq_{n-1} \langle a|e^{-iH\epsilon}|q_{n-1}\rangle \langle q_{n-1}|e^{-iH\epsilon}|q_{n-2}\rangle \cdots \langle q_2|e^{-iH\epsilon}|q_1\rangle \langle q_1|e^{-iH\epsilon}|b\rangle.$$

Here, the intermediate states q_i are taken to be position eigenstates, that are a complete set, and in the limit $\epsilon \to 0$ this expression is a path integral. $A(t)$ is a sum of all the contributions that correspond to possible choices of the values of q_n at the nth slice, so it is a sum over *stories*. For each story, we can also assign the quantum particle an instantaneous velocity by setting

$$q_{n+1} - q_n = \epsilon \dot{q}.$$

Note, however, that many stories are not plausible. A small ϵ does not guarantee that neighboring q values are close. Like the diffusion equation, the SE can take the particle to the Andromeda galaxy in any short time (albeit with a very small amplitude), so that the paths that interfere in giving the final answer need not be (and are not) regular and physically reasonable. Paths involving velocities exceeding c must be included. This type of monster multiple integrals was introduced by N. Wiener in the theory of the Brownian motion.

We can make contact with the classical Lagrangian formalism as follows. The Hamiltonian can be written as the sum $H = T + V$ of a kinetic energy $T = T(p)$, where p are the canonical momenta, plus a potential energy $V(q)$. To proceed, we write $\exp(-iH\epsilon) = \exp(-iT\epsilon)\exp(-iV\epsilon) + O(\epsilon^2)$ and then neglect $O(\epsilon^2)$ since $\epsilon \to 0$ at the end. Then, introducing a set of momentum eigenstates, we write the n-th step as:

$$\langle q_{n+1}| \exp(-iT\epsilon) \exp(-iV\epsilon)|q_n\rangle = \int dp_n \langle q_{n+1}|p_n\rangle \langle p_n| \exp(-iT\epsilon) \exp(-iV\epsilon)|q_n\rangle.$$

Now, using $\langle q|p\rangle = e^{ipq}$, we let the kinetic energy act on momentum eigenstates and the potential energy act on position eigenstates. One finds that $\langle q_{n+1}| \exp(-iT\epsilon) \exp(-iV\epsilon)|q_n\rangle = \int dp_n e^{ip_n(q_{n+1}-q_n)} e^{-i\epsilon H(p,q)}$.

Putting all together,

$$A(t) = \prod_n \int \frac{dq(t_n)dp(t_n)}{2\pi} \exp\left\{i\left[(p_n(q_{n+1}-q_n) - H(p_n,q_n)\epsilon\right]\right\}. \quad (15.15)$$

Note that the product produces a summation at the exponent:

$$A(t) = \left[\prod_n \left(\int \frac{dq(t_n)dp(t_n)}{2\pi}\right)\right] \exp[i \sum_j (p_j(q_{j+1}-q_j - H(p_j,q_j)\epsilon]. \quad (15.16)$$

Introducing a path-integral notation for the infinite-dimensional integration, and the velocity, with $\epsilon \to d\tau$, we obtain:

$$A(t) = \int Dq D\left(\frac{p}{2\pi}\right) \exp[i \int_t^{t'} d\tau (p\dot{q}(\tau) - H(p, q, \tau)].$$ (15.17)

Under some conditions one can work out the momentum integrals. In particular, if the Hamiltonian is $H = \frac{p^2}{2m} + V(q)$, the exponent in the integrand can be cast in the form $-\frac{(p-\dot{q})^2}{2m} + L(q, \dot{q})$ where of course $L(q, \dot{q}) = \frac{p^2}{2m} - V(q)$ is the Lagrangian. Then the integrations are all of the Gaussian type $\int_{-\infty}^{\infty} dx \exp(-\frac{ax^2}{2}) = \sqrt{\frac{2\pi}{a}}$, for $a > 0$ and

$$A(t) = \int Dq \exp[i \int_t^{t'} d\tau L(q, \dot{q})].$$ (15.18)

Consider two configurations $q(t_1)$ and $q(t_2)$ of a system. For each virtual path connecting them, one can calculate the action integral. The Euler–Lagrange equations select the allowed classical evolutions of the system. In Quantum Theory, each virtual path contributes to the wave function the amount:

$$\Delta \Psi_{path} \propto \exp\left[\frac{2\pi i S}{h}\right].$$

For a particle going from $x_1(t_1)$ to $x_2(t_2)$ in the laboratory, one must consider all trajectories, in order to get the exact result. For motions involving actions $\gg h$ the paths that make the action stationary give a dominant contribution. In the continuum limit and in 1 dimension, the above amplitude $A(t)$ is the amplitude $P(x', t', x, t)$ that the particle is in x at time t if it is in x' at time t'. Assuming, for simplicity, a time-independent Hamiltonian, and setting $\hbar = 1$,

$$P(x', t', x, t) = \langle x' | e^{-iH(t-t')} | x \rangle.$$ (15.19)

The corresponding Green's function, defined as

$$G(x', t', x, t) = i P(x', t', x, t)\theta(t - t'),$$ (15.20)

satisfies

$$(H - i\frac{\partial}{\partial t})G = \delta(x - x')\delta(t - t').$$ (15.21)

In the case of a free particle, one finds (reinserting $\hbar$)

$$P(x', t', x, t) = \left(\frac{m}{2\pi i\hbar(t - t')}\right)^{\frac{1}{2}} \exp\left(\frac{im(x - x')^2}{2\hbar(t' - t)}\right).$$ (15.22)

Problem 28 Obtain Equation (15.23) by the path integral method.

Solution 28 Start directly from the above $A(t)$. All the integrals involved are Gaussian.

All this is immediately extended to 3 dimensions. Then,

$$P(x', t', x, t) = \left(\frac{m}{2\pi i\hbar(t - t')}\right)^{\frac{3}{2}} \exp\left(\frac{im(x - x')^2}{2\hbar(t' - t)}\right). \tag{15.23}$$

This formalism is well-suited to find simple approximations to problems like the diffraction by a double slit, where the dominating paths are evident (Fig. 15.1). An electron with energy E is emitted from G at time $t = 0$; the amplitude to arrive at time t_1 in slit $S1$ and hits the detector D at time T is $G(s1, t_1, G, 0)G(D, T, S1, t1)$, and the total amplitude to go to point D is $\langle D, T|G, 0\rangle = \int_{-\infty}^{\infty} G(S1, t_1, G, 0)G(D, T, S1, t1)dt_1$. This integral is a convolution. Since $G(S1, t_1, G, 0) = \frac{i}{\hbar}\theta(t_1)(\frac{m}{2\pi i\hbar t_1})^{\frac{3}{2}}$ $\exp(\frac{ima^2}{2\hbar t_1})$ has the transform $\frac{-1}{4\pi}\frac{e^{ika}}{a}$, where $k = \sqrt{\frac{2mE}{\hbar}}$ and the transform of $G(D, T, S1, t_1$ is $\frac{-1}{4\pi}\frac{e^{ikb}}{b}$, it is easy to transform back the product of the transforms with the result that the amplitude to go through $S1$ and arrive at time T is

$$\langle D, T|G, 0\rangle_{S1} = \frac{Z}{T^{\frac{3}{2}}}\frac{a+b}{ab}e^{\frac{im(a+b)^2}{2\hbar T}}. \tag{15.24}$$

In the same way, the amplitude to go through $S2$ is

$$\langle D, T|G, 0\rangle = \frac{Z}{T^{\frac{3}{2}}}\frac{c+d}{cd}e^{\frac{im(c+d)^2}{2\hbar T}}. \tag{15.25}$$

These amplituds must be summed and interfere, giving us

$$\langle D, t|G, 0\rangle = |\langle D, T|G, 0\rangle_{S1} + \langle D, T|G, 0\rangle_{S2}|^2. \tag{15.26}$$

If a, b, c, d are about the same, the probability is proportional to $|\exp(\frac{imL_1^2}{2\hbar T}) + \exp(\frac{imL_2^2}{2\hbar T})$ with $L_1 = a + b, L_2 = c + d$. Since $m\frac{L_1^2 - L_2^2}{T} = 2p(L_1 - L_2)$, where p is

Fig. 15.1 The geometry of the double slit experiment and of the Bohm–Aharonov effect

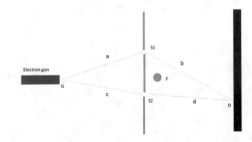

the momentum, the probability turns out to be proportional to $1 + \cos(\frac{p}{\hbar})(L_1 - L_2)$. In the presence of a magnetic flux F the Lagrangian has an additional term $-\frac{e}{c}A.v$, where v is the velocity. Now,

$$\langle D, T | G, 0 \rangle = \frac{Z}{T^{\frac{3}{2}}} \frac{c+d}{cd} e^{\frac{im(c+d)^2}{2\hbar T}} \exp\left(\frac{-ie}{c\hbar} \int_{C1} dx.A\right) \exp\left(\frac{-ie}{c\hbar} \int_{C2} dx.A\right).$$

$$(15.27)$$

where $C1$ is a path through G,S1 and D and C2 is a path through G, S2 and D. This goes like $1 + \cos[\frac{m}{2\hbar T}(L_1^2 - L_2^2) - \frac{e\Phi}{c\hbar}]$. Thus, the Bohm–Aharonov effect is borne out by this formalism.

Chapter 16
The Quantum Harmonic Oscillator

This is not just another one-dimensional example. It is a
fundamental piece of the general theory.

The oscillator Hamiltonian in the coordinate representation is:

$$\hat{H} = \frac{p^2}{2m} + \frac{1}{2}m\omega^2 x^2. \tag{16.1}$$

We know (Eq. 2.103) that by a canonical trasformation it can be cast in the simpler form

$$\tilde{H} = A\omega, \tag{16.2}$$

where A is the amplitude of the oscillation. In Classical Mechanics, the amplitude of the oscillation is arbitrary, but in Quantum Mechanics, there is a natural **scale** of energies given by $E \sim \hbar\omega$. Setting

$$m\omega^2 x_0^2 \sim \hbar\omega,$$

we realize that there is also a characteristic length

$$x_0 = \sqrt{\frac{\hbar}{m\omega}}.$$

© Springer International Publishing AG, part of Springer Nature 2018
M. Cini, *Elements of Classical and Quantum Physics*,
UNITEXT for Physics, https://doi.org/10.1007/978-3-319-71330-4_16

16.1   Coordinate Representation

At the moment, it is not clear how to deal with the form (16.2) but from (16.1) the
Schrödinger Equation (S.E.) for the stationary states $\hat{H}\psi(x) = E\psi(x)$ gives us

$$\frac{d^2}{dx^2}\psi = \frac{m^2\omega^2}{\hbar^2}x^2\psi - \frac{2mE}{\hbar^2}\psi, \tag{16.3}$$

that is,

$$\frac{d^2}{dx^2}\psi = \frac{x^2}{x_0^4}\psi - \frac{2mE}{\hbar^2}\psi.$$

Introducing the dimensionless length q by $x = x_0 q$,

$$\frac{1}{x_0^2}\frac{d^2}{dq^2}\psi = \frac{q^2}{x_0^2}\psi - \frac{2mE}{\hbar^2}\psi, \tag{16.4}$$

the S.E. becomes

$$\boxed{\frac{d^2\psi}{dq^2} = (q^2 - 2\epsilon)\psi,} \tag{16.5}$$

where

$$\epsilon = \frac{E}{\hbar\omega}. \tag{16.6}$$

is the dimensionless energy. Even in the Hamiltonian (16.1) one can do a similar
simplification. We eliminate m in favor of x_0, by $m = \frac{\hbar}{\omega x_0^2}$, thus

$$\hat{H} = \frac{\hbar\omega}{2}\left[-\frac{d^2}{dq^2} + q^2\right], \tag{16.7}$$

which agrees with (16.5). By inspection,

$$\psi_0(q) = \frac{1}{\sqrt[4]{\pi}}e^{-\frac{q^2}{2}} \tag{16.8}$$

is a solution of the S.E. and is normalised, i.e., $\int_{-\infty}^{\infty}\psi(x)^2 dx = 1$. In terms of x,

$$\psi_0(x) = C e^{-\frac{x^2}{2x_0^2}},$$

with

$$C = \left(\frac{m\omega}{\pi\hbar}\right)^{\frac{1}{4}}$$

up to a phase factor.

This simple solution is the ground state, as we shall see. To find the general solution, we set

$$\psi(q) = h(q)e^{-\frac{q^2}{2}};$$

then, by substituting in (16.5) we find:

$$h'' - 2qh' + (2\epsilon - 1)h(q) = 0. \tag{16.9}$$

By inspection, $h(q) \propto q$ is a solution with $\epsilon = \frac{3}{2}$, and one can verify easily that $h(q) \propto 4q^2 - 2$ is another solution, with $\epsilon = \frac{5}{2}$. This suggests that a systematic search for polynomial solutions is in order.

Polynomial Solutions

Setting

$$h(q) = a_0 + a_1 q + a_2 q^2 + a_3 q^3 + \cdots + a_N q^N = \sum_{j=0}^{N} a_j q^j,$$

and putting into (16.9) we get the *recurrence relation*:

$$(j + 1)(j + 2)a_{j+2} - 2ja_j + (2\epsilon - 1)a_j = 0, \quad j = 0, \ldots, N - 2.$$

Hence, by

$$a_{j+2} = \frac{2j + 1 - 2\epsilon}{(j + 1)(j + 2)} a_j \quad j = 0, \ldots, N - 2. \tag{16.10}$$

we can find all the even terms from a_0 and all the odd ones from a_1. To systematically find all the polynomial solutions, let us put in (16.10) the condition that $a_{N+2} = 0$; this gives us a solution of degree N;

The condition $2N + 1 - 2\epsilon_N = 0$ with (16.6) yields the eigenvalues[1]

$$\boxed{E_N = (N + \tfrac{1}{2})\hbar\omega.} \tag{16.11}$$

The Gaussian solution deduced above was the ground state $N = 0$; the lowest possible energy is called zero point energy; classically, the oscillator could be fixed at

---

[1] The recurrence formula (16.10) also gives us a transcendental solution, but this is not acceptable for a wave function. For $j \to \infty$, (16.10) becomes $a_{j+2} \sim \frac{2}{j}a_j$; this is solved by $a_j \sim \frac{C}{(\frac{j}{2})!}$, with some constant C. In the even case, $h = \sum_{j=2k} \frac{C}{(\frac{j}{2})!}q^j = C\sum_k \frac{1}{k!}q^{2k} = Ce^{q^2}$; this asymptotic behavior at large q leads to $\psi \to \infty$ for $x \to \infty$, and such a solution cannot be normalized. In the odd case, $h = \sum_{j=2k-1} \frac{C}{(\frac{j}{2})!}q^j = C\sum_k \frac{1}{(k-\frac{1}{2})!}q^{2k-1}$, and since $h > C\sum_k \frac{1}{k!}q^{2k-1} \sim \frac{C}{q}e^{q^2}$, even this solution cannot be normalized.

the origin, but this would violate the uncertainty principle. The characteristic energy $\hbar\omega$ is the difference $\hbar\omega$ between consecutive energy levels.

Hermite Polynomials

Setting $2N + 1 - 2\epsilon = 0$ in (16.9), we find the *Hermite equation*

$$h_N''(q) - 2qh_N' + 2Nh_N(q) = 0; \tag{16.12}$$

the (16.10) relations become

$$a_{j+2} = \frac{2j - 2N}{(j + 1)(j + 2)}a_j, \quad j = 0, \ldots, N - 2,$$

and the solutions are Hermite polynomials $h_N(q)$; these are orthogonal in the sense that

$$\int_{-\infty}^{\infty} h_m(x)h_n(x)e^{-x^2}dx = 0, \quad m \neq n.$$

The first few are: $h_0 = 1$, $\quad h_1(q) = 2q$, $\quad h_2(q) = 4q^2 - 2$, $\quad h_3(q) = 8q^3 - 12q$, $\quad h_4(q) = 16q^4 - 48q^2 + 12$.

I report without proof the following interesting formulas:

$$\frac{dH_n(x)}{dx} = 2nH_{n-1}(x) \tag{16.13}$$

$$H_{n+1}(x) = 2xH_n(x) - 2nH_{n-1}(x). \tag{16.14}$$

In addition, the Rodrigues formula holds:

$$H_n(x) = (-1)^n \exp(x^2) \left(\frac{d}{dx}\right)^n \exp(-x^2). \tag{16.15}$$

This formula is useful to work out the normalization condition. First, one integrates by parts (16.1) n times to show that

$$\int_{-\infty}^{\infty} e^{-x^2} H_n(x)^2 dx = \int_{-\infty}^{\infty} e^{-x^2} \frac{d^n H_n(x)}{dx^n} dx.$$

Then from (16.13) one obtains that $\frac{d^n H_n(x)}{dx^n} = 2^n n!$ So, the normalized wave function of the nth level of the harmonic oscillator is:

$$\psi_n(x) = \sqrt[4]{\frac{m\omega}{\pi\hbar}} \frac{\exp[-\frac{m\omega x^2}{2\hbar}]}{\sqrt{2^n n!}} H_n\left(x\sqrt{\frac{m\omega}{\hbar}}\right). \tag{16.16}$$

16.2 Number Representation and Coherent States

It is important to solve again the quantum oscillator by the neat and powerful *operator method* which is basic in field theory. One would be tempted to factor the Hamiltonian

$$H = \frac{\hbar\omega}{2}\left[-\frac{d^2}{dq^2} + q^2\right] = ?? = \frac{\hbar\omega}{2}\left(\frac{d}{dq} + q\right)\left(-\frac{d}{dq} + q\right), \qquad (16.17)$$

but this is wrong! $\frac{d}{dq}$ and q do not commute, since

$$\left[\frac{d}{dq}, q\right]_{-} = 1.$$

However, we can fix it. Introducing the annihilation operator

$$a = \frac{1}{\sqrt{2}}\left(q + \frac{d}{dq}\right), \qquad (16.18)$$

and its Hermitean conjugate, called creation operator

$$a^{\dagger} = \frac{1}{\sqrt{2}}\left(q - \frac{d}{dq}\right), \qquad (16.19)$$

we obtain

$$H = \hbar\omega\left(a^{\dagger}a + \frac{1}{2}\right). \qquad (16.20)$$

Still, the operators do not commute, since

$$\boxed{[a, a^{\dagger}]_{-} = 1,} \qquad (16.21)$$

but now the Schrödinger equation takes the form

$$\left(a^{\dagger}a + \frac{1}{2}\right)\psi = \epsilon\psi. \qquad (16.22)$$

If ψ is a solution, $a^{\dagger}\psi$ is also a solution, with eigenvalue $\epsilon + 1$, since the commutation rule yields $a^{\dagger}aa^{\dagger} = a^{\dagger}a^{\dagger}a + a^{\dagger}$, and so

$$\left(a^{\dagger}a + \frac{1}{2}\right)a^{\dagger}\psi = \left(a^{\dagger}[1 + a^{\dagger}a] + a^{\dagger}\frac{1}{2}\right)\psi$$

$$= a^{\dagger}(a^{\dagger}a + 1/2 + 1)\psi = a^{\dagger}(\epsilon + 1)\psi = (\epsilon + 1)a^{\dagger}\psi.$$

Moreover, $a\psi$ is another solution, with energy eigenvalue $\epsilon - 1$; to see that, in

$$\left(a^\dagger a + \frac{1}{2}\right) a\psi = \left(a^\dagger a a + \frac{1}{2} a\right) \psi,$$

we put a in evidence on the left with the help of the commutation rule:

$$= \left(a a^\dagger a - a + \frac{1}{2} a\right) \psi = a \left(a^\dagger a + \frac{1}{2} - 1\right) \psi = (\epsilon - 1) a\psi.$$

Evidently, $a^\dagger$ creates excitations and a destroys them; in field theories like Quantum Electrodynamics (QED) the oscillator is not observable at all. We observe, create, and absorb the excitations, which are photons! No excitations can be destroyed in the ground state. Accordingly, the ground state is given by $a\psi_0 = 0$. We can verify that this is the same ground state as the one in Eq. (16.42). Indeed, in x_0 units,

$$0 = \left(x + i\frac{p}{\hbar}\right) \psi_0 = \left(x + \frac{d}{dx}\right) \psi_0.$$

Integrating, we recover $\psi_0 = e^{-\frac{x^2}{2}}$.

In the original variables,

$$a^\dagger = \frac{1}{\sqrt{2}} \left[\frac{x}{x_0} - \frac{i x_0 p}{\hbar}\right],$$

$$a = \frac{1}{\sqrt{2}} \left[\frac{x}{x_0} + \frac{i x_0 p}{\hbar}\right].$$

(16.23)

The inverse transformation is:

$$x = \frac{x_0}{\sqrt{2}} (a + a^\dagger),$$

$$p = \frac{-i\hbar}{x_0 \sqrt{2}} (a - a^\dagger).$$

(16.24)

One can check that $[p, x]_- = -i\hbar$.

All the eigenstates can be determined in this formalism by

$$a^\dagger \psi_n = u_n \psi_{n+1}, \quad a\psi_n = v_n \psi_{n-1},$$

but we must still find the constants u_n, v_n. To this end, we write the normalization condition $\langle \psi_{n+1} | \psi_{n+1} \rangle = 1$; we obtain $u_n^2 = \langle a^\dagger \psi_n | a^\dagger \psi_n \rangle = \langle \psi_n | a a^\dagger \psi_n \rangle = (1 + n)$. So,

$$\psi_{n+1} = \frac{1}{\sqrt{n+1}} a^\dagger \psi_n,$$

(16.25)

that is,

$$|\psi_n\rangle = \frac{1}{\sqrt{n!}}(a^\dagger)^n|\psi_0\rangle. \tag{16.26}$$

Similarly, setting $a\psi_n = v_n\psi_{n-1}$, we obtain from $v_n^2 = \langle a\psi_n|a\psi_n\rangle$ that

$$a\psi_n = \sqrt{n}\psi_{n-1}. \tag{16.27}$$

Thus, $\hat{x}\psi_n = \frac{x_0}{\sqrt{2}}(\sqrt{n}\psi_{n-1} + \sqrt{n+1}\psi_{n+1})$ and the matrix of $\hat{x}$ has elements

$$x_{mn} = \frac{x_0}{\sqrt{2}}(\sqrt{n}\delta_{m,n-1} + \sqrt{n+1}\delta_{m,n+1}). \tag{16.28}$$

Similarly,

$$p_{mn} = i\hbar\frac{1}{x_0\sqrt{2}}(-\sqrt{n}\delta_{m,n-1} + \sqrt{n+1}\delta_{m,n+1}).$$

We know that
$$([\hat{p},\hat{x}])_{mn} = (-i\hbar)_{mn} = -i\hbar\delta_{mn}.$$

We can verify that, through matrix multiplication, using the Einstein convention (sum over repeated indices),

$$([\hat{p},\hat{x}])_{mn} \equiv (px - xp)_{mn} \equiv p_{mk}x_{kn} - x_{mk}p_{kn} = -i\hbar\delta_{mn}. \tag{16.29}$$

Similarly, one can build the matrices of $\hat{x}^2, \hat{p}^2$ and the Hamiltonian in diagonal form:

$$H_{mn} = \left(\hat{n} + \frac{1}{2}\right)\hbar\omega\delta_{mn}.$$

It is now a simple exercise to find the eigenstates of the non-Hermitean annihilation operator. Using $a|n\rangle = \sqrt{n}|n-1\rangle$, one finds that $a|\psi_\lambda\rangle = \lambda|\psi_\lambda\rangle$ is solved by the normalized state

$$\psi_\lambda\rangle = \exp\left(-\frac{\lambda}{2}\right)\exp[\lambda a^\dagger]|0\rangle = \exp\left(-\frac{\lambda}{2}\right)\sum_{n=0}^{\infty}\frac{(\lambda a^\dagger)^n}{n!}|0\rangle.$$

Letting the creator operators act, we find:

$$|\psi_\lambda\rangle = \exp\left(-\frac{\lambda}{2}\right)\sum_{n=0}^{\infty}\frac{\lambda^n}{\sqrt{n!}}|n\rangle. \tag{16.30}$$

Actually ψ_λ is called a *coherent state* and is very interesting for many purposes. To start with, this solves the shifted oscillator. Suppose we replace the annihilation operator a by $s = a + \lambda$, where λ is a c-number (that is, not an operator). The consequence is that $s + s^\dagger = \sqrt{2}(q + \frac{\lambda + \lambda^*}{\sqrt{2}})$, that is, $\frac{\lambda + \lambda^*}{\sqrt{2}}$ is a shift of the origin; but since $[s, s^\dagger]_- = 1$, the new operators are canonically equivalent to the old ones. The canonically conjugated Hamiltonian is

$$\tilde{H} = \hbar\omega\left[s^\dagger s - \lambda^* s - \lambda s^\dagger + \lambda^2 + \frac{1}{2}\right] \tag{16.31}$$

and obviously has the same eigenvalues as H. However, the ground state $\tilde{\psi}$ of $\tilde{H}$ is not found by $s\tilde{\psi} = 0$ but by $(s - \lambda)\tilde{\psi} = 0$. Thus the ground state (i.e., vacuum) is a coherent state if the Hamiltonian undergoes a shift.

Problem 29   An oscillator is prepared in a state proportional to $|\varphi\rangle = (a^\dagger + a^{\dagger 2})|0\rangle$. Calculate $\langle x \rangle$.

Solution 29   The normalized state is $|\phi\rangle = \frac{1}{\sqrt{3}}(a^\dagger + a^{\dagger 2})|0\rangle$. Therefore, $\langle\phi|\hat{x}|\phi\rangle = \frac{1}{3}\frac{x_0}{\sqrt{2}}\langle 0|(a + a^2)(a + a^\dagger)(a^\dagger + (a^\dagger)^2)|0\rangle = \frac{4}{3}\frac{x_0}{\sqrt{2}}$.

The coherent state ψ_λ has further points of interest. The wave packet (16.30) may be rewritten more explicitly in terms of the harmonic oscillator wave functions as

$$\psi_\lambda(x, t) = \exp\left(-\frac{\lambda}{2}\right)\sum_{n=0}^{\infty}\frac{\lambda^n}{\sqrt{n!}}\psi_n(x, t); \tag{16.32}$$

and the sum can be worked out thanks to the identity

$$\sum_{n=0}^{\infty} H_n(z)\frac{t^n}{n!} = \exp(z^2 - (x - t)^2). \tag{16.33}$$

One obtains

$$\psi_\lambda(x, t) = \frac{1}{\sqrt{x_0\sqrt{\pi}}}\exp\left(-\frac{1}{2}\left(\frac{x}{x_0} - \lambda\sqrt{2}\right)^2\right)\exp\left(-\frac{i}{2}\varphi(t)\right), \tag{16.34}$$

where $\varphi(t)$ is a real phase. Thus,

$$|\psi(x, t)|^2 = \frac{1}{x_0\sqrt{\pi}}\exp\left[-\left(\frac{x - x_0\lambda\sqrt{2}\cos(\omega_0 t)}{x_0}\right)^2\right] \tag{16.35}$$

keeps its shape during the evolution and oscillates in a way resembling a classical oscillator. That is the reason why this is called a coherent state: it keeps oscillating in analogy with the classical oscillator. In the next section it will allow for a quantum field that is as close as possible to its classical approximation. The main source of coherent light is the Laser (see Sect. 25.4.1).

16.3 Photons

Consider the modes of the electromagnetic field in a box. In the Coulomb gauge $\phi = 0$, $div A = 0$, the vector potential corresponding to wave vector k and frequency ω_k is $u_k(r)e^{-i\omega_k t} + c.c.$, where $u_k(r)$ satisfies

$$\left(\nabla^2 + \frac{\omega^2}{c^2}\right)u_k(r) = 0$$

(i.e., the wave equation) and $\nabla u_k(r) = 0$ (the Coulomb gauge). Consider the electric field amplitude of any monochromatic solution. It can be written as

$$E(t) = \frac{a(t) + a^*(t)}{2}, \tag{16.36}$$

where the complex amplitude $a(t)$ is proportional to $\exp(-i\omega t)$. We can treat the monochromatic solution as a harmonic oscillator by setting $q = \frac{a(0)+a^*(0)}{2}$, $p = \frac{a(0)-a^*(0)}{2i}$; so, $E(t) = q\cos(\omega t) + p\sin(\omega t)$. Here, q and p are the so called quadrature components of the field. We have rewritten the monochromatic solution in terms of a harmonic oscillator that oscillates in amplitude space rather than in real space and has no elastic constant and no mass. Up to now the treatment is classical, but we are ready to quantize the field by imposing $[a, a^\dagger]_- = 1$. The quantum expression is

$$A(r, t) = \sum_k \sqrt{\frac{\hbar}{2\omega_k\epsilon_0}}[a_k u_k(r)e^{-i\omega_k t} + a_k^\dagger u_k^*(r)e^{i\omega_k t}], \tag{16.37}$$

where ϵ_0 is the electric permittivity of free space and $a_k^\dagger$, a_k creation and annihilation operators of free space, which merely replace dimensionless amplitudes in what would be a classical expression; the pre-factor is chosen in such a way that the energy $E = \int[\frac{\epsilon_0}{2}E^2 + \frac{B^2}{2\mu_0}]d^3r$ of a quantum is $\hbar\omega$, as it should be. Note that the oscillation amplitude of the field corresponding to the x_0 of the mechanical oscillator is thereby specified. Here μ_0 is the magnetic susceptivity and B the magnetic field; the magnetic term is equal to the electric one in vacuo. A rigorous derivation of the above is outside the scope of this book, but may be found in any book about Quantum Optics. Many consequences are relevant to the appreciation of Quantum Mechanics, however. The corresponding electric field is obtained by $E = -\frac{1}{c}\frac{\partial A}{\partial t}$,

$$E(r, t) = \sum_k \sqrt{\frac{\hbar\omega_k}{2\epsilon_0}}[a_k u_k(r)e^{-i\omega_k t} - a_k^\dagger u_k^*(r)e^{i\omega_k t}]. \tag{16.38}$$

First of all, we see that the field is an operator, as every observable should be. Moreover, it is evident that the vacuum average of each k component $E(k) =$

$\sqrt{\frac{\hbar\omega_k}{2\epsilon_0}}[a_k\boldsymbol{u}_k(\boldsymbol{r})e^{-i\omega_k t} - a_k^{\dagger}\boldsymbol{u}_k^*(\boldsymbol{r})e^{i\omega_k t}]$ vanishes, but $\boldsymbol{E}(\boldsymbol{k})^2$ does not. Physically this means that if a field component is measured it is found to oscillate around zero.

16.3.1 Zero Point Energy and Casimir Effect

The physical vacuum contains the zero-point oscillation and the zero-point energy of all normal modes of the electromagnetic field (and of any other physical field). The quantum vacuum is full of stuff in turmoil, and in any cube centimeter, there is infinite energy. This sounds amazing, but does not cause a difficulty like the ultraviolet catastrophe because it involves an infinite energy difference between the physical vacuum and an absolute void that does not exist in Nature. Then, one could believe that this is just formal. However, the zero point fields exist: while the average electric field is indeed zero, the average square electric field can be measured. These vacuum fields produce observable consequences. In Schrödinger's theory, as we shall see, the 2s and 2p levels of H are degenerate. The relativistic Dirac's theory removes some of this degeneracy, but $^2S_{1/2}$ and $^2P_{1/2}$ levels remain degenerate. The Lamb shift of atomic $^2S_{1/2}$ and $^2P_{1/2}$ levels was measured in 1947 with great precision as a splitting of about 1000 MHz. This shift can be calculated in QED (Quantum Electrodynamics) with similarly great precision. This is due to the fact that the atomic electrons are acted upon by the vacuum field in addition to the nuclear electric field, and the vacuum field couples differently with the two states.

Another important consequence is known as the Casimir effect. While we cannot eliminate the zero point fields, we can change the boundary conditions for the field, e.g., by placing two mirrors facing each other at a small distance. This produces an attractive potential that goes with the inverse fourth power of the distance between the mirrors.

Consider a pillbox with reflecting walls; let the basis be square with side L and the thickness be s. The modes with $k = (\frac{\pi a}{s}, \frac{\pi b}{s}, \frac{\pi c}{s})$ with integer a, b and c contribute $\hbar c k$ to the vacuum energy twice (because of the two polarizations). The sum $U(s)$ diverges. In reality, the reflecting walls can reflect only for $\hbar c k < \omega_p$, where ω_p is the plasma energy of the material, of order $10\,\text{eV}$, so the sum diverges to an even worse degree. But we are interested in the finite difference due to the presence of the walls. A serious calculation must take into account the dielectric properties of the material and of its surface. A model calculation can hardly be realistic, but it can show the finiteness of the effect and its s dependence. Using a cutoff length α to avoid the divergence,

$$U(s) = \pi\hbar c \sum_{a,b,c}^{\infty} \lambda(a, b, c) \exp(-\alpha\lambda(a, b, c)), \qquad (16.39)$$

where $\lambda(a, b, c) = \sqrt{\frac{a^2}{s^2} + \frac{b^2}{L^2} + \frac{c^2}{L^2}}$ is an inverse length. One wishes eventually to remove the cutoff with $\alpha \to 0$. It is evident that the result must be proportional to the area L^2 and for dimensional reasons. The summation gives us

$$U(s) = \frac{\hbar c \pi^2 L^2}{2} \left(\frac{d}{d\alpha}\right)^2 \left(\frac{1}{\alpha} \frac{1}{e^{\frac{\alpha}{s}}} - 1\right). \tag{16.40}$$

Now, if a cavity with a width W is divided in two halves by a mirror, the vacuum energy is $2U(\frac{W}{2})$. This diverges. Instead, if the mirror is at a distance s from one wall and $W - s$ from the other, the vacuum energy is $U(s) + U(W - s)$. All these energies diverge, but the energy cost of shifting the mirror from a distance s to the middle of the cavity is finite. It is the Casimir energy

$$C(s) = U(s) + U(W - s) - 2U(\frac{W}{2}) \sim \frac{\pi^2 \hbar c}{720 s^3}. \tag{16.41}$$

More precisely, the attractive force per unit area (1 cm^2) between perfect conductors at distance s is calculated to be $\frac{0.013}{s^4}$ dyne, and the effect has been confirmed experimentally. The numerical values in the above derivation cannot be taken too seriously. By varying the nature of the surfaces and angles, one can control the Casimir force, which in some cases, can also be repulsive, and perhaps one day, this will be used to operate nano-engines. The vacuum energy due to all possible fields is also suspected to be at the origin of the Λ term which produces the dark energy (Sect. 8.12), but no quantitative theory is available to get from the Planck length $\lambda_P = \sqrt{\frac{\hbar G}{c^3}} \sim 1.6 \quad 10^{-35} \text{ m}$ and the Planck mass $M_P \sim 2.2 \quad 10^{-8} \text{ kg}$ to the estimated energy density of dark energy $10^{-29} \frac{g}{cm^3}$.

16.3.2 Parametric Down-Conversion and Squeezed Light

In the above discussion of the harmonic oscillator, I have emphasized that in the quantum case, there is a fundamental length x_0, which characterizes the eigenstate wave functions in both real and momentum space. However, it is possible to manipulate the oscillator by putting it in a nonlinear medium in such a way that the wave functions are scaled. For instance, the wave function of the squeezed vacuum with squeezing parameter R is obtained from the oscillator ground state (16.8) by scaling

$$\psi_R(q) = \frac{\sqrt{R}}{\sqrt[4]{\pi}} e^{-\frac{(Rq)^2}{2}}. \tag{16.42}$$

It is, of course, a normalized solution, i.e., $\int_{-\infty}^{\infty} \psi_R(x)^2 dx = 1$. The scaling modifies the variances, with $\Delta q^2 = \frac{1}{2R^2}$, $\Delta p^2 = \frac{R^2}{2}$, in such a way that the uncertainty in one of the quadratures (x or p) is decreased but the uncertainty in the other quadrature is increased so that the Heisenberg uncertainty principle is preserved.

It is possible to squeeze the position of an oscillator by producing the state $|\psi\rangle = |0\rangle - \frac{s}{\sqrt{2}}|2\rangle$. In this state, $\langle\psi|x^2|\psi\rangle = \frac{1}{2} - s$, so the spread in x is reduced.

With two modes a and b the vacuum

$$\Psi_{00} = \frac{1}{\sqrt{\pi}} \exp\left(-\frac{q_a^2 + q_b^2}{2}\right) = \frac{1}{\sqrt{\pi}} \exp\left(-\frac{(q_a + q_b)^2}{4}\right) \exp\left(-\frac{(q_a - q_b)^2}{4}\right)$$

can be squeezed to become

$$\Psi_{00sq} = \frac{1}{\sqrt{\pi}} \exp\left(-\frac{(q_a + q_b)^2}{4R^2}\right) \exp\left(-\frac{R^2(q_a - q_b)^2}{4}\right),$$

where R is called the squeezing parameter. One way to do that is by using Parametric Down Conversion. This is a nonlinear optical phenomenon occurring in crystals like beta Barium Borate under intense laser light. A fraction $f \ll 1$ of the photons split into pairs of photons having half frequency, and thus half of the energy and momentum each. The two photons may have the same polarization (Type I correlation) or perpendicular polarizations (Type II). Let us consider the latter case. The half frequency mode oscillators are no longer in the vacuum state but in a state

$$|S\rangle = |00\rangle + f|11\rangle.$$

Let a and b denote the two polarizations and q_a and q_b denote the dimensionless coordinates of the oscillators $q_a = \frac{a+a^\dagger}{\sqrt{2}}, q_b = \frac{b+b^\dagger}{\sqrt{2}}$. The variance of $q_a - q_b$ is given by $\langle S|(q_a - q_b)^2|S\rangle$. Expanding, one finds that the only terms that contribute are those in $aa^\dagger + a^\dagger a + bb^\dagger + b^\dagger b - 2ab - 2a^\dagger b^\dagger$ and the variance turns out to be $1 - 2f + O(f^2)$.

Similarly, the variance of $p_a - p_b$ is given by $\langle S|(p_a - p_b)^2|S\rangle = 1 + 2f + O(f^2)$, as one can verify by setting $p_a = \frac{-i(a-a^\dagger)}{\sqrt{2}}, p_b = \frac{-i(b-b^\dagger)}{\sqrt{2}}$ and working out the calculation in the same way. This calculation implies an increased correlation between the coordinates and a decreased correlation between the momenta. This implies that the wave function of the two mode vacuum state has become a *squeezed vacuum*.

Squeezed coherent states have also been obtained, e.g., by nonlinear optics techniques. Squeezed states have important applications in precision measurements (clocks, interferometers) and in gravitational wave detectors.

Chapter 17
Stationary States of One Particle in 3 Dimensions

> *The partial differential equations are quite a bit harder to solve than the ordinary ones, unless the symmetry allows us to separate the variables. Fortunately, among the most interesting stationary problems, there are some that can be solved analytically.*

17.1 Separation of Variables in Cartesian Coordinates

The 3-dimensional plane wave is the product of one-dimensional plane waves and the kinetic energy is the sum of the contributions of motions along x, y, z. More generally, the problem is separable into Cartesian coordinates if the potential energy is of the form

$$V(x, y, z) = U_x(x) + U_y(y) + U_z(z), \tag{17.1}$$

where $U_x(x), U_y(y)$ e $U_z(z)$ are arbitrary functions. In this case, the motions along the three directions are independent. In the Schrödinger stationary state equation

$$\left[\frac{\partial^2}{\partial x^2} + \frac{\partial^2}{\partial y^2} + \frac{\partial^2}{\partial z^2} \right] \Psi(x, y, z) + V(x, y, z)\Psi(x, y, z) = E\Psi(x, y, z),$$

one puts the factored solution

$$\Psi(x, y, z) = X(x)Y(y)Z(z); \tag{17.2}$$

dividing by $\Psi(x, y, z)$,

© Springer International Publishing AG, part of Springer Nature 2018
M. Cini, *Elements of Classical and Quantum Physics*,
UNITEXT for Physics, https://doi.org/10.1007/978-3-319-71330-4_17

$$\frac{1}{X(x)}\frac{\delta^2 X}{\delta x^2} + U_x(x) + \frac{1}{Y(y)}\frac{\delta^2 Y}{\delta y^2} + U_z(z) + \frac{1}{Z(z)}\frac{\delta^2 Z}{\delta z^2} + U_z(z) = E.$$

Therefore,

$$\begin{cases} \dfrac{1}{X(x)}\dfrac{\delta^2 X}{\delta x^2} + U_x(x) = \epsilon_x \Rightarrow \dfrac{\delta^2 X}{\delta x^2} + U_x(x)X(x) = \epsilon_x X(x), \\[2mm] \dfrac{1}{Y(y)}\dfrac{\delta^2 Y}{\delta y^2} + U_y(y) = \epsilon_y \Rightarrow \dfrac{\delta^2 Y}{\delta y^2} + U_y(y)Y(y) = \epsilon_y Y(y), \\[2mm] \dfrac{1}{Z(z)}\dfrac{\delta^2 Z}{\delta z^2} + U_z(z) = \epsilon_z \Rightarrow \dfrac{\delta^2 Z}{\delta z^2} + U_z(z)Z(z) = \epsilon_z Z(z), \\[2mm] \epsilon_x + \epsilon_y + \epsilon_z = E, \end{cases}$$

and the problem is broken. In this way, one gets special solutions, but since they form a complete set (Second Postulate) the most general solution can be expanded as a convergent series of those special solutions.

Box

Consider a particle confined in a potential

$$V(x, y, z) = \begin{cases} 0, & -\frac{L_x}{2} < x < \frac{L_x}{2}, \; -\frac{L_y}{2} < y < \frac{L_y}{2}, \; -\frac{L_z}{2} < z < \frac{L_z}{2}, \\ \infty & \text{otherwise.} \end{cases}$$

V is of the form (17.1). Separating the variables, one finds the eigenfunctions

$$\psi_{n_x n_y n_z}(x, y, z) = u_{n_x}(x)u_{n_y}(y)u_{n_z}(z)$$

and the energy eigenvalues

$$E_{n_x n_y n_z}(x, y, z) = \epsilon_{n_x} + \epsilon_{n_y} + \epsilon_{n_z},$$

with

$$u_{n_x} = \sqrt{\frac{2}{L_x}} \sin\left[\frac{\pi n_x}{L_x}\left(x + \frac{L_x}{2}\right)\right],$$
$$\epsilon_{n_x} = \hbar^2 \frac{\pi^2 n_x^2}{2mL_x^2},$$

etc. The motions along x, y and z are independent; while classically one multiplies the probabilities, here the amplitudes are multiplied. It ia *statistical independence*.

In the cubic case $L_x = L_y = L_z = L$, many levels are *degenerate*, that is, several choices of the qquantum numbers give the same energy. For instance, $E_{511} = E_{151} = E_{115} = E_{333} = 27\hbar^2 \frac{\pi^2}{2mL^2}$. This is a very important and general remark. More symmetry generally implies more degeneracy, when there are several non-commuting symmetry operations that commute with the Hamiltonian. This idea is fully developed by Group Theory.

Oscillator in 3d

Similarly, one solves the problem

$$H = \frac{p^2}{2m} + \frac{1}{2}m\left[\omega_x^2 x^2 + \omega_y^2 y^2 + \omega_z^2 z^2\right],$$

still with $E_{n_x n_y n_z}(x, y, z) = \epsilon_{n_x} + \epsilon_{n_y} + \epsilon_{n_z}$ and high degeneracy in the isotropic case $\omega_x = \omega_y = \omega_z = \omega$.

Landau Levels

The Hamiltonian of a charged particle, subject to the Lorentz force, is

$$\hat{H} = \frac{(\vec{p} - \frac{e}{c}\vec{A})^2}{2m}. \tag{17.3}$$

We put the field along the z axis; we use the Landau gauge[1]

$$\vec{A} = (-y, 0, 0)B.$$

So,

$$\hat{H} = \frac{(p_x + \frac{eBy}{c})^2 + p_y^2 + p_z^2}{2m}.$$

Here, p_x and p_z are conserved, and we can take the wave function in the form

$$\psi = e^{i[p_x x + p_z z]}\chi(y);$$

along z, the particle is free. Take, for simplicity, $p_z = 0$ since the free motion is uninteresting, and $p_x = 0$ ince this simply involves a shift in the origin of y; the motion along y is quantized[2] and harmonic with frequency $\omega_L = \frac{eB}{mc}$. The quantized levels are called *Landau levels*.

17.2   Separation of Variables in Spherical Coordinates

When the potential energy depends only on the distance from a point, which is conveniently taken as the origin (central problems) the S.E. is separable in spherical coordinates; the transformations from Cartesian to spherical coordinates and back are:

---

[1]Lev Davidovic Landau (Baku 1908- Moscow 1968), was the most important Soviet physicist. He won the Nobel prize in 1962 for his works on superfluid He.

[2]The special role of the y direction is due to the gauge; by a gauge transformation we can rotate the pair x, y as we like.

$$\begin{cases} x = r\sin\theta\cos\phi, \\ y = r\sin\theta\sin\phi, \\ z = r\cos\theta, \end{cases} \quad \begin{cases} r = \sqrt{x^2 + y^2 + z^2}, \\ \theta = \arccos(\frac{z}{\sqrt{x^2+y^2+z^2}}), \\ \phi = \arctan(\frac{y}{x}). \end{cases} \tag{17.4}$$

By using the chain rule and elementary, lengthy algebra (which is best done by computer) one finds the momentum components

$$\begin{cases} ip_x = \sin(\theta)\cos(\phi)\dfrac{\partial}{\partial r} + \dfrac{\cos(\phi)\cos(\theta)}{r}\dfrac{\partial}{\partial\theta} - \dfrac{\sin(\phi)}{r\sin(\theta)}\dfrac{\partial}{\partial\phi} \\[2mm] ip_y = \sin(\theta)\sin(\phi)\dfrac{\partial}{\partial r} + \dfrac{\sin(\phi)\cos(\theta)}{r}\dfrac{\partial}{\partial\theta} + \dfrac{\cos(\phi)}{r\sin(\theta)}\dfrac{\partial}{\partial\phi} \\[2mm] ip_z = \cos(\theta)\dfrac{\partial}{\partial r} - \dfrac{\sin(\theta)}{r}\dfrac{\partial}{\partial\theta}. \end{cases} \tag{17.5}$$

We know already that $L_z = xp_y - yp_x = -i\frac{\partial}{\partial\phi}$. Moreover, we find

$$L_\pm \equiv L_x \pm iL_y = e^{\pm i\phi}\left[\frac{\partial}{\partial\theta} \pm i\cot(\theta)\frac{\partial}{\partial\phi}\right]$$

and

$$-L^2 = \frac{1}{\sin(\theta)}\frac{\partial}{\partial\theta}\sin(\theta)\frac{\partial}{\partial\theta} + \frac{1}{\sin(\theta)^2}\frac{\partial^2}{\partial\phi^2}.$$

Finally, the Laplacian in spherical coordinates is:

$$\nabla^2 = \frac{1}{r^2}\frac{\partial}{\partial r}\left(r^2\frac{\partial}{\partial r}\right) + \frac{1}{r^2\sin(\theta)}\frac{\partial}{\partial\theta}\left(\sin(\theta)\frac{\partial}{\partial\theta}\right) + \frac{1}{r^2\sin^2(\theta)}\frac{\partial^2}{\partial\phi^2}. \tag{17.6}$$

If $V = V(r)$, the stationary state equation

$$\left[-\frac{\hbar^2}{2m}\nabla^2 + V(r)\right]\Psi(r, \theta, \phi) = E\Psi(r, \theta, \phi)$$

splits in a radial equation and an angular one by putting

$$\Psi(r, \theta, \phi) = R(r)Y_{lm}(\theta, \phi).$$

Let $|\theta, \phi\rangle$ denote the amplitude that the system has a sharp orientation in space in the direction with well-defined angles θ, ϕ. This may be the angular factor of a wave function of a particle which separates in spherical coordinates. The *Spherical Harmonics*

$$Y_{lm}(\theta, \phi) = \langle\theta\phi|l, m\rangle$$

satisfy the angular equation

$$\frac{1}{\sin(\theta)}\frac{\partial}{\partial\theta}\left(\sin(\theta)\frac{\partial Y}{\partial\theta}\right) + \frac{1}{\sin^2(\theta)}\frac{\partial^2 Y}{\partial\phi^2} = -\lambda Y, \tag{17.7}$$

while the *radial function* is obtained by solving

$$\frac{d}{dR}\left(r^2\frac{dR}{dr}\right) - \frac{2mr^2}{\hbar^2}[V(r) - E]R = \lambda R. \tag{17.8}$$

17.2.1 Spherical Harmonics

The angular equation (17.7) separates if we put

$$Y_{lm}(\theta,\phi) = \Theta_{lm}(\theta)\Phi_m(\phi), \tag{17.9}$$

where

$$\frac{d^2\Phi_m}{d\phi^2} = -m^2\Phi_m \;\Rightarrow\; \Phi_m = e^{im\phi}, \text{ integer m.} \tag{17.10}$$

We have already encountered the eigenfunctions of L_z. The Θ equation, known as the Legendre equation, reads as:

$$\sin(\theta)\frac{d}{d\theta}\left(\sin(\theta)\frac{d\Theta_{lm}}{d\theta}\right) + \{\lambda\sin^2(\theta) - m^2\}\Theta_{lm} = 0. \tag{17.11}$$

where we know from Sect. 13.1 that $\lambda = l(l+1)$. For $l = m = 0$, this is simply $\frac{\partial}{\partial\theta}(\sin(\theta)\frac{\partial}{\partial\theta})\Theta(\theta) = 0$ and is solved by $\Theta = 1$. For $l = 1$, $\lambda = 2$, one readily finds that for $m = 0$, $\Theta = \cos(\theta)$, and for $m = \pm1$, $\Theta = \sin(\theta)$. In general, it is elementary to find that the solution is a polynomial in $\cos(\theta)$ for $m = 0$, otherwise it is a polynomial in $\cos(\theta)$ and $\sin/\theta$). Indeed, the general solution is $\Theta(\theta) = P_l^m(\cos(\theta))$, where

$$P_l^m(x) = (-1)^m(1-x^2)^{\frac{m}{2}}\frac{d^m}{dx^m}P_l(x) \tag{17.12}$$

is an *associated Legendre polynomial*. The Legendre polynomials $P_l(x)$ satisfy

$$\frac{d}{dx}[(1-x^2)\frac{d}{dx}P_m(x)] + m(m+1)P_m(x) = 0. \tag{17.13}$$

In particular, $P_0(x) = 1$, $P_1(x) = x$, $P_2(x) = \frac{3x^2-1}{2}$, $P_3(x) = \frac{5x^3-3x}{2}$. They can be computed from the Rodriguez formula

$$P_n(x) = \frac{1}{2^n n!}\frac{d^n}{dx^n}(x^2-1)^n.$$

Then, $Y_{lm}(\theta, \phi) = \langle \theta, \phi | l, m \rangle$ is the amplitude in the direction defined by θ and ϕ of an eigenfunction of L^2 with eigenvalue $\hbar^2 l(l+1)$ and of L_z with eigenvalue $m\hbar$.

The spherical harmonics are tabulated in terms of associated Legendre functions, which are also easily available.

l	m	Y_{lm}	$Y_{lm}r^l$
0	0	$\frac{1}{\sqrt{4\pi}}$	$\frac{1}{\sqrt{4\pi}}$
1	0	$\sqrt{\frac{3}{4\pi}}\cos\theta$	$\sqrt{\frac{3}{4\pi}}\cos\theta z$
1	±1	$\pm\sqrt{\frac{3}{8\pi}}\sin\theta e^{\pm i\phi}$	$\pm\sqrt{\frac{3}{8\pi}}(x\pm iy)$
2	0	$\sqrt{\frac{5}{16\pi}}(3\cos^2\theta-1)$	$\sqrt{\frac{5}{16\pi}}(3z^2-r^2)$
2	±1	$\pm\sqrt{\frac{15}{8\pi}}\cos\theta\sin\theta e^{\pm i\phi}$	$\pm\sqrt{\frac{15}{8\pi}}z(x\pm iy)$
2	±2	$\sqrt{\frac{15}{32\pi}}\sin^2\theta e^{\pm 2i\phi}$	$\sqrt{\frac{15}{32\pi}}(x\pm iy)^2$

The closure relation is[3]

$$\sum_{k=0}^{\infty}\sum_{m=-k}^{k} Y_{km}^*(\theta_1,\phi_1)Y_{km}(\theta_2,\phi_2) = \delta(\Omega_1-\Omega_2)$$
$$= \frac{\delta(\theta_1-\theta_2)\delta(\phi_1-\phi_2)}{|\sin(\theta_1)|}.$$

This means that any good $f(\theta,\phi)$ can be developed:

$$f(\theta,\phi) = \sum_{k=0}^{\infty}\sum_{m=-k}^{k}\int Y_{km}^*(\theta_1,\phi_1)f(\theta_1,\phi_1)d\Omega_1 Y_{km}(\theta,\phi). \qquad (17.14)$$

Parity is the operation $P : (x,y,z) \to (-x,-y,-z)$. In order to change the sign of $z = r\cos(\theta)$, one can do $\theta \to \pi-\theta$, which implies $\sin(\theta) \to \sin(\theta)$; to change the sign of x and y, one can do $\phi \to \phi+\pi$. So,

$$P : \phi \to \phi+\pi, \quad \theta \to \pi-\theta.$$

Under the action of P, $Y_{ll} = e^{li\phi}(\sin^l(\theta))$ takes a factor $e^{il\pi} = (-1)^l$. This result cannot depend on m, since $L_\pm$, like the components of the pseudovector angular momentum, are even under parity.

---

[3]Recall that

$$\delta(g(x)) = \frac{\sum_\alpha \delta(x-x_\alpha)}{\left|\frac{dg}{dx}\right|}.$$

17.3 Central Field

In a central potential $V(r)$, we use spherical coordinates and seek special solutions of the form

$$\psi(r, \theta, \phi) = R(r)Y_{lm}(\theta, \phi);$$

then,

$$-\nabla^2\psi = \left[\frac{-1}{r^2}\frac{\partial}{\partial r}\left(r^2\frac{\partial}{\partial r}\right) + \frac{L^2}{\hbar^2 r^2}\right]\psi$$

$$= \left[\frac{-1}{r^2}\frac{\partial}{\partial r}\left(r^2\frac{\partial}{\partial r}\right) + \frac{l(l+1)}{r^2}\right]\psi.$$

Removing the spherical harmonic, we are left with the *radial equation*

$$\frac{-1}{r^2}\frac{\partial}{\partial r}\left(r^2\frac{\partial R}{\partial r}\right) + \frac{l(l+1)}{r^2}R + \frac{2mV(r)}{\hbar^2}R = k^2R(r), \qquad (17.15)$$

where

$$k^2 = \frac{2mE}{\hbar^2}.$$

We gain some simplification by setting

$$R = \frac{u}{r};$$

then,

$$\frac{\hbar^2}{2m}\left[\frac{-1}{r}\frac{d^2u}{dr^2} + \frac{l(l+1)}{r^2}\frac{u}{r}\right] + V(r)\frac{u}{r} = E\frac{u}{r},$$

that is,

$$-\frac{\hbar^2}{2m}\frac{d^2u}{dr^2} + \frac{\hbar^2}{2m}\frac{l(l+1)}{r^2}u(r) + V(r)u(r) = Eu(r).$$

This is a 1d problem with a modified potential. The term $l(l+1)$ is a centrifugal potential and forces the wave function to vanish at the centre unless $l = 0$. It is convenient to write

$$-\frac{d^2u}{dr^2} + \frac{l(l+1)}{r^2}u(r) + \frac{2m}{\hbar^2}V(r)u(r) = k^2u(r).$$

Free Particle Again!

For $V = 0$ we get

$$\frac{d^2u}{dr^2} = \frac{l(l+1)}{r^2}u(r) - k^2u(r).$$

Equation (17.15) with $\rho = kr$ is the *spherical Bessel equation*

$$2\rho R' + \rho^2 R'' + (\rho^2 - l(l+1))R = 0,$$

and the solutions are

$$j_l(\rho) = \rho^l \left(-\frac{1}{\rho}\frac{d}{d\rho}\right)^l \frac{\sin(\rho)}{\rho}.$$

The free particle eigenstates of H with diagonal L^2 and L_z are quite different from the De Broglie eigenstates of H and momentum. For $l = 0$,

$$u_{l=0}(r) = \sin(kr) \Longrightarrow R_{l=0}(r) = \frac{\sin(kr)}{r}.$$

The alternative solution $u_{l=0}(r) = \cos(kr)$ must be discarded, because $R_{l=0}(r)$ would blow up at the origin. We can expand the plane wave in spherical solutions.
   This is simplest for e^{ikz}, since we need only harmonics with $m = 0$, $Y_{l0}(\theta, \phi) = \sqrt{\frac{2l+1}{4\pi}} P_l(\cos(\theta))$, with P_l Legendre polynomials. One can show that

$$e^{ikz} = \sum_{l=0}^{\infty}(-i)^l(2l+1)P_l(\cos(\theta))(\frac{r}{k})^l(\frac{1}{r}\frac{d}{dr})^l \frac{\sin(kr)}{r}.$$

Spherical Potential Well

By imposing boundary conditions for $r = R_0$, one finds the energy eigenvalues for a particle in a spherical potential well

$$V(r) = \begin{cases} V = 0 & r < a, \\ V = \infty & r > a. \end{cases}$$

In the case $l = 0$, one finds that

$$E_{n0} = \frac{\hbar^2\pi^2}{2ma^2}n^2, \; n = 1, 2, 3, \ldots$$

For general l the eigenvalues E_{nl} are

$$E_{nl} = \frac{\hbar^2 x_{nl}^2}{2ma^2},$$

where x_{nl} is the n-th root of $j_l(x) = 0$ and j_l is the spherical Bessel function.

17.4 The Hydrogenoid Atom

We now solve the Schrödinger equation for the bound states of the H atom and one-electron atoms with the nuclear charge Ze. It is fair to mention, however, that these results were first obtained by Pauli[4] in a masterpiece of a paper before the formulation of the Schrödinger equation, by using the conservation of the quantum Runge–Lenz vector

$$\mathbf{R} = \frac{\mathbf{p} \wedge \mathbf{L} - \mathbf{L} \wedge \mathbf{p}}{2m} - k\frac{\mathbf{r}}{r}, \tag{17.16}$$

where k is the coefficient of $\frac{1}{r}$ in the Hamiltonian. It differs from the classical expression (17.16) in order to preserve the property that it commutes with the Hamiltonian. For $V(r) = -\frac{Ze^2}{r}$, the radial equation reads as

$$-\frac{d^2u}{dr^2} + \frac{l(l+1)}{r^2}u - \frac{2m}{\hbar^2}\frac{Ze^2}{r}u = \frac{2mE}{\hbar^2}u.$$

The classical Kepler problem has no length scale, and the orbit can have any size. In Quantum Mechanics, it is the existence of $\hbar$ that determines the atomic radius. Since $\frac{d^2u}{dr^2}$ is an inverse square length, the characteristic size is

$$a_0 = \frac{\hbar^2}{me^2Z} = \frac{a_B}{Z},$$

where a_B is the so-called *Bohr radius*; we must solve

$$-\frac{d^2u}{dr^2} + \frac{l(l+1)}{r^2}u - \frac{2}{a_0 r}u = \frac{2mE}{\hbar^2}u.$$

It turns out that $a_B = 0.529$ Angström, where 1 Angström $=10^{-8}$ cm. It is convenient to put the equation in a dimensionless form by setting

$$\rho = \frac{r}{a_0}, \quad \epsilon = \frac{2ma_0^2 E}{\hbar^2} < 0 \tag{17.17}$$

and multiplying by a_0^2. The dimensionless form is

$$u'' + \epsilon u(\rho) = [-\frac{2}{\rho} + \frac{l(l+1)}{\rho^2}]u(\rho). \tag{17.18}$$

---

[4]W. Pauli, Z. Phys. 36, 336 (1926).

Special Case

The case[5] $l = 0$ is simplest and for $\rho \to \infty$, by approximating by $u'' + \epsilon u(\rho) \sim 0$, one finds the trend $u \to e^{-\rho\sqrt{-\epsilon}}$: this prompts the trial solution

$$u = \rho e^{-\lambda\rho}, \quad \lambda = \sqrt{-\epsilon}.$$

It turns out that

$$u' = (1 - \lambda\rho)e^{-\lambda\rho},$$
$$u'' = (-2\lambda + \lambda^2\rho)e^{-\lambda\rho} = (\tfrac{-2\lambda}{\rho} + \lambda^2)\rho e^{-\lambda\rho}.$$

So, $\epsilon = -1$. These results concern the ground state, as we shall see, and lead to

$$R(r) \propto e^{-\frac{r}{a_0}} = e^{-\frac{Zr}{a_B}}, \quad E = -\frac{\hbar^2}{2\,ma_0^2} = -\frac{1}{2}\frac{me^4 Z^2}{\hbar^2}.$$

Note the strong Z- dependence. Moreover, $E = -\frac{Ze^2}{2a_0}$.

Normalization

Since $\frac{2}{\sqrt{a_0^3}} e^{-\frac{r}{a_0}}$ is normalized with

$$\int_0^\infty dr\, r^2 |R_{1,0}|^2 = 1$$

and $Y_{00} = \frac{1}{\sqrt{4\pi}}$,

$$\psi = \frac{1}{\sqrt{\pi a_0^3}} e^{-\frac{r}{a}}.$$

Excited Bound States

From (17.18), we see that even if $l > 0$, the asymptotic trend for $\rho \to \infty$, remains $u \to e^{-\rho\sqrt{-\epsilon}}$; the short distance behavior changes at short distances from the centre because of the centrifugal barrier, and becomes

$$u'' \sim \frac{l(l+1)}{\rho^2} u(\rho), \quad \rho \to 0.$$

Therefore, $u \to \rho^{l+1}$. Then, we set
$\lambda = \sqrt{-\epsilon}$, with

$$u = e^{-\lambda\rho}\rho^{l+1}\left[c_0 + c_1\rho + \ldots + c_{n_r}\rho^{n_r}\right]$$
$$= e^{-\lambda\rho}\left[c_0\rho^{l+1} + c_1\rho^{l+2} + \ldots + c_{n_r}\rho^{l+1+n_r}\right],$$

---

[5]The states with $l = 0, 1, 2, 3, \ldots$, are called s, p, d, f, g, h, i, etc.,and so on, in alphabetic order.

that is,

$$u(\rho) = e^{-\lambda\rho} \sum_{\nu=0}^{n_r} c_\nu \rho^{\nu+l+1}. \tag{17.19}$$

Therefore, the solution is of the form

$$u = e^{-\lambda\rho} \rho^{l+1} f^{(n_r,l)}(\rho), \tag{17.20}$$

where n_r is the *radial quantum number*. We shall find that $n_r \to \infty$ cannot be accepted, and $f^{(n_r,l)}(\rho)$ is a polynomial.

Then, differentiating (17.19),

$$u' = e^{-\lambda\rho} \sum_{\nu=0}^{n_r} c_\nu [-\lambda\rho^{\nu+l+1} + (\nu+l+1)\rho^{\nu+l}].$$

Differentiating again,

$$u'' = e^{-\lambda\rho} \sum_{\nu=0}^{n_r} c_\nu \{-\lambda[-\lambda\rho^{\nu+l+1} + (\nu+l+1)\rho^{\nu+l}] \\ -\lambda(\nu+l+1)\rho^{\nu+l} + (\nu+l+1)(\nu+l)\rho^{\nu+l-1}\}.$$

Let us simplify:

$$u'' = e^{-\lambda\rho} \sum_{\nu=0}^{n_r} c_\nu \{\lambda^2 \rho^{\nu+l+1} - 2\lambda(\nu+l+1)\rho^{\nu+l} + (\nu+l)(\nu+l+1)\rho^{\nu+l-1}\}.$$

The first term is $\lambda^2 \sum c_\nu \rho^{\nu+l+1} e^{-\lambda\rho} = -\epsilon u$. Therefore,

$$u'' + \epsilon u = e^{-\lambda\rho} \left\{ -2\lambda \sum_{\nu=0}^{n_r} c_\nu (\nu+l+1)\rho^{\nu+l} + \sum_{\nu=0}^{n_r} c_\nu (\nu+l)(\nu+l+1)\rho^{\nu+l-1} \right\};$$

we shift the second sum by renaming $\nu \to \nu+1 : \sum_{\nu=a}^{b} f(\nu) = \sum_{\nu+1=a}^{b} f(\nu+1) = \sum_{\nu=a-1}^{b-1} f(\nu+1)$. To avoid complications with the first and last terms, we put $c_\nu = 0$ for $\nu < 0$ and for $\nu > n_r$. So,

$$u'' + \epsilon u = e^{-\lambda\rho} \left\{ -2\lambda \sum_{\nu=0}^{n_r} c_\nu (\nu+l+1)\rho^{\nu+l} \right. \\ \left. + \sum_{\nu=-1}^{n_r-1} c_{\nu+1}(\nu+l+1)(\nu+l+2)\rho^{\nu+l} \right\}.$$

Collecting equal powers ρ with

$$u'' + \epsilon u = e^{-\lambda\rho} \left\{ \sum_{\nu=-1}^{n_r} \rho^{\nu+l} \left[-2\lambda c_\nu (\nu + l + 1) \right. \right.$$
$$\left. \left. + c_{\nu+1}(\nu + l + 1)(\nu + l + 2)\right] \right\}. \tag{17.21}$$

But (17.18) implies that $u'' + \epsilon u = [-\frac{2}{\rho} + \frac{l(l+1)}{\rho^2}]u(\rho)$; inserting (17.19) and expanding, the rhs reads as:

$$\frac{-2}{\rho}u + \frac{l(l+1)}{\rho^2}u = e^{-\lambda\rho} \left[-2 \sum_{\nu=0}^{n_r} c_\nu \rho^{\nu+l} + l(l+1) \sum_{\nu=0}^{n_r} c_\nu \rho^{\nu+l-1} \right].$$

Now, shifting the sum again (with $c_{-1} = 0$), one finds

$$e^{-\lambda\rho} \left[-2 \sum_{\nu=0}^{n_r} c_\nu \rho^{\nu+l} + l(l+1) \sum_{\nu=-1}^{n_r-1} c_{\nu+1} \rho^{\nu+l} \right]$$
$$= e^{-\lambda\rho} \sum_{\nu=-1}^{n_r} \rho^{\nu+l} \left[-2c_\nu + l(l+1)c_{\nu+1} \right].$$

Equating with (17.21),

$$-2\lambda c_\nu (\nu + l + 1) + c_{\nu+1}(\nu + l + 1)(\nu + l + 2) = -2c_\nu + l(l+1)c_{\nu+1},$$

and, collecting terms,

$$c_{\nu+1}[(\nu + l + 1)(\nu + l + 2) - l(l+1)] + 2c_\nu[1 - \lambda(\nu + l + 1)] = 0.$$

We need to simplify the coefficient of $c_{\nu+1}$. Set $a = \nu + 1$; we are left with

$$(\nu + l + 1)(\nu + l + 2) - l(l+1) = (a + l)(a + l + 1) - l(l+1)$$
$$= a^2 + a(l+1) + al + l(l+1) - l(l+1) = a(a + 2l + 1) = (\nu + 1)(\nu + 2 + 2l).$$

Now solve for $c_{\nu+1}$:

$$c_{\nu+1} = 2\frac{\lambda(\nu + l + 1) - 1}{(\nu + 1)(\nu + 2l + 2)}c_\nu. \tag{17.22}$$

This recurrence relation like (16.10), for $\nu \to \infty$, gives an exponential series $f^{(n_r,l)}(\rho) \sim e^{2\lambda\rho}$. The series must terminate, otherwise the exponential growth would take us outside the space of square integrable functions. So, we must find $c_{n_r+1} = 0$ for $\nu = n_r$, where the *radial quantum number* n_r is the degree of the polynomial. The condition is

$$\lambda n = 1, \tag{17.23}$$

where

$$n = n_r + l + 1 \tag{17.24}$$

defines the so-called *principal quantum number* n. The hydrogenic wave function is labelled by n and by the angular quantum numbers:

$$\psi_{n,l,m}(r, \theta, \phi) = R_{n,l}(r)Y_{l,m}(\theta, \phi).$$

The energy eigenvalues are $\epsilon = -\frac{1}{n^2}$, or, in view of (17.17),

$$E_n = -\frac{Z^2 e^2}{2 a_B n^2} = -\frac{13.59 Z^2}{n^2} eV. \tag{17.25}$$

The energies depend exclusively on n; therefore, $2s$ and $2p$ are degenerate, $3s$, $3p$, $3d$ are degenerate, and so on. The low entries of the energy level scheme are: $1s$, $2s$, $2p$, $3s$, $3p$, $3d$, ... as follows.

shell	n	level	angular momentum l	allowed m
K	1	1s	0	$m = 0$
L	2	2s	0	$m = 0$
		2p	1	$m = -1, 0, 1$
M	3	3s	0	$m = 0$
		3p	1	$m = -1, 0, 1$
		3d	2	$m = -2, -1, 0, 1, 2$
N	4	4s	0	$m = 0$
		4p	1	$m = -1, 0, 1$
		4d	2	$m = -2, -1, 0, 1, 2$
		4f	3	$m = -3, -2, -1, 0, 1, 2, 3$
...	...	...	...	...

These levels agree with experiment, apart from corrections that are small for Hydrogen. Relativistic corrections (spin-orbit interaction, dependence of mass from velocity, etc.) have a relative size $Z\alpha^2$, where Z is the atomic number and α is the fine structure constant. Other corrections arise from the electron spin magnetic moment, nuclear spin and quadrupole moment, and Quantum Electrodynamics. . Spectroscopically, the photon frequencies that are emitted and absorbed are

$$\nu_{mn} = R\left(\frac{1}{n^2} - \frac{1}{m^2}\right),$$

where R is the Rydberg constant. Emission occurs in the decay $m \to n$, while the opposite transition leads to absorption. The series of lines with $n = 2$ was discovered by Balmer in 1885, and starts in the visible spectrum, with the line H_α with ν_{23} in the red, the blue line H_β with ν_{24}, and the violet H_γ with ν_{25}; the series continues in the ultraviolet. Later, the ultraviolet Lyman series $n = 1$; then the infrared series with $m = 3, 4, 5$; however the explanation remained a mystery before the advent of Quantum Mechanics.[6] The explanation is:

---

[6] An ad hoc model by N. Bohr in 1913 gave the same levels but nothing else.

$$\hbar\omega_{mn} = E_m - E_n = 13,59\text{eV}\left(\frac{1}{n^2} - \frac{1}{m^2}\right).$$

Eliminating λ in (17.22) by the condition $\lambda n = 1$, one finds the recurrence relation with the coefficients of $f^{(n_r,l)}$:

$$c_{\nu+1} = \frac{2}{n}\frac{\nu + l + 1 - n}{(\nu + 1)(\nu + 2l + 2)}c_\nu. \tag{17.26}$$

This may be simplified by removing the factor $\frac{2}{n}$. To this end, we change scale. If a function $L(x) = \sum a_\nu x^\nu$ is defined by a recurrence relation $\frac{a_{\nu+1}}{a_\nu} = \xi(\nu)$, we can change scale by setting $L(x) \to L(sx)$, where s is arbitrary. Then,

$$L(sx) = \sum a'_\nu x^\nu, \quad a'_\nu = a_\nu s^\nu \implies \frac{a'_{\nu+1}}{a'_\nu} = s\,\xi(\nu).$$

Therefore, we set $f^{(n_r,l)}(\rho) = L(\frac{2\rho}{n})$, and $L_{n_r}^{2l+1}$ has a simplified recurrence relation

$$a_{\nu+1} = \frac{\nu + l + 1 - n}{(\nu + 1)(\nu + 2l + 2)}a_\nu. \tag{17.27}$$

This defines[7] the *associated Laguerre polynomials*. Now, from (11.12) one obtains

$$\begin{aligned} u(\rho) &= e^{-\lambda\rho}\rho^{l+1}L_{n_r}^{2l+1}(\tfrac{2\rho}{n}); \quad \lambda = 1/n, \\ R_{nl}(\rho) &= e^{-\frac{\rho}{n}}\rho^l L_{n_r}^{2l+1}(\tfrac{2\rho}{n}). \end{aligned} \tag{17.28}$$

Radial Functions

We can derive the radial functions from (17.27) by recalling that $\nu \le n_r$ and that $n = n_r + l + 1$. For $n = 1$, $n_r = l = 0$ e $\nu = 0$. Putting $a_0 = 1$ one gets $a_1 = 0$.

For $n = 2$, there are two cases: $n_r = 1, l = 0$, $n_r = 0, l = 1$.

For $n_r = 1, l = 0$, (17.27) becomes $a_{\nu+1} = \frac{\nu-1}{(\nu+1)(\nu+2)}a_\nu$. For $\nu = 0$, $a_1 = -a_0/2$; for $\nu = 1$ $a_2 = 0$. Therefore,

$$L_1^1(\rho) = 1 - \frac{1}{2}\rho,$$

---

[7]The *associated Laguerre polynomials*

$$L_{q-p}^p(x) = (-1)^p \left(\frac{d}{dx}\right)^p L_q(x)$$

are defined in terms of the *Laguerre polynomials*

$$L_q(x) = e^x \left(\frac{d}{dx}\right)^q \left(e^{-x}x^q\right).$$

and, up to a normalization constant,

$$R_{20}(\rho) = e^{-\frac{\rho}{2}}(1 - \frac{1}{2}\rho).$$

For $n_r = 0, l = 1$, there is only $\nu = 0$; Equation (17.27) $a_{\nu+1} = \frac{\nu}{(\nu+1)(\nu+4)}a_\nu$, yields $a_1 = 0, l = 1$ and so

$$R_{21}(\rho) = e^{-\frac{\rho}{2}}\rho.$$

Here, I report few radial functions normalized with:

$$\int_0^\infty |R_{n,l}(r)|^2 r^2 dr = 1.$$

$$R_{1,0} = \frac{2}{\sqrt{a_0^3}}e^{-\rho}$$

$$R_{2,0} = \frac{1}{\sqrt{2a_0^3}}(1 - \frac{\rho}{2})e^{-\frac{\rho}{2}}$$

$$R_{2,1} = \frac{1}{\sqrt{24a_0^3}}\rho e^{-\frac{\rho}{2}}$$

$$R_{3,0} = \frac{2}{\sqrt{27a_0^3}}(1 - \frac{2\rho}{3} + \frac{2\rho^2}{27})e^{-\frac{\rho}{3}}$$

$$R_{3,1} = \frac{8}{27\sqrt{6a_0^3}}(1 - \frac{\rho}{6})\rho e^{-\frac{\rho}{3}}$$

$$R_{3,2} = \frac{4}{81\sqrt{30a_0^3}}\rho^2 e^{-\frac{\rho}{3}}.$$

In general, $R_{nl}(r)$ has $n - 1$ nodes.

17.4.1 Coulomb Wave Functions

The continuum (positive energy) solutions called Coulomb waves are of interest in electron-nucleus scattering, and also in the description of the final state of photo-electrons (i.e., electrons ejected by an atom that absorbs a photon) and Auger electrons, i.e. electrons emitted by an atom while another electron jumps to a lower empty level. There is no special interest here to go into the solution of the radial equation in terms of confluent hypergeometric functions. However the interesting thing is that for large ρ, the wave function does not approach a free electron function, but gets a phase that depends on distance at all distances. The reason is that the Coulomb potential is long range.

17.4.2 Atomic Shells

The states with $n = 1, 2, 3, 4, \ldots$ form the so-called *atomic shells* $K, L, M,$ $N, O, P \ldots$ which are observed *grosso modo* in all atoms. Actually, the description of many-electron atoms is much more involved, but to some extent one can understand the observed charge densities and binding energies in terms of atomic electrons moving in an effective field, due to the field of the nucleus plus the mean field of the others. This is the idea behind the Hartree-Fok method, see Chap. 21. Since $n_r \geq 0, l \in (0, n-1)$. Shell K has one orbital, shell L has 4, shell M 9. How many in shell n? Since $\sum_{n=1}^{P} = \frac{p(p+1)}{2}$, the number of orbitals is $\sum_{l=0}^{n-1}(2l+1) = n^2$.

Chapter 18
Spin and Magnetic Field

An electron has radius 0 (as far as we know), a quantized angular momentum (spin) $\frac{1}{2}\hbar$ and a magnetic moment. Let us discover how.

18.1 Magnetic and Angular Moment in Classical Physics

Classically, a point charge that circulates on a ring of radius r produces a current $i = \frac{ev}{2\pi r}$, which causes a magnetic dipole moment $\overrightarrow{\mu} = \frac{i}{c}S\overrightarrow{n}$, $S = \pi r^2$, in obvious notation;

$$\overrightarrow{\mu} = \frac{ev}{2c}r\,\overrightarrow{n} = \frac{e}{2c}\overrightarrow{r}\wedge\overrightarrow{v} = \frac{e}{2mc}\overrightarrow{L}. \tag{18.1}$$

The so-called gyromagnetic factor $\frac{e}{2mc}$ changes an angular momentum to a magnetic moment. The two are always parallel, in all theories. The magnetic dipole produces a magnetic field. We can take for the vector potential the expression

$$\overrightarrow{A} = \frac{\overrightarrow{\mu}\wedge\overrightarrow{r}}{r^3}.$$

In an external magnetic field $\overrightarrow{B}$, the dipole has energy

$$E = -\overrightarrow{\mu}\cdot\overrightarrow{B}. \tag{18.2}$$

When a particle has a dipole moment, the angle θ between the magnetic moment and B is a constant of the motion (since it takes energy to modify θ).

According to Eq. (18.2), a dipole in a space-dependent field feels a force

$$\overrightarrow{F} = -\overrightarrow{\nabla}E = \overrightarrow{\nabla}[\overrightarrow{\mu}\cdot\overrightarrow{B}]. \tag{18.3}$$

© Springer International Publishing AG, part of Springer Nature 2018
M. Cini, *Elements of Classical and Quantum Physics*,
UNITEXT for Physics, https://doi.org/10.1007/978-3-319-71330-4_18

Measurement of Microscopic Magnetic Moments

Suppose we have a beam of unknown magnetic moments μ, with no charge, and known mass m. One can measure μ by letting the beam cross an inhomogeneous field in a short time Δt and measuring the deflection. (The electric charge would be deflected by the Lorentz force, which is much larger than the force on the dipole, and this is the reason for excluding it).

If l is the width of the inhomogeneous field region where $\frac{\partial B}{\partial z} \neq 0$ and the magnetic moment has a known velocity v, it spends a time $\Delta t = \frac{l}{v}$ in the field. Then,

$$m\Delta v \approx F\Delta t, \text{ quindi } \Delta v \approx \mu_z \frac{\partial B_z}{\partial z} \frac{l}{mv},$$

and the deflection must be

$$\Delta\theta \approx \frac{\Delta v}{v} \approx \mu_z \frac{\partial B_z}{\partial z} \frac{l}{mv^2}. \tag{18.4}$$

18.2 Stern–Gerlach Experiment

In 1922, Stern e Gerlach conducted one of the most important and astonishing experiments in history. They produced an atomic beam of Ag and sent it through an inhomogeneous magnetic field. Their question was: do the electronic currents in the atom produce a magnetic moment? Ag had been chosen because it was monovalent, so it was possible that the only chemically active electron circulating in a larger orbit played a role in the experiment.[1] The beam was obtained from metal vapor produced in a crucible; the whole apparatus operated in vacuo and Ag was known to remain atomic. The beam was collimated and in the absence of a field, formed a spot on a photographic plate. In Fig. 18.1, a schematic drawing of the geometry is shown. The shape of the magnet was optimized in order to produce an inhomogeneous field, such that the trajectory depended on the orientation of the atomic dipole.

The plan of the experiment was to achieve a measurement of the dipole μ: it was expected that the beam should broaden into a cone, due to the random orientation of the dipoles in the crucible and in the beam; from the angle θ at the vertex of the cone one could deduce μ according to Eq. (18.4). Stern and Gerlach had planned their experiment very well, from the classical point of view, and the atoms did show a moment of the expected order of magnitude. But the beam does not broaden; it splits into two, equally intense branches, one with the moment parallel to the field and the other with opposite moment.

Each branch consists of atoms with the same moment orientation, and the same μ_z. If one of these beams is sent to a second apparatus oriented like the previous one,

---

[1]The Bohr model predicted orbits, and classically there is no other way than a current to produce a magnetic moment; but Nature had a surprise in store.

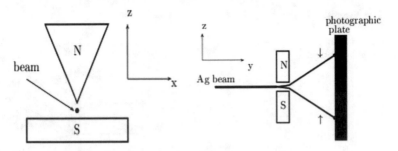

Fig. 18.1 Schematic sketch of the Stern–Gerlach apparatus. Left: seen from the crucible, along the y axis; the magnetic field is strongly inhomogeneous due to the pointed shape of the North pole. Right: seen from the direction of the x axis

it can be deflected, but does not split again. This is expected, since all the atoms are in the same state, see the same field, and there is no reason for a splitting. But, if the second apparatus is oriented differently, a new splitting occurs. Why? This fact cannot be understood in terms of classical magnetic moments. There is, however, an optical analogue (see the next subsection).

It was understood that the magnetic moment of Ag is due to the electronic spin,[2] the one of the chemically active optical electron. Summarizing the results of the experiment:

1. The Ag atom has a magnetic moment μ and the values of μ_z along any direction are quantized.
2. Since $\vec{\mu} \propto \vec{L}$, the values of L_z are quantized, too.
3. There are $2l + 1$ possible values of m; the experiment with Ag shows that the values are 2, so the angular momentum is $\frac{1}{2}$.
4. The magnetic moment is found either up or down in any direction one measures it.
5. The outcome of the experiment cannot be understood in terms of classical point masses carrying a magnetic moment. Indeed, the direction of the magnetic moments is not a property of the Ag atoms, but is determined by the field.

This shows essential features of Quantum Mechanics. The interpretation is:

1. Every angular momentum due to some rotation in space (orbital angular momentum) is associated with $\vec{L}$ and has integer eigenvalues; besides, there is spin, which has no classical counterpart. In general, we shall denote angular momenta of any origin by $\vec{j}$ and the component by m_j.
2. The Ag atoms do not have a sharply defined component μ_z of the magnetic moment when they form the beam, but only when one measures μ_z.

---

[2]Most nuclei have spins and magnetic moments; the nuclear magnetic moments, however, are negligible compared to the electronic ones. This is related to the proton and neutron masses, which are much heavier than the electron.

Fig. 18.2 A second
Stern–Gerlach apparatus
splits the beam again if its
N-S direction is different

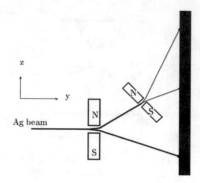

3. The act of measurement produces a collapse of the wave function and two branches
   with well-defined μ_z; μ_x and μ_y are not defined.
4. Any measurement along a different direction produces a new collapse of the wave
   function (see Fig. 18.2).

18.2.1   Electromagnetic Spin

Nobody had imagined the electron spin before the discovery, but there is an analogy
with the polarization of electromagnetic waves which was known from Maxwell's
equations. An electromagnetic wave carries momentum and therefore also orbital
angular momentum. The orbital angular momentum is

$$L = \frac{1}{4\pi} \int d^3 r\, r \wedge [E \wedge B]$$

in Gaussian units. Besides the orbital angular momentum, the wave also carries spin
angular momentum if it is left or right circularly polarized, like for instance a wave
propagating long z with $E_x = E_0 \cos(\omega t)$, $E_y = \pm E_0 \sin(\omega t)$. Even in this case no
object turns, and the angular momentum is intrinsic. Since the photon does not couple to magnetic fields, we must observe its spin through the angular momentum
that photons can exchange with matter. A left polarized photon carries an angular
momentum $\hbar$. Beth[3] reports a measurement of the spin using a special film that converts a righ polarized wave to left polarized. The film was suspended to a quartz fiber
and exposed to left polarized radiation, and the angular momentum was mechanically measured. Spin $S = 1$ implies three states with M = $-1, 0, 1$, but this rule
holds in the rest reference of the particle. The photon has no rest reference and
only M = $-1, 1$. One can perform an optical experiment which bears some analogy
the Stern–Gerach experiment using a Wollaston prism, invented by William Hyde

---

[3]R.A. Beth Phys. Rev. 5, 115 (1936).

Wollaston, that separates an unpolarized beam in two orthogonal linearly polarized outgoing beams. This is based on the property of birefringence of some anisotropic crystals like calcite whose refraction index depends on the polarization and on the direction of propagation. The analogy would be perfect if the experiment separated left and right polarization.

18.3 Angular Momentum Matrices

If one knows that an electron is in $\overrightarrow{x}$, the information is not complete. It has two orthogonal ways to be there. The two ways correspond to the two components of $S_z = \pm\frac{\hbar}{2}$, where z is the direction privileged by the experiment. No motion of particles can realize a fractional angular momentum. No spherical harmonic is relevant to it. Actually spin is intrinsic angular momentum, intrinsic like mass or charge, with a z component that can be exchanged with the outside world in fixed amounts, while z is in a direction that can be chosen at will. There is actually a vague analogy with rotating rigid bodies in classical mechanics.

For the orbital angular momentum $\overrightarrow{L}$, and for integer spins, we know the representations (13.11)–(13.13)

$$\langle l_1 m_1 | L^2 | l_2 m_2 \rangle = \hbar^2 l_1 (l_1 + 1)\delta_{l_1,l_2}\delta_{m_1,m_2},$$

$$\langle l_1 m_1 | L_z | l_2 m_2 \rangle = \hbar m_1 \delta_{l_1,l_2}\delta_{m_1,m_2},$$

$$\langle l_1 m_1 | L_\pm | l_2 m_2 \rangle = \hbar\sqrt{l_1(l_1 + 1) - m_2(m_2 \pm 1)}\delta_{l_1,l_2}\delta_{m_1\pm1,m_2},$$

with integer l, m. None of these is suitable for half-integer spin; we need the spinor representations,[4] which correspond to half integer angular momenta $(\frac{1}{2}, \frac{3}{2}, \frac{5}{2} \cdots)$. We shall deal with spin $\frac{1}{2}$ in detail. Compound systems like nuclei and atoms may have semi-integer spins, which arises as a combination of the spins of the constituent particles (electrons and quarks). For the electron spin degree of freedom, we need a formalism without the derivatives $(\frac{\partial}{\partial r}, \frac{\partial}{\partial \theta}, \frac{\partial}{\partial \phi})$. The direction of the magnetic field is the quantization axis and will be taken as the z axis. For spin $\frac{1}{2}$, there are just two orthogonal states $|\alpha\rangle = |S_z = \frac{1}{2}\rangle$ and $|\beta\rangle = |S_z = -\frac{1}{2}\rangle$. These are called conventionally up and down spin and will be taken as the basis of the spin Hilbert space.[5]

The matrix representation of states and operators is achieved as in Sect. 13.1.2. Letting $S_z = \frac{\hbar}{2}\sigma_z$,

---

[4]The following theory is due to Wolfgang Pauli (1900 Vienna - 1958 Zurich) one of the most important theoreticians of the twentieth century.

[5]In general the system has many other degrees of freedom and the full Hilbert space is the tensor product of the spin Hilbert space times the Hilbert space arising from the other degrees of freedom. See Sect. 12.6.

$$|\alpha\rangle \equiv |\tfrac{1}{2}\rangle \rightarrow \begin{pmatrix} 1 \\ 0 \end{pmatrix}; \quad |\beta\rangle \equiv |-\tfrac{1}{2}\rangle \rightarrow \begin{pmatrix} 0 \\ 1 \end{pmatrix}; \quad \sigma_z = \begin{pmatrix} 1 & 0 \\ 0 & -1 \end{pmatrix}.$$

The two-component wave functions, called *spinors*, are written like complex column vectors $\psi = \begin{pmatrix} a \\ b \end{pmatrix}$; the Hermitean conjugate is

$$\psi^\dagger = \begin{pmatrix} a \\ b \end{pmatrix}^\dagger = \left(a^*\ b^* \right).$$

Introducing shift operators

$$S_+ = \hbar \begin{pmatrix} 0 & 1 \\ 0 & 0 \end{pmatrix}, \qquad\qquad S_- = \hbar \begin{pmatrix} 0 & 0 \\ 1 & 0 \end{pmatrix}, \qquad\qquad (18.5)$$

one can switch spin:

$$S_+|\alpha\rangle = 0, \quad S_-|\alpha\rangle = \hbar|\beta\rangle, \ S_-|\beta\rangle = 0, \quad S_+|\beta\rangle = \hbar|\alpha\rangle.$$

By analogy with the orbital angular momentum, from the shift operators $S_\pm$, one gains the x and y components:

$$S_+ = S_x + iS_y$$
$$S_- = S_x - iS_y.$$

Here is the spin representation in terms of Pauli matrices:

$$\vec{S} = \hbar \frac{1}{2} \vec{\sigma},$$

with

$$\sigma_x = \begin{pmatrix} 0 & 1 \\ 1 & 0 \end{pmatrix}, \qquad \sigma_y = \begin{pmatrix} 0 & -i \\ i & 0 \end{pmatrix}, \qquad \sigma_z = \begin{pmatrix} 1 & 0 \\ 0 & -1 \end{pmatrix}.$$

It is easy to check that the Pauli matrices have square $\begin{pmatrix} 1 & 0 \\ 0 & 1 \end{pmatrix}$ and anticommute,

$$[\sigma_i, \sigma_j]_+ = 2\delta_{ij},$$

where the indices run over the 3 components. The identification of $\vec{S}$ with angular momentum is legitimate, since

$$\vec{S} \wedge \vec{S} = i\hbar \vec{S},$$

which holds, since

$$[\sigma_x, \sigma_y]_- = 2i\sigma_z,$$

and so on. Moreover,

$$S^2 = \frac{3}{4}\hbar^2 = \frac{1}{2}\left(\frac{1}{2} + 1\right)\hbar^2.$$

An electron can go from α and β or back by adsorbing or creating a spin-1 photon in magnetic resonance experiments. It is a magnetic dipole transition. If the electron spin is localized over an atom, the wavelength of the photon is much larger; in such cases, the orbital angular momentum of the photon is practically zero and the photon spin changes by 1.

18.3.1 For the Electron Spin, Every Direction Is OK

The Stern–Gerlach experiment shows that the spin component is $\pm\frac{1}{2}\hbar$ in any direction, and the analogy with the spin of a tennis ball is misleading. The Pauli formalism describes this state of affairs well.

The (Hermitean) matrix that represents the spin in the direction $\vec{n} = (\sin\theta\cos\phi, \sin\theta\sin\phi, \cos\theta)$ is

$$\vec{S} \cdot \vec{n} = \frac{1}{2}\hbar \begin{pmatrix} \cos\theta & \sin\theta e^{-i\phi} \\ \sin\theta e^{i\phi} & -\cos\theta \end{pmatrix}.$$

The eigenvectors are

$$|\uparrow, \vec{n}\rangle = \begin{pmatrix} \cos\frac{\theta}{2} \\ \sin\frac{\theta}{2}e^{i\phi} \end{pmatrix}, \tag{18.6}$$

$$|\downarrow, \vec{n}\rangle = \begin{pmatrix} -\sin\frac{\theta}{2}e^{-i\phi} \\ \cos\frac{\theta}{2} \end{pmatrix}. \tag{18.7}$$

For example, the physical meaning of the upper component $\cos\frac{\theta}{2}$ of $|\uparrow, \vec{n}\rangle$ is the following. Suppose a beam of particles is prepared in the state $|\uparrow, \vec{n}\rangle$, by a Stern–Gerlach experiment with the inhomogeneous field along the $\vec{n}$ axis. Then, suppose the z component of the spin is measured, maybe using a new Stern–Gerlach apparatus with the field along z. Then, $\cos\frac{\theta}{2} = \langle\alpha|\uparrow, \vec{n}\rangle$ is the amplitude of up spin.

Problem 30 Verify the eigenvectors.

Solution 30

$$\begin{pmatrix} \cos\theta & \sin\theta e^{-i\phi} \\ \sin\theta e^{i\phi} & -\cos\theta \end{pmatrix} \begin{pmatrix} \cos\frac{\theta}{2} \\ \sin\frac{\theta}{2} e^{i\phi} \end{pmatrix} = \begin{pmatrix} \cos(\theta)\cos(\frac{\theta}{2}) + \sin(\theta)\sin(\frac{\theta}{2}) \\ \sin(\theta)\cos(\frac{\theta}{2})e^{i\phi} - \cos(\theta)\sin(\frac{\theta}{2})e^{i\phi} \end{pmatrix}.$$

Substituting the identities

$$\cos(\theta) = \cos\left(\frac{\theta}{2}\right)^2 - \sin\left(\frac{\theta}{2}\right)^2, \quad \sin(\theta) = 2\cos\left(\frac{\theta}{2}\right)\sin\left(\frac{\theta}{2}\right)$$

one finds

$$= \begin{pmatrix} \cos(\frac{\theta}{2})\left[\cos(\frac{\theta}{2})^2 - \sin(\frac{\theta}{2})^2 + 2\sin(\frac{\theta}{2})^2\right] \\ \sin\frac{\theta}{2}e^{i\phi}\left[2\cos(\frac{\theta}{2})^2 - \cos(\frac{\theta}{2})^2 + \sin(\frac{\theta}{2})^2\right] \end{pmatrix} = \begin{pmatrix} \cos(\frac{\theta}{2}) \\ \sin(\frac{\theta}{2})e^{i\phi} \end{pmatrix}.$$

The True Origin of Electron Spin

There is no premonition of spin in the postulates. The Pauli theory looks like a brilliant patch; in fact it is *phenomenological*. This is how one describes theories that properly explain some phenomena, but also show the need for a broader and more satisfactory framework. Actually in the Schrödinger theory one starts with the classically known observables and Pauli has shown that the novel concept of spin fits well in the theory. A much more satisfactory formulation arises from the relativistic formulation. For the electron, this is Dirac's theory.

18.3.2 How to Rotate a Spin 1/2

How to rotate a spinor like, for instance $| \uparrow, \vec{n} \rangle = \begin{pmatrix} \cos\frac{\theta}{2} \\ \sin\frac{\theta}{2}e^{i\phi} \end{pmatrix}$ by an angle $\Delta\phi$ around z axis? One could quite reasonably argue simply that $\phi \to \phi + \Delta\phi$; well, this is *almost true*, but there is a subtle, striking complication. In Sect. 13.2, we determined how to achieve the rotation of an orbital acting not on the argument x of $\psi(x)$ but on ψ. This requires a unitary operator $R_\phi = e^{-i\frac{\Delta\phi \cdot L}{\hbar}}$. By analogy, the generator of the infinitesimal rotations of spin is $\vec{S} = \frac{1}{2}\hbar\vec{\sigma}$ and the operator that does the rotation is

$$R_\phi = e^{-i\Delta\phi \cdot \frac{S}{\hbar}}. \tag{18.8}$$

At the exponent, we find a 2×2 matrix, which is easy to diagonalize and exponentiate. The operator (18.8) that rotates around the z axis reads as

$$R_{\Delta\phi} = e^{-i\frac{\Delta\phi}{2}\sigma_z} = e^{-i\frac{\Delta\phi}{2}\begin{pmatrix} 1 & 0 \\ 0 & -1 \end{pmatrix}}.$$

Diagonal matrices are exponentiated immediately:

$$\begin{pmatrix} \mu & 0 \\ 0 & \nu \end{pmatrix}^2 = \begin{pmatrix} \mu^2 & 0 \\ 0 & \nu^2 \end{pmatrix},$$

and so

$$e^{\begin{pmatrix} \mu & 0 \\ 0 & \nu \end{pmatrix} t} = \begin{pmatrix} e^{\mu t} & 0 \\ 0 & e^{\nu t} \end{pmatrix},$$

and also

$$R_{\Delta\phi} = \begin{pmatrix} e^{-i\frac{\Delta\phi}{2}} & 0 \\ 0 & e^{i\frac{\Delta\phi}{2}} \end{pmatrix}.$$

Therefore,

$$R_{\Delta\phi} | \uparrow, \overrightarrow{n} \rangle = \begin{pmatrix} e^{-i\frac{\Delta\phi}{2}} \cos\frac{\theta}{2} \\ e^{i\frac{\Delta\phi}{2}} \sin\frac{\theta}{2} e^{i\phi} \end{pmatrix} = e^{-i\frac{\Delta\phi}{2}} \begin{pmatrix} \cos\frac{\theta}{2} \\ \sin\frac{\theta}{2} e^{i[\phi+\Delta\phi]} \end{pmatrix} :$$

increments ϕ, but also gives an overall phase to the spinor. The implication is that $R_{2\pi} = -1$: a rotation by 2π changes the sign of the spinor! This fundamental property of half-integer spin particles is rooted in the mathematics. The set of all the matrices (18.8) with the rule of matrix multiplication constitutes a mathematical structure called the SU(2) Group (S stands for special, and means that all the matrices have a unit determinant). The O(3) Group made with the matrices (13.21) does almost the same job. In both cases, all the rotations are represented, but in $SU(2)$ there are two operators for each rotation. $SU(2)$ is called the covering Group of $O(3)$. The (-1) factor does not change the state, but if one could superpose two beams then the (-1) factor affecting one of them would lead to destructive interference. Such an experiment has not yet been reported.

18.3.3 Spin Magnetic Moment and Pauli Equation

The classical equation (18.1) associates a magnetic moment with the angular momentum. Since within the electron, there are no currents, we cannot compute the electron moment classically. In Dirac's Relativistic Quantum Theory one finds that

$$\overrightarrow{\mu} = \frac{e}{2mc} g \overrightarrow{S}, \tag{18.9}$$

where m is the electron mass and $g = 2$. Actually, for the electron this quantity has been measured with great accuracy, with the result that $g \approx 2.0023193043617$. The small discrepancy is due to the fact that a real electron is the source of a field, not considered in Dirac's theory, and as such, it interacts with virtual electron-positron pairs. Quantum Electrodynamics is a field theory, taking into account the interaction, and its result agrees extremely well with the data. For the proton, which is about 1800 times more massive and has internal structure and strong interactions, $g \approx 5$, and μ is much smaller. The neutron has μ comparable to that of the proton.

In a magnetic field B, the dipole energy is given by (18.2) like in Classical Physics. The Schrödinger-Pauli equation reads as

$$i\hbar \frac{\partial \psi}{\partial t} = H\psi. \tag{18.10}$$

For a particle with charge e and spin 1/2 in a magnetic field,

$$H = \begin{pmatrix} \frac{(\vec{p} - \frac{e}{c}\vec{A})^2}{2m} & 0 \\ 0 & \frac{(\vec{p} - \frac{e}{c}\vec{A})^2}{2m} \end{pmatrix} - \frac{e}{2mc} g \, \vec{S} \cdot \vec{B} =$$

$$\begin{pmatrix} \frac{(\vec{p} - \frac{e}{c}\vec{A})^2}{2m} - \frac{e\hbar}{2mc}\frac{g}{2} B & 0 \\ 0 & \frac{(\vec{p} - \frac{e}{c}\vec{A})^2}{2m} + \frac{e\hbar}{2mc}\frac{g}{2} B \end{pmatrix}.$$

The magnetic term $\frac{e}{2mc} g \, \vec{S} \cdot \vec{B}$, alone has eigenstates $|\alpha\rangle$ with eigenvalue $\mu_B B$ and $|\beta\rangle$ with $-\mu_B B$, where μ_B is the *Bohr-Procopiu magneton*; for $e < 0$ the solutions are exchanged; so, for an electron in a magnetic field parallel to the z axis, $|\beta\rangle$ is the ground state. Approximating g with 2,

$$\mu_B = \frac{e\hbar}{2mc}.$$

The two levels are separated by the *Larmor frequency*,

$$\omega_L = \frac{eB}{mc}.$$

Including the kinetic term only, one finds the Landau levels (Sect. 17.1); the spin term replaces each level by two sub-levels; since $g \sim 2$, the spin down adds $\frac{eB}{2mc}$ to the energy and the spin up subtracts the same. Thus the two magnetic levels are separated by $\sim \hbar \omega_L$.

The direct transition between the spin sublevels can be done by absorbing or emitting a photon, in a magnetic dipole transition; this is EPR (electron spin resonance), or NMR (nuclear spin resonance). For a proton the ground state is α, and the frequency is three orders of magnitude lower.

Problem 31 Rotate the spinor $|\alpha\rangle$ around the x (a) by an angle $\varphi = \pi/2$ and (b) by an angle $\varphi = \pi$. Interpret the results physically.

Solution 31 The rotation operator by φ around x is

$$R_\varphi = \exp\left[\frac{-i\varphi}{2}\begin{pmatrix} 0 & 1 \\ 1 & 0 \end{pmatrix}\right].$$

Separating odd and even powers in the expansion of $e^{ix} = \cos(x) + i\sin(x)$, one finds

$$R_\varphi = \sum_{\text{evenn}} \frac{1}{n!}\left(\frac{-i\varphi}{2}\right)^n + \sigma_x \sum_{\text{oddn}} \frac{1}{n!}\left(\frac{-i\varphi}{2}\right)^n.$$

$$\sum_{\text{npari}} \frac{(ix)^n}{n!} = Re\,e^{ix} = \cos(x) \qquad \sum_{\text{ndispari}} \frac{(ix)^n}{n!} = i\,Im\,e^{ix} = i\sin(x).$$

Therefore,

$$R_\varphi = \begin{pmatrix} \cos\left(\frac{\varphi}{2}\right) & -i\sin\left(\frac{\varphi}{2}\right) \\ -i\sin\left(\frac{\varphi}{2}\right) & \cos\left(\frac{\varphi}{2}\right) \end{pmatrix}.$$

Note that $R_\varphi^\dagger = R_\varphi^{-1}$. One finds: (a)

$$R_{\pi/2}|\alpha\rangle = \frac{1}{\sqrt{2}}\begin{pmatrix} 1 \\ -i \end{pmatrix}.$$

This is an eigenstate (spin down) of S_y. (b)
$R_\pi|\alpha\rangle = -i|\beta\rangle$. This is in line with one's intuition.

Problem 32 Verify the identity

$$\vec{j}(1)\cdot\vec{j}(2) = j_z(1)j_z(2) + \frac{j_+(1)j_-(2) + j_-(1)j_+(2)}{2},$$

where $\vec{j}(1), \vec{j}(2)$ are angular momenta of different particles with $[\vec{j}(1), \vec{j}(2)]_- = 0$.

Solution 32 Just substitute the definitions.

18.4 Quantum 'Sum' of Angular Momenta

An electron in a central field has orbital angular momentum and spin. This holds for each of the 92 electrons in a Uranium atom. We must also consider the spin of the nucleus, which results from the spins of protons and neutrons and their orbital angular momenta.

The total angular momentum of any isolated system is particularly interesting, since it is always conserved. The isotropy of space implies that the Hamiltonian must

commute with all rotations, and with their generators. The nuclear angular momentum would be conserved separately, if we could neglect the magnetic dipole coupling
with the electron angular momentum; the electron orbital momentum is coupled to
the electron spin momentum by the relativistic spin-orbit interaction arising in Dirac's
theory. When relativistic effects are small, the second-order corrections in v/c are
enough. The spin-orbit interaction

$$H' = -\frac{e\hbar}{4m^2c^2}\sigma.(E \wedge p)$$

is prominent among the corrections arising from Dirac's theory since it can lift
degeneracies. In central problems, the spin-orbit interaction is proportional to $\vec{l} \cdot \vec{s}$
for each electron, and there are similar spin-spin, orbit-orbit and spin-other orbit
interactions. These terms arise in the *Breit approximation* that includes the lowest-
order relativistic corrections to the Coulomb interaction. The latter is strictly correct
only in the case of static charges, and since electrons in atoms do move, the Breit
approximation should replace the Coulomb interaction, at least for heavy elements.

Consider a system made up of two parts with angular momenta j_1 and j_2 that
depend on different variables and therefore commute:

$$[j_1, j_2]_- = 0;$$

the angular momentum operator of the system is the sum

$$j = j_1 + j_2,$$

as in Classical Physics. It is really an angular momentum, since trivially,

$$j \wedge j = i\hbar j.$$

We need a basis of eigenstates for each:

$$j_1^2|j_1m_1\rangle = j_1(j_1 + 1)\hbar^2|j_1m_1\rangle, \ j_{1z}|j_1m_1\rangle = m_1\hbar|j_1m_1\rangle$$
$$j_2^2|j_2m_2\rangle = j_2(j_2 + 1)\hbar^2|j_2m_2\rangle, \ j_{2z}|j_2m_2\rangle = m_2\hbar|j_2m_2\rangle.$$

The direct product basis is a basis for the compound system:

$$|j_1m_1j_2m_2\rangle \equiv |j_1m_1\rangle|j_2m_2\rangle. \tag{18.11}$$

For all the $(2j_1 + 1)(2j_2 + 1)$ basis vectors, we already know the eigenvalue m_j
of j_z:

$$j_z|j_1m_1j_2m_2\rangle = (m_1 + m_2)\hbar|j_1m_1j_2m_2\rangle = m_j\hbar|j_1m_1j_2m_2\rangle.$$

Table 18.1 Spin states of H

Stati	m_j	j
$|\alpha(1)\alpha(2)\rangle$	1	1
$|\alpha(1)\beta(2)\rangle, \ |\beta(1)\alpha(2)\rangle$	0	Mixed 1, 0
$|\beta(1)\beta(2)\rangle$	−1	1

We cannot (generally) assign j, m_1 and m_2. In fact,

$$j^2 = j_1^2 + j_2^2 + 2j_1 \cdot j_2;$$

the squares do commute with the components, but $j_1 \cdot j_2$ fails to comute:

$$[j^2, j_{1z}]_- = 2[j_1 \cdot j_2, j_{1z}]_- = 2[j_{1x}j_{2x} + j_{1y}j_{2y}, j_{1z}]_- \qquad (18.12)$$

does not vanish.

Problem 33 Compute the commutator (18.12).

Solution 33 One finds $2i\hbar(j_1 \wedge j_2)_z$.

Sum of Two Spins 1/2

In the ground state of the H atom,[6] the orbital angular momentum is $l = 0$, but there are two 1/2 spins to sum, one electron and a proton spin. There is a magnetic dipole-dipole interaction, and neither angular momentum is conserved, while the resultant gives good quantum numbers j, m_j. Let us denote by $\alpha(1)$, $\beta(1)$ the eigenstates of the electron S_z and by $\alpha(2)$, $\beta(2)$ those of the nucleus; the direct product of the two bases gives the 4 states in Table 18.1:

We can label all 4 states with the total angular momentum component m_j by simply summing the individual components, since the operators commute; we can also put a total $j = 1$ label on the first and last lines. Indeeed, $j = 0$ is excluded because it must give $m_j = 0$, while higher values are excluded since they should also give $m_j > 1$ components that do not exist. The 4-dimensional Hilbert space has a three-dimensional subspace $j = 1$, called the *triplet*, and the component $|j = 1, m_j = 0\rangle$, called the *singlet*, must be a combination of the two states with $m_j = 0$. To separate the two components, one can start from $|\alpha(1)\alpha(2)\rangle$ and lower the z component by acting with $j^- = S^-(1) + S^-(2)$.

In the example of Table 18.1, using the shift operators (18.5) $S^+ = \hbar \begin{pmatrix} 0 & 1 \\ 0 & 0 \end{pmatrix}$, $S^- = \hbar \begin{pmatrix} 0 & 0 \\ 1 & 0 \end{pmatrix}$, the required shift operator is $j^- = S_-(1) + S_-(2)$.

---

[6] Specifically, I refer to Protium ^1H; the Deuterium ^2H nucleus has a neutron and a proton and spin 1, and the unstable Tritium ^2H has spin 1/2, too.

Since $S^-(1)|\alpha(1)\alpha(2)\rangle = |\beta(1)\alpha(2)\rangle$, $S^-(2)|\alpha(1)\alpha(2)\rangle = |\alpha(1)\beta(2)\rangle$, we obtain at once the $m_j = 0$ component of the triplet

$$|j = 1, m_j = 0\rangle = \frac{\alpha(1)\beta(2)\rangle + \beta(1)\alpha(2)\rangle}{\sqrt{2}}. \tag{18.13}$$

The orthogonal combination must be the singlet

$$|j = 0, m_j = 0\rangle = \frac{\alpha(1)\beta(2)\rangle - \beta(1)\alpha(2)\rangle}{\sqrt{2}}. \tag{18.14}$$

To verify that these are indeed eigenstates of $j^2 = S_1^2 + S_2^2 + 2S_1S_2$, one can start with $S_1^2 = S_2^2 = \frac{3}{4}\hbar^2$ and note that $S_1S_2 = S_{1z}S_{2z} + S_{1x}S_{2x} + S_{1y}S_{2y}$; hence $j^2 = \frac{3}{2}\hbar^2 + 2[S_{1z}S_{2z} + \frac{1}{2}(S_1^+S_2^- + S_1^-S_2^+)]$. This makes the verification immediate. The separation between the singlet ground state and the triplet excited state is small $(5.9 \times 10^{-6}$ eV$)$ and belongs to the so-called hyperfine structure of H. It corresponds to a 21 cm wavelength line (at 1420 MHz) and it is very important in Astrophysics, since it is weakly absorbed by dust clouds, and so allows us to see the atomic H clouds throughout the Milky Way and beyond.

Clebsh-Gordan Coefficients

Suppose we have a system consisting of two parts with angular momenta j_1 and j_2, which would be conserved in the absence of a mutual interaction. From the separate bases $|j_1, m_1\rangle$ and $|j_2, m_2\rangle$, one obtains the so-called *direct product basis* for the system,

$$\{|j_1 m_1 j_2 m_2\rangle\}, \qquad m_1 \in (-j_1, j_1), m_2 \in (-j_2, j_2).$$

But since the total angular momentum is conserved, we may wish to go to a new basis $|J, M_j\rangle$ where $\hat{J} = \hat{j}_1 + \hat{j}_2$ and M is the eigenvalue of $J_z = j_{1z} + j_{2z}$. The maximum value of M_j is $M_j^{max} = j_1 + j_2$ and the minimum is $M_j^{min} = -M_j^{max}$. Therefore, the space spanned by the direct product basis includes the space spanned by the basis of $J = j_1 + j_2$, and this is the maximum J that can occur (since larger J would imply larger M_j). Consequently, by taking linear combinations of the product basis vectors, one can form a basis for the subspace $J = j_1 + j_2$; the number of linear combinations for this J is $2(j_1 + j_2) + 1$. As we know, these states are labelled by J and M_j, but not all of them have sharp m_1 and m_2. In fact only the maximum M_j and the minimum M_j have such sharp values. Instead, $M_j = j_1 + j_2 - 1$ corresponds to two different members of the direct product basis, namely, $|j_1, m_1 = j_1, j_2, m_2 = j_2 - 1\rangle$ and $|j_1 m_1 = j_1 - 1, j_2 m_2 = j_2\rangle$. From these vectors, we know how to make one linear combination belonging to $J = j_1 + j_2$: we must just apply the shift operator $\hat{J}^- = \hat{j}_1^- + \hat{j}_2^-$ to $|J, M = J\rangle$, getting $|J, M = J - 1\rangle$. The orthogonal combination must then belong to $J = j_1 + j_2 - 1$ and $M_j = j_1 + j_2 - 1$. Now, $J = j_1 + j_2 - 1$ must occur only once. In this way, one finds all the values from $j = j_1 + j_2$ to $j = |j_1 - j_2|$ and all appear once. Indeed, one can prove the following identity:

$$\sum_{j=|j_1-j_2|}^{j_1+j_2} (2j+1) = (2j_1+1)(2j_2+1). \tag{18.15}$$

Thus, in general one can go from the direct product basis to the total angular momentum basis $|JM_j\rangle$ by a unitary transformation

$$|jm_j\rangle = \sum_{m_1=-j_1}^{j_1} \sum_{m_2=-j_2}^{j_2} |j_1m_1j_2m_2\rangle\langle j_1m_1j_2m_2|jm_j\rangle. \tag{18.16}$$

One can do the reverse transformation by the inverse unitary transformation. The *Clebsch Gordan coefficients* $\langle j_1m_1j_2m_2|jm_j\rangle$ can be found by using the shift operator and orthogonalization as we have just seen. This is just the procedure that we used to derive equations (18.13), (18.14). There are extensive tables of Clebsch Gordan coefficients, that are also readily obtained by computer systems like Mathematica. Here is another example.

Sum of L = 1 and a Spin 1/2

To find the angular momentum of an electron in a $2p$ orbital, we must combine $l = 1$ with the spin ($3 \times 2 = 6$ states). One can denote the states by $|m_l, m_s\rangle$.

The maximum and the minimum m_j must belong to $j = \frac{3}{2}$ (it cannot be less, because $m_j = \frac{3}{2}$ exists, but it cannot be more, because otherwise higher m_j should appear). But $j = \frac{3}{2}$ means 4 states, and the other 2 must belong to $j = \frac{1}{2}$.

We know that (Table 18.2)

$$|\frac{3}{2}, \frac{3}{2}\rangle = |1, 1\rangle|\frac{1}{2}\rangle.$$

The right combination for $j = \frac{3}{2}$, $m_j = \frac{1}{2}$ can be found by acting on the state $m_j = \frac{3}{2}$ with

$$j_- = L_- + s_-.$$

Using the general formula $L_\pm|l, m\rangle = \sqrt{l(l+1) - m(m \pm 1)}\hbar|l, m \pm 1\rangle$, and normalizing,

Table 18.2 Angular momentum of a 2p electron

Stati	m_j	j
$|1, \frac{1}{2}\rangle$	$\frac{3}{2}$	$\frac{3}{2}$
$|1, -\frac{1}{2}\rangle,\ |0, \frac{1}{2}\rangle$	$\frac{1}{2}$	Mixed $\frac{3}{2}, \frac{1}{2}$
$|0, -\frac{1}{2}\rangle,\ |-1, \frac{1}{2}\rangle$	$-\frac{1}{2}$	Mixed $\frac{3}{2}, \frac{1}{2}$
$|-1 -\frac{1}{2}\rangle$	$-\frac{3}{2}$	$\frac{3}{2}$

$$|\frac{3}{2}, \frac{1}{2}\rangle = \sqrt{\frac{2}{3}}|1, 0\rangle|\frac{1}{2}\rangle + \frac{1}{\sqrt{3}}|1, 1\rangle|\frac{-1}{2}\rangle. \tag{18.17}$$

The orthogonal combination is

$$|\frac{1}{2}, \frac{1}{2}\rangle = \frac{1}{\sqrt{3}}|1, 0\rangle|\frac{1}{2}\rangle - \sqrt{\frac{2}{3}}|1, 1\rangle|\frac{-1}{2}\rangle. \tag{18.18}$$

From (18.17) we deduce that

$$\langle 1, 0, \frac{1}{2}, \frac{1}{2}|\frac{3}{2}, \frac{1}{2}\rangle = \sqrt{\frac{2}{3}}, \quad \langle 1, 1, \frac{1}{2}, -\frac{1}{2}|\frac{3}{2}, \frac{1}{2}\rangle = \sqrt{\frac{1}{3}},$$

and from (18.18) it follows that

$$\langle 1, 0, \frac{1}{2}, \frac{1}{2}|\frac{1}{2}, \frac{1}{2}\rangle = \sqrt{\frac{1}{3}}, \quad \langle 1, 1, \frac{1}{2}, -\frac{1}{2}|\frac{1}{2}, \frac{1}{2}\rangle = -\sqrt{\frac{2}{3}}.$$

Thus, one can do very well without a table of Clebsch–Gordan coefficients.

Action of j^2 on the LS Basis

In order to evaluate the matrix elements and to verify if a wave function is eigenstate of j^2, one can combine the identity $j^2 = L^2 + S^2 + 2L \cdot S$ with

$$\vec{L} \cdot \vec{S} = L_z S_z + \frac{1}{2}(L_+ S_- + L_- S_+),$$

recalling that (Tables 18.3 and 18.4)

$$\langle l_1 m_1|L_\pm|l_2 m_2\rangle = \hbar\sqrt{l_1(l_1 + 1) - m_2(m_2 \pm 1)}\delta_{l_1,l_2}\delta_{m_1\pm 1,m_2}.$$

When summing more than two angular momenta, one can choose an arbitrary order and combine the first two, then the resulting sums with the third, and so on.

Table 18.3 The Clebsch–Gordan coefficients for $j_1 = j_2 = \frac{1}{2}$

m_1	m_2	j	$\langle j_1 j_2 m_1 m_2|jm = m_1 + m_2\rangle$
$\frac{1}{2}$	$\frac{1}{2}$	1	1
$\frac{1}{2}$	$-\frac{1}{2}$	1	$\frac{1}{\sqrt{2}}$
$\frac{1}{2}$	$-\frac{1}{2}$	0	$\frac{1}{\sqrt{2}}$
$-\frac{1}{2}$	$\frac{1}{2}$	1	$\frac{1}{\sqrt{2}}$
$-\frac{1}{2}$	$\frac{1}{2}$	0	$-\frac{1}{\sqrt{2}}$
$-\frac{1}{2}$	$-\frac{1}{2}$	1	1

Table 18.4 The i Clebsch–Gordan coefficients for $j_1 = 1$, $j_2 = \frac{1}{2}$

m_1	m_2	j	$\langle j_1 j_2 m_1 m_2 \lvert jm = m_1 + m_2 \rangle$
1	$\frac{1}{2}$	$\frac{3}{2}$	1
1	$-\frac{1}{2}$	$\frac{3}{2}$	$\sqrt{\frac{1}{3}}$
1	$-\frac{1}{2}$	$\frac{1}{2}$	$\sqrt{\frac{2}{3}}$
0	$\frac{1}{2}$	$\frac{3}{2}$	$\sqrt{\frac{2}{3}}$
0	$\frac{1}{2}$	$\frac{1}{2}$	$-\sqrt{\frac{1}{3}}$
0	$-\frac{1}{2}$	$\frac{3}{2}$	$\sqrt{\frac{2}{3}}$
0	$-\frac{1}{2}$	$\frac{1}{2}$	$\sqrt{\frac{1}{3}}$
-1	$\frac{1}{2}$	$\frac{3}{2}$	$\sqrt{\frac{1}{3}}$
-1	$\frac{1}{2}$	$\frac{1}{2}$	$-\sqrt{\frac{2}{3}}$
-1	$-\frac{1}{2}$	$\frac{3}{2}$	1

The final pattern of possible total angular momenta does not depend on the arbitrary order; in fact, it is an objective physical reality. However, when several multiplets with the same j appear, their bases can get mixed by changing the order.

18.5 Photon Spin and Quantum Cryptography

There is an obvious difference between electron spin and photon spin: the photon, unlike the electron, is a vector particle. In the case of a photon going up along z, we can take an orthogonal basis $(\lvert x \rangle, \lvert y \rangle)$ of linearly polarized states. The spin states $\sigma_z = \pm 1$ correspond to the combinations $\frac{\lvert x \rangle \pm i \lvert y \rangle}{\sqrt{2}}$. A field polarized at 45° is like a vector that points at 45° $\frac{\lvert x \rangle + \lvert y \rangle}{\sqrt{2}}$ and is an equal mixture of both polarizations. This quantum property has an important application in cryptography.

There is information that must remain secret for a long time, like passwords, bank coordinates, or the controls of a commercial satellite-not to mention classified military and security files. For thousands of years, there has been a competition between those who wanted to encrypt secrets with increasingly ingenious codes and those who wanted to decipher them; the Allies made great progress after breaking the code of the Enigma machines of the Wehrmacht in World War II. With the advent of Quantum Mechanics, the competition is now being won by those who want to keep secrets. Various devices to achieve cryptography using the uncertainty principle have come to the industrial level and are sold by various commercial companies. Here is the principle of operation.

The basic problem of cryptography is well known: two people, let us say Alice and Bob, need to exchange secret information, using a public channel such as a telephone line or radio; people who have interest in the breaking the code can receive signals, but cannot possibly understand them without a cryptographic key, which may be represented by a secret number. If the number has many digits, the secret can be effectively kept. But first, Bob and Alice must share this number. The essential problem that Alice and Bob must solve is this: how to exchange the key while avoiding it also being received by the eavesdropper who has access to the public channel.

Charles Bennet of IBM ed Gilles Brassard of Montreal University developed a method based on Quantum Mechanics to do this in a safe way. The signal is transmitted as a succession of bits that can be 1 or 0. First, Alice and Bob agree that a photon polarized "horizontally" is 1 and one polarized "vertically" means 0. In order to transmit and receive photons, both have parallel polarizing filters that can emit the light and read the polarizations. The filters act like projectors $|x\rangle\langle x|$, which beep if the signal is polarized along x. However, for the present purpose, Alice and Bob also need a second filter rotated 45° compared to the first. So, indicated with a dash of the polarization plane, the bit 1 can travel as − or \, while bit 0 can travel as | or /.

If the filter used to receive is the same as that used to transmit, the bit is read without errors, but if, for example, a signal emitted as | is picked up with the inclined filter, half of the time, it ill be interpreted as −.

The transmission of the key is done as follows. Alice sends a random bit sequence, even with randomly choosing the filter for sending them, and keeping track each time about what she does. For each incoming bit, Bob randomly chooses which filter to use to read it and makes a note of the filter used for the reading. After the end of the photon transmission, Bob communicates to Alice, through the public channel, the sequence of filters used in the readings (obviously without communicating the sequence of readings). Next, Alice can reveal, using the same channel, which filters were used correctly. Then, deleting the photons read by the wrong filter, both Alice and Bob have obtained a sequence of bits that constitute the secret key.

A possible eavesdropper can receive the photons, but Quantum Mechanics prevents him from doing what classically would be possible, i.e., he cannot observe the state of polarization without the risk of changing it. Choosing between − and | means choosing between two orthogonal states, like the states α and β of a spin. When the eavesdropper receives the photon with the wrong filter, he not only reads the polarization incorrectly, but can also change the polarization. This is similar to the effect of the second Stern–Gerlach filter in Fig. 18.2. The errors introduced in this way will eventually reveal his presence. Indeed, randomly choosing a small number of bits among those making up the message, Alice and Bob can verify on the public channel whether they arrived unchanged or not; in the latter case, the presence of the eavesdropper will be unmasked.

18.5.1 Rabi Model

As an example of the nontrivial behavior of quantum time-dependent systems, consider the Rabi model, which is fundamental in Quantum Optics.[7] It represents a spin in a magnetic field or a two-level atom. Suppose that it is initially in the ground state $|g\rangle$; then, it starts interacting with a radiation field whose frequency ω is comparable to the separation ω_0 between the ground state level E_g and the excited state E_e. The difference $\Delta = \omega - \omega_0$ is called detuning. The Rabi Hamiltonian is:

$$H = -\frac{\hbar\omega_0}{2}\sigma_z - \tau\cos(\omega t)\sigma_x. \tag{18.19}$$

Perhaps, one might expect that the probability P of finding the atom in the excited state $|e\rangle$ should vanish through energy conservation whenever $\Delta \neq 0$. However this is not true, because H depends on the time, and both levels become dressed with light. Let $E_e = \frac{\hbar\omega_0}{2} = -E_g$, and let us write the wave function as

$$\psi(t) = a_g|g\rangle\exp\left(\frac{-iE_g t}{\hbar}\right) + a_e|e\rangle\exp\left(\frac{-iE_e t}{\hbar}\right); \tag{18.20}$$

the Schrödinger equation yields

$$\begin{aligned}
\dot{a}_g &= \tfrac{i}{\hbar}\tau\cos(\omega t)\exp\left(\tfrac{-i\omega_0 t}{\hbar}\right)a_e \\
\dot{a}_e &= \tfrac{i}{\hbar}\tau\cos(\omega t)\exp\left(\tfrac{i\omega_0 t}{\hbar}\right)a_g.
\end{aligned} \tag{18.21}$$

This is readily solved exactly, but to allow for a simple analytic formula, we can neglect the quickly rotating terms in $\omega + \omega_0$, while keeping the slow ones in $\omega - \omega_0$. This is known as the Rotating Wave Approximation. Then, we obtain the easily solvable equations

$$\begin{aligned}
\dot{a}_g &= \tfrac{i}{2\hbar}\tau\exp(\tfrac{i(\omega-\omega_0)t}{\hbar})a_e \\
\dot{a}_e &= \tfrac{i}{2\hbar}\tau\exp(\tfrac{-i(\omega-\omega_0)t}{\hbar})a_g.
\end{aligned} \tag{18.22}$$

Now it is elementary to find that

$$\alpha_e(t) = \frac{i\tau}{\hbar\Omega}\exp\left(\frac{it}{2}\Delta\right)\sin\left(\frac{\Omega t}{2}\right); \tag{18.23}$$

the Rabi frequency Ω defined by

$$\hbar\Omega = \sqrt{\tau^2 + (\hbar\Delta)^2}. \tag{18.24}$$

---

[7]Isidor Isaac Rabi (1898–1988) was a U.S. Physicist, who received the Nobel Prize in 1944.

For $\Delta = 0$, the atom jumps up and down with the frequency $\frac{\tau}{\hbar}$. Increasing the detuning makes the Rabi oscillations more frequent, while the system spends less and less time in the excited state.

Chapter 19
Systems of Particles

The study of many-body systems requires new concepts in dramatic contrast with Classical Physics. Two quantum particles can be in an entangled state, and like particles must be entangled because of a permutation symmetry of the wave function.

19.1 N-Particle Systems

We need to extend the quantum mechanical theory to the case of N particles; separately, the particles would be described by a Schrödinger equation or the Pauli equation, depending on their spins. As in the classical case, the Hamiltonian of the system will be the sum of those of the particles plus an interaction term (possibly). For independent particles (no interaction), the Hamiltonian is, in obvious notation

$$\hat{H}(1, 2, \ldots, N) = \sum_{n}^{N} \hat{h}(n),$$

where $h(n)$ describes particle n. The wave function $\Psi(1, 2, \ldots, N)$ depends on all the orbital and spin degrees of freedom of the particles, so it is a spinor in the spin space of each particle. Consequently, the scalar product of two wave functions Ψ and Φ can be written symbolically as

$$\langle \Phi | \Psi \rangle = \int \Phi^* \Psi \, d1 \ldots dN, \tag{19.1}$$

where, besides the integration over the coordinates, one must understand the scalar products of the corresponding spins, for particles that possess one. In this way, one can define normalized states. To describe the state of the system, one needs a set

© Springer International Publishing AG, part of Springer Nature 2018
M. Cini, *Elements of Classical and Quantum Physics*,
UNITEXT for Physics, https://doi.org/10.1007/978-3-319-71330-4_19

of single particle states $\{A, B, C \ldots\}$ where all the particles of all kinds can be accommodated. The many-body wave function ψ is the amplitude of having particle 1 in some state A, particle 2 in some B, and so on. The Schrödinger equation for N $= 2$ particles reads as:

$$i\hbar \frac{\partial \Psi(1, 2)}{\partial t} = H(1, 2)\Psi(1, 2);$$

if H is time-independent, the time dependence of Ψ factors as usual, and

$$H(1, 2)\Psi(1, 2) = E\Psi(1, 2). \tag{19.2}$$

Here, we find a new example of separation of variables: if

$$h(1)\psi_a(1) = \epsilon_a \psi_a(1),$$
$$h(2)\psi_b(2) = \epsilon_b \psi_b(2),$$

where $\{\psi_a\}$ and $\{\psi_b\}$ are complete set of solutions for one particle, then $\psi_{ab}(1, 2) = \psi_a(1)\psi_b(2)$ solves (19.2) with $E = E_{ab} = \epsilon_a + \epsilon_b$. These are special solutions, but the most general solution can be expanded on the set of those functions. Up to this point, the extension to several particles seems to lead to mathematical complications but no new ideas. However, we need only look to the next section.

19.2 Pauli Principle: Two Identical Particles

For identical particles, a new ingredient comes into play. Consider, for instance, two electrons, and let A denote the spinor of electron 1 and B denote the spinor of electron 2. Let the operator $P(i, j)$ exchange two electrons (coordinates and spins), putting 1 in B and 2 in A. It is clear that the physical state does not change. There can exist no way to distinguish one electron from another. They are unlike billiard balls, which can be marked with a chalk. In the wave function they are marked 1 and 2, so $P(1, 2)$ does a mathematical operation. Since $P(1, 2)^2 = 1$, $P(1, 2)$ has eigenvalues of ± 1, one could argue that sometimes we get 1 and sometimes -1. That's reasonable. But instead, we get always -1, when exchanging two electrons, or two protons, or two identical particles having half integer spins. These are called Fermions. On the other hand, the Bosons (integer spin particles) always have $P(1, 2) = 1$. In summary:

$$P(1, 2)\Psi(1, 2) = \Psi(2, 1) = \begin{cases} -\Psi(1, 2) & \text{Fermions,} \\ \Psi(1, 2) & \text{Bosons.} \end{cases}$$

At the fundamental level, Bosons are related to fields and Fermions to matter. Pions (spin 0), Photons and the vector particles $W^\pm$, Z^0 (spin 1) are Bosons, like Gravitons (spin 2, provided they exist). Electrons, Neutrinos, Quarks and nucleons are 1/2 spin

Fermions. On the other hand, composite systems like nuclei and atoms may have integer or half-integer spins in their ground states and will behave accordingly at low enough energy. The terms come from the names of Enrico Fermi[1] and Satyendranath Bose.[2] The distinction between Bosons and Fermions is relevant to angular momentum, that is, to the way the wave function of a single particle transforms under rotation. The Pauli principle says that the permutation symmetry of a many-particle wave function (also called the *statistics* of the particle) depends on its spin. Why this connection? This is a new law of Nature that cannot be deduced from the postulates of Quantum Mechanics alone. However, in relativistic Quantum Mechanics the wave equations depend on spin: one has the Klein-Gordon equation for spin 0, the Dirac equation for spin 1/2, for the massive particles of spin 1, the Proca equation, and so on. All those equations are untenable as single-particle equations, because of negative-energy solutions and other severe reasons. They can only stand as field equations, and the quanta of these fields are the particles. Then, the requirement that the field have a ground state (the vacuum state) is mandatory, and Fermion fields have one only if their wave functions is antisymmetric, while Bose fields require total permutation symmetry. In *relativistic field theory*, this is the spin-statistics theorem (Pauli 1940), but un the non-relativistic theory, it must be taken as a new principle.

Let $\phi_1(x_1)$, $\phi_2(x_2)$ denote two wave functions for a boson field. They could represent wave functions of α particles or the wave vectors of two normal modes of the electromagnetic field. A two-boson wave function can be taken in the form

$$\Phi_{ab}(1, 2) = \frac{1}{\sqrt{2}} [\phi_a(1)\phi_b(2) + \phi_a(2)\phi_b(1)]. \qquad (19.3)$$

One can represent the same two-body state in second quantization in the following form:

$$|ab\rangle = a_1^\dagger a_2^\dagger |vac\rangle. \qquad (19.4)$$

Here, $a_1^\dagger$, $a_2^\dagger$ are the creation operators of the harmonic oscillators whose quanta have the space-time dependence $\phi_a(x, t)$ and $\phi_b(x, t)$), while $|vac\rangle$ stands for the ground state of all the oscillators. The presence of a boson of each type is equivalent to the state with both modes in the first excited state. Thus, Eqs. (19.3) and (19.4) contain the same information, and the quantum states built through the two methods are in complete one-to-one correspondence if one takes that all the operators of different modes commute ($[a_1^\dagger, a_2^\dagger]_- = 0$, $[a_1^\dagger, a_2]_- = 0$ and the like). The second quantization, invented by Dirac, is a clever notation, and in addition, it is designed to

---

[1] Enrico Fermi (Rome 1901–Chicago 1954) was one of the great scientists of the twentieth century, both for theoretical and experimental achievements. Among his contributions: the Fermi statistics, the theory of the Weak Interactions, and the discovery of the Δ^{++} resonance. He was awarded the Nobel price in 1938 and he took advantage of the trip to Stockholm to quit fascist Italy for America, where he became a citizen and built the first nuclear reactor.

[2] Satyendranath Bose, Indian physicist (Calcutta 1894–1974), developed the statistics of integer spin particles. His work was completed in cooperation with Albert Einstein.

handle the transitions between states with different numbers of bosons more easily.
For each mode, the known commutation rules $[a, a^\dagger]_- = 1$ apply.

Using two orthogonal spin-orbitals $\psi_1(\mathbf{x}_1)$, $\psi_2(\mathbf{x}_2)$, one must write a two-electron
wave function in agreement with the Pauli principle. The antisymmetric wave func-
tion, or *Slater determinant,* setting a Fermion in each of the two spinors reads as:

$$\Psi_{ab}(1,2) = \frac{1}{\sqrt{2}}[\psi_a(1)\psi_b(2) - \psi_a(2)\psi_b(1)]$$

$$= \frac{1}{\sqrt{2!}}\begin{vmatrix} \psi_a(1) & \psi_b(2) \\ \psi_a(1) & \psi_b(2) \end{vmatrix} \equiv |\psi_a\psi_b|. \tag{19.5}$$

$\Psi(1,2)$ vanishes if $\psi_1 = \psi_2$; one cannot put two electrons in the same spin-orbital.
This rule is the Pauli principle. N electrons occupy N distinct spin-orbitals. In the
ground state, the lowest N spin-orbitals are occupied (the *aufbau* principle).. This is
the ground configurations of atoms and molecules.[3]

Since many one-body states are degenerate, the ground state configuration built
in this way is not unique in general. A common misconception leads some people
to *believe* in the existence of individual spin-orbitals where the electrons belong in
the many electron states. Equation (19.5) appears to say that the two electrons are in
single-particle states ψ_a and ψ_b and the two-particle state is the Slater determinant.
However, any unitary transformation, like $\psi_a = \alpha\xi_a + \beta\xi_b$, $\psi_b = \beta\xi_a - \alpha\xi_b$ with
$\alpha^2 + \beta^2 = 1$, gives[4] the same determinant, but with different orbitals. While the Slater
determinant represents a possible wave function of a physical two-particle state, there
is no way to assign the individual electrons unique spin-orbitals in the physical state.
This remark extends to any many-body state and is one of the manifestations of
entanglement (see Chap. 27).

In summary, the Pauli theorem says that for two particles, the admissible product
functions are:

$$\begin{cases} \psi_a(1)\psi_b(2) & \text{different particles,} \\ \psi_a(1)\psi_b(2) - \psi_a(2)\psi_b(1) & \text{identical fermions,} \\ \psi_a(1)\psi_b(2) + \psi_a(2)\psi_b(1) & \text{identical bosons.} \end{cases}$$

For N identical Bosons, on the other hand, one must symmetrize. Instead of
determinants one writes *permanents.* In both cases, the deviations from classical
predictions are dramatic.

---

[3]This simple rule holds as long as we can neglect the Coulomb interactions, which can mix different
configurations. The antisymmetry principle always holds.
[4]As a consequence of $det(XY) = det(X)det(Y)$.

19.2.1 Simple Examples

In atoms and in condensed matter, the electron-electron interactions are far from negligible, but the determinantal wave functions are useful in approximate methods (Hartree-Fock and various Density Functional schemes) in which the interactions are accounted for by means of an effective potential or mean field. Many qualitative facts can be understood in terms of an effective central field $V(r)$. Changing $V(r)$, the radial wave functions and the level energies change, but the multiplet structure depends mainly on symmetry and is not much affected.

So, He has a $1s^2$ configuration, with two opposite spin electrons in a 1 s level with a modified nuclear charge Z, Li has $1s^2 2s$, Be has $1s^2 2s^2$, B $1s^2 2s^2 2p$, and so on. This simple rule breaks down for heavy elements as a result of many complications including relativistic effects.

Singlet and Triplet States of He

Let $a(\vec{r})$ denote the 1 s orbital in an independent particle approximation (central field) for the He atom. The Hamiltonian is, in obvious notation,

$$H = \sum_{i=1}^{2} \left[\frac{p_i^2}{2m} - \frac{2e^2}{r_i} \right] + \frac{e^2}{r_{12}}. \tag{19.6}$$

According to the Pauli principle, we can put the two electrons in the spin-orbital states $\psi_1 = a(\vec{r})\alpha$, $\psi_2 = a(\vec{r})\beta$; a more expressive notation is $\psi_1 = a(\vec{r}) \uparrow$, $\psi_2 = a(\vec{r}) \downarrow$. The ground state reads as:

$$\Phi(1, 2) = |a(\vec{r}) \uparrow a(\vec{r}) \downarrow |,$$

that is,

$$\Phi(1, 2) = \frac{1}{\sqrt{2}} [\psi_1(1)\psi_2(2) - \psi_1(2)\psi_2(1)]$$
$$= a(\vec{r}_1)a(\vec{r}_2) \frac{[\alpha(1)\beta(2) - \alpha(2)\beta_2(1)]}{\sqrt{2}} = a(\vec{r}_1)a(\vec{r}_2)\chi_S.$$

Here,

$$\chi_S = \frac{\alpha(1)\beta(2) - \beta(1)\alpha(2)}{\sqrt{2}}$$

is the spin singlet, such that

$$S^2 \chi_S \equiv (S(1) + S(2))^2 \chi_S = 0.$$

It is no accident that we got a spin eigenstate, since the Hamiltonian H commutes with the spin operators. Since weak spin-orbit coupling terms are neglected, H is actually spin-independent. Therefore, we can label all the energy eigenstates by the eigenvalues S, m_S of S^2, S_z. We agree with the Pauli principle, since χ_S is odd under

$P(1, 2)$, and the orbital function is even under the exchange of r_1 and r_2; the total wave function is odd.

The triplet ($S = 1$) combinations are even:

$$\chi_{M_S}^T = \begin{cases} \chi_{M_S=1}^T = \alpha(1)\alpha(2), \\[2mm] \chi_{M_S=0}^T = \frac{\alpha(1)\beta(2)+\beta(1)\alpha(2)}{\sqrt{2}}, \\[2mm] \chi_{M_S=-1}^T = \beta(1)\beta(2). \end{cases}$$

They admit odd orbital functions. There is no such function with ony one orbital. Letting b denote the 2 s orbital of He, one can model the excited states of the $1s2s$ configuration:

$$D_{\uparrow\uparrow} = |a \uparrow b \uparrow |, \; D_{\uparrow\downarrow} = |a \uparrow b \downarrow |,$$
$$D_{\downarrow\uparrow} = |a \downarrow b \uparrow |, \; D_{\downarrow\downarrow} = |a \downarrow b \downarrow |.$$

For example, with the notation $a(1) \equiv a(\vec{r}_1)$, $b(1) \equiv b(\vec{r}_1)$, and so on,

$$D_{\uparrow\downarrow} = a(1)b(2)\alpha(1)\beta(2) - a(2)b(1)\alpha(2)\beta(1)$$

is correctly antisymmetric, but it is no singlet and no triplet. Since H is spin-independent, we have the right to demand spin eigenstates. Indeed,

$$\Psi_{S=0} = \frac{D_{\uparrow\downarrow} + D_{\downarrow\uparrow}}{\sqrt{2}} = \frac{a(1)b(2) + b(1)a(2)}{\sqrt{2}} \chi^S. \tag{19.7}$$

and

$$\Psi_{S=1} = \frac{a(1)b(2) - b(1)a(2)}{\sqrt{2}} \chi_{M_S}^T; \tag{19.8}$$

$M_S = 1$ is just $D_{\uparrow\uparrow}$, $M_S = -1$ is $D_{\downarrow\downarrow}$. These are single Slater determinants; for $M_S = 0$, one needs the combination $\frac{D_{\uparrow\downarrow}+D_{\downarrow\uparrow}}{\sqrt{2}}$.

The Exchange Interaction

Let us compute the expectation value of the Coulomb interaction $\frac{e^2}{r_{12}} \equiv \frac{e^2}{|r_1-r_2|}$ over the two-electron eigenstates of spin. The spin averages yield 1.

For the triplets, the results are independent of M_s. One finds that:

$$\langle \Psi_{\uparrow\uparrow} | \frac{e^2}{r_{12}} | \Psi_{\uparrow\uparrow} \rangle = \frac{1}{2} \langle a(1)b(2) - b(1)a(2) | \frac{1}{r_{12}} | a(1)b(2) - b(1)a(2) \rangle$$

$$= \frac{1}{2} [\langle a(1)b(2) | \frac{1}{r_{12}} | a(1)b(2) \rangle - \langle b(1)a(2) | \frac{1}{r_{12}} | a(1)b(2) \rangle \tag{19.9}$$

$$- \langle a(1)b(2) | \frac{1}{r_{12}} | b(1)a(2) \rangle + \langle b(1)a(\vec{r}_2) | \frac{1}{r_{12}} | b(1)a(2) \rangle];$$

since 1 and 2 are dummy indices,

$$\left\langle \Psi_{\uparrow\uparrow} \left| \frac{e^2}{r_{12}} \right| \Psi_{\uparrow\uparrow} \right\rangle = (a(1)b(2) - b(1)a(2)| \frac{1}{r_{12}} |a(1)b(2)). \qquad (19.10)$$

The first term

$$C = \langle a(1)b(2)| \frac{1}{r_{12}} |a(1)b(2)\rangle = \int d^3 r_1 d^3 r_2 \frac{|a(r_1)|^2 |b(r_2)|^2}{r_{12}} \qquad (19.11)$$

has a clear electrostatic meaning and is called the *Coulomb term*; the second contribution

$$-E_x = -(b(1)a(2)| \frac{1}{r_{12}} |a(1)b(2))$$

is the *exchange term*. E_x is always positive, so its contribution is negative for the triplet; it turns out to be and positive for the singlet, as one can verify by working out the expectation value over $\Psi_{\uparrow\downarrow}$. The metastable triplet $1s2s\,^3S$ of He is below the singlet $1s2s\,^1S$ by an amount $2E_x$. This prediction is in fair agreement with the spectroscopic evidence.

As we see in Eq. (19.8), the triplet wave function is antisymmetric when one exchanges $\vec{r}_1$ with $\vec{r}_2$ and the amplitude gets small at small distances. By this simple mechanism, high spin leads to lower repulsion. According to Hund's first rule, the ground state of any atom has the maximum possible spin compatible with its electronic configuration. This is the basic reason for the existence of magnetism in solids. The exchange term plays a key role in explaining the cohesion of solids and the covalent bond in general.

19.3 Many-Electron States and Second Quantization

The determinantal wave functions are a basis for expanding all totally antisymmetric Fermion wave functions. For N independent electrons, $H = \sum_i^N h(i)$, and Schrödinger's equation $H\Psi = E\Psi$ is separable. One can solve in terms of the single-particle solutions of $h(i)\psi_\mu(i) = \epsilon_\mu \psi_\mu(i)$ by setting

$$\Psi(1, 2, \ldots, N) = \hat{A} \prod_i \psi(i),$$

with $E = \sum_\mu \epsilon(\mu)$; here $\hat{A}$ is the anti-symmetrizer that converts the product to a Slater determinant. In general, one can show that in order to normalize the wave function, one must multiply the N electron anti-symmetrized product by $\frac{1}{\sqrt{N!}}$. A convenient notation for the normalized wave function is:

$$\Phi(1, 2, \ldots N) = |\psi_1 \psi_2 \ldots \psi_N|.$$

The N electron states must be a linear combinations of Slater determinants made with spin-orbitals.

Consider the set of orthonormal spin-orbitals that are eigenspinors of some one-body Hamiltonian h such that $h\psi_\mu = \epsilon_\mu \psi_\mu$. For 3 noninteracting electrons, a normalized eigenfunction of the total Hamiltonian $H = h(a) + h(b) + h(c)$ with energy $E_{abc} = \epsilon_a + \epsilon_b + \epsilon_c$ is of the form

$$\Phi_{a,b,c}(1,2,3) = \frac{1}{\sqrt{3!}} det \begin{pmatrix} \psi_a(1) & \psi_a(2) & \psi_a(3) \\ \psi_b(1) & \psi_b(2) & \psi_b(3) \\ \psi_c(1) & \psi_c(2) & \psi_c(3) \end{pmatrix}. \tag{19.12}$$

This means $\Phi_{a,b,c}(1,2,3)\sqrt{3!} = \psi_a(1)\psi_b(2)\psi_c(3) + \psi_a(2)\psi_b(3)\psi_c(1) + \psi_a(3)\psi_b(1)\psi_c(2) - \psi_a(1)\psi_b(3)\psi_c(2) - \psi_a(3)\psi_b(2)\psi_c(1) - \psi_a(2)\psi_b(1)\psi_c(3)$. One-body operators depend on the coordinates (and spin) of one electron at a time, that is, are of the form $\hat{F} = \hat{f}(1) + \hat{f}(2) + \hat{f}(3)$. One can easily verify that the expectation values

$$\langle \Phi_{a,b,c} | \hat{F} | \Phi_{a,b,c} \rangle = \langle \psi_a | \hat{f} | \psi_a \rangle + \langle \psi_b | \hat{f} | \psi_b \rangle + \langle \psi_c | \hat{f} | \psi_c \rangle. \tag{19.13}$$

Thus, the expectation values of momentum, angular momentum, kinetic energy and the like are the sum of those of the spin-orbitals. The charge density is

$$\langle \Phi_{a,b,c} | \hat{\rho}(\boldsymbol{x}, t) | \Phi_{a,b,c} \rangle = |\psi_a(\boldsymbol{x}, t)|^2 + |\psi_b(\boldsymbol{x}, t)|^2| + \psi_c(\boldsymbol{x}, t)|^2. \tag{19.14}$$

Here, $|.|^2$ denotes the square norm of spinors. One can show that the same rule applies for any number N of electrons.

One can verify easily for 3 electrons and also prove in general for any number of electrons that all the out-of-diagonal elements between determinants of one-body operators vanish if the states differ by two or more spin-orbitals. If the states differ by one spin-orbital, one obtains the matrix element between the different orbitals. For instance, for 3 electrons,

$$\langle \Phi_{abc} | \boldsymbol{p}(1) + \boldsymbol{p}(2) + \boldsymbol{p}(3) | \Phi_{dbc} \rangle = \langle \psi_a(1) | \boldsymbol{p}(1) | \psi_d(1) \rangle. \tag{19.15}$$

For a prototype two-body operator, I introduce the Coulomb interaction, which I write in obvious notation:

$$H_C = \sum_{i<j} \frac{1}{r_{ij}} = \frac{1}{r_{12}} + \frac{1}{r_{13}} + \frac{1}{r_{23}}.$$

Developing the expectation value $\langle \Phi_{abc} | H_C | \Phi_{abc} \rangle$ with the determinant (19.12), one gets $6*3*6 = 108$ terms.

So, I introduce a lighter notation, writing $a(1)$ instead of $\psi_a(1)$ and the like. Then I observe that the 6 terms a rising from the development of left determinant are permutations of 1, 2 and 3. We can apply the inverse permutation to all the terms. From odd permutations we collect a minus sign from both determinants. So one can write

$$\langle \Phi_{abc}|H_C|\Phi_{abc}\rangle =$$

$$\left\langle a(1)b(2)c(3) \left| \frac{1}{r_{12}} + \frac{1}{r_{13}} + \frac{1}{r_3} \right| \det \begin{pmatrix} a(1)\ a(2)\ a(3) \\ b(1)\ b(2)\ b(3) \\ c(1)\ c(2)\ c(3) \end{pmatrix} \right\rangle, \qquad (19.16)$$

and these are 18 terms. Consider the terms in $\frac{1}{r_{12}}$. Since there is no operator acting on 3, only the terms with $c(3)$ multiplied by $c(3)$ survive. Therefore, the contribution is

$$\left\langle a(1)b(2)|\frac{1}{r_{12}}| \begin{pmatrix} a(1)\ b(1) \\ a(2)\ b(2) \end{pmatrix} \right\rangle.$$

Hence, the result is:

$$\langle \Phi_{abc}|H_C|\Phi_{abc}\rangle = \left\langle a(1)b(2)\left|\frac{1}{r_{12}}\right| \begin{pmatrix} a(1)\ b(1) \\ a(2)\ b(2) \end{pmatrix}\right\rangle + \langle a(1)c(3) \left|\frac{1}{r_{13}}\right| \begin{pmatrix} a(1)\ c(1) \\ a(2)\ c(2) \end{pmatrix}\right\rangle$$

$$+ \langle b(2)c(3) \left|\frac{1}{r_{23}}\right| \begin{pmatrix} b(2)\ c(2) \\ b(3)\ c(3) \end{pmatrix}\right\rangle.$$

This can be developed and simplified to read as

$$\langle \Phi_{abc}|H_C|\Phi_{abc}\rangle = C - E, \qquad (19.17)$$

where the Coulomb term is

$$C = \langle a(1)b(2)\left|\frac{1}{r_{12}}\right| a(1)b(2)\rangle + \langle a(1)c(3)\left|\frac{1}{r_{13}}\right| a(1)c(3)\rangle\rangle + \langle b(2)c(3)\left|\frac{1}{r_{23}}\right| b(2)c(3)\rangle; \qquad (19.18)$$

the exchange term is

$$E = \langle a(1)b(2)\left|\frac{1}{r_{12}}\right| a(2)b(1)\rangle + \langle a(1)c(3)\left|\frac{1}{r_{13}}\right| a(3)c(1)\rangle\rangle + \langle b(2)c(3)\left|\frac{1}{r_{23}}\right| b(3)c(2)\rangle; \qquad (19.19)$$

and it vanishes for antiparallel spins, since the matrix elements imply spin scalar products. The treatment extends directly to a many-body system where the sums extend to all pairs (Fig. 19.1).

Next, we consider the off-diagonal terms in a many-body system. If just one spin orbital differs from bra and ket, the treatment is similar. Suppose there is a different spin-orbital d in the ket. Then, we must consider

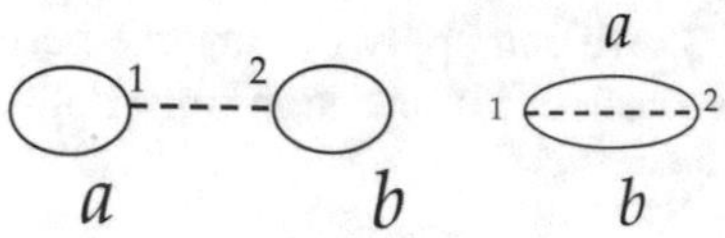

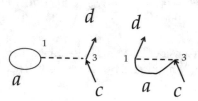

Fig. 19.1 Graphical representation of the direct term (19.18) (left) and exchange term (19.19); the sum over pairs of spin-orbitals is implied

Fig. 19.2 Graphical representation of the direct (left) and exchange terms in the Coulomb scattering $ab \rightarrow ac$

$$\langle \Phi_{abc} | H_C | \Phi_{abd} \rangle =$$

$$\left\langle a(1)b(2)c(3) \left| \frac{1}{r_{12}} + \frac{1}{r_{13}} + \frac{1}{r_3} \right| det \begin{pmatrix} a(1) \ a(2) \ a(3) \\ b(1) \ b(2) \ b(3) \\ d(1) \ d(2) \ d(3) \end{pmatrix} \right\rangle. \tag{19.20}$$

The result vanishes if the spin of d is opposite to the spin of c, since H_C commutes with the z component of the total spin. Therefore, assume that the spins of c and d are the same, but $\langle c|d \rangle = 0$. The terms in $\frac{1}{r_{12}}$ vanishes, but $\frac{1}{r_{13}}$ and $\frac{1}{r_{23}}$ do contribute. The result is:

$$\langle \Phi_{abc} | H_C | \Phi_{abd} \rangle = \sum_k^{occ} \left[\langle k(1)c(2) | \frac{1}{r_{12}} | k(1)d(2) \rangle - \langle k(1)c(2) | \frac{1}{r_{12}} | k(1)d(2) \rangle \right], \tag{19.21}$$

where $\sum_k^{occ}$ sums over the occupied states and the second term is the exchange contribution (for parallel spins) (Fig. 19.2). If both spin-orbitals are different in the ket, the calculation is easy:

$$\langle \Phi_{abc} | H_C | \Phi_{aed} \rangle =$$

$$\left\langle a(1)b(2)c(3) \left| \frac{1}{r_{12}} + \frac{1}{r_{13}} + \frac{1}{r_3} \right| det \begin{pmatrix} a(1) \ a(2) \ a(3) \\ e(1) \ e(2) \ e(3) \\ d(1) \ d(2) \ d(3) \end{pmatrix} \right\rangle, \tag{19.22}$$

but $\frac{1}{r_{12}}$ and $\frac{1}{r_{13}}$ yield nothing by orthogonality, and one is left with

Fig. 19.3 Graphical
representation of the direct
(left) and exchange terms
representing the Coulomb
scattering $ab \rightarrow de$

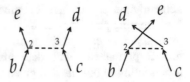

$$\langle \Phi_{abc}|H_C|\Phi_{aed}\rangle = \langle b(2)c(3)\left|\frac{1}{r_{23}}\right|e(2)d(3)\rangle - \langle b(2)c(3)\left|\frac{1}{r}\right|d(2)e(3)\rangle, \quad (19.23)$$

where the second term is of the exchange type (Fig. 19.3).

As a result of a rather involved calculation, we obtained a simple rule that extends directly to any number of electrons.

19.4 Hellmann–Feynman Theorem

One of the few exact results for interacting many-body problems runs as follows. Let $H = H(\lambda)$ denote a many- body Hamiltonian which depends on a parameter λ and E denote an eigenstate of H. Then $E(\lambda) = \langle \Psi|H(\lambda)|\Psi\rangle$ and a small $d\lambda$ produces a first-order variation $d\Psi$, which is orthogonal to Ψ. Therefore,

$$\frac{dE}{d\lambda} = \left\langle \Psi|\frac{dH}{d\lambda}|\Psi\right\rangle. \quad (19.24)$$

19.5 Second Quantization for Fermions

In Sect. 16.2 the creator and annihilation operators for Boson excitations of a 1d field, namely the harmonic oscillator, were presented. In a similar way, the second quantization formalism for Fermions allows us to deal with phenomena where particles are created or destroyed. One such phenomenon is the decay of the neutron into a proton, an electron and an antineutrino

$$n \rightarrow p + e + \bar{\nu};$$

in pair creation, an electron and a positron are created by γ rays, and many more reactions are known in elementary particle physics. In condensed matter physics, at much lower energies, the formalism is still useful. The promotion of an electron from a filled to an empty level can be described as the creation of an electron-hole pair. In scattering processes, when all the particles are conserved, one can proceed with Slater determinants in first quantization; however, second quantization formalism is generally preferred, since it is much more agreable to work with; the picture is that the incoming particle is destroyed and the outgoing electron is created.

In order to change the formalism from Bosons to Fermions, we must simply replace permanents with determinants. In place of an N-times excited oscillator representing N bosons in a given mode, we now consider N-fermion determinants $|u_1 u_2 \ldots u_N|$, where the spin-orbitals are chosen from a *complete orthonormal* set $\{w_i\}$. The index i can be discrete or continuous, but implies a *fixed ordering* of the complete set. In this way, one can convene, e.g., that in $|u_1 u_2 \ldots u_N|$, the indices $1 \ldots N$ are in increasing order, thereby avoiding multiple counting of the same state. The zero-particles or *vacuum* state $|vac\rangle$ replaces the oscillator ground state. For the determinants, it is generally preferable to use a compact notation like $|u_m u_n|$, rather than the explicit $\frac{1}{\sqrt{2}} Det \begin{pmatrix} u_m(1) & u_m(2) \\ u_n(1) & u_m(2), \end{pmatrix}$ which contains the same information. In second quantization, we aim for a unified description of all the Hilbert spaces with any number of particles, starting with the no-particle state that is the vacuum or $|vac\rangle$; the one-body state with one particle in spin-orbital u_k is denoted by $c_k^\dagger|vac\rangle$ and is described as the effect of the creation operator $c_k^\dagger$ over the vacuum. Similarly, the 2-body determinantal state $|u_m u_n|$ becomes $c_m^\dagger c_n^\dagger|vac\rangle$, and the 3-body determinant $|u_m u_n u_p|$ is written as $c_m^\dagger c_n^\dagger c_p^\dagger|vac\rangle$.

Since a determinant is odd when columns are exchanged, we want an anticommutation rule

$$[c_m^\dagger, c_n^\dagger]_+ \equiv c_m^\dagger c_n^\dagger + c_n^\dagger c_m^\dagger = 0. \tag{19.25}$$

It follows that the square of a creation operator vanishes.

Adding a particle to any state cannot lead to the vacuum state,

$$\langle vac|c_m^\dagger = 0. \tag{19.26}$$

By definition,

$$c_m^\dagger \{c_n^\dagger c_r^\dagger|vac\rangle\} = c_m^\dagger c_n^\dagger c_r^\dagger|vac\rangle. \tag{19.27}$$

The notation suggests that $c_m^\dagger$ is the Hermitean conjugate of c_m; this is called an annihilation operator. Taking the conjugate of (19.27)

$$\{\langle vac|c_r c_n\}c_m = \langle vac|c_r c_n c_m \tag{19.28}$$

and taking the scalar product with $c_m^\dagger c_n^\dagger c_r^\dagger|vac\rangle$, we deduce that

$$\{\langle vac|c_r c_n\}c_m \mid c_m^\dagger c_n^\dagger c_r^\dagger|vac\rangle = 1. \tag{19.29}$$

If we now consider c_m as acting on the right, we see that it is changing the 3-body state $c_m^\dagger c_n^\dagger c_r^\dagger|vac\rangle$ into the 2-body one $c_n^\dagger c_r^\dagger|vac\rangle$. Thus, annihilation operator is a well deserved name: an annihilation operator c_m for a fermion in the spin-orbital state u_m removes the leftmost state in the determinant, leaving a $N - 1$ state determinant:

$$c_1 |u_1 u_2 \ldots u_N| = |u_2 \ldots u_N| \tag{19.30}$$

and

$$c_m |vac\rangle = 0. \tag{19.31}$$

It obeys the conjugate of the anticommutation rules (19.25), namely,

$$[c_m, c_n]_+ \equiv c_m c_n + c_n c_m = 0, \quad c_m^2 = 0. \tag{19.32}$$

Next, consider

$$c_n c_m^\dagger c_n^\dagger c_r^\dagger |vac\rangle, \quad n, m, r \ \text{ all different.} \tag{19.33}$$

Since the creation operators anticommute, we get

$$-c_n c_n^\dagger c_m^\dagger c_r^\dagger |vac\rangle = c_m^\dagger c_r^\dagger |vac\rangle,$$

since the m state is created at the leftmost place in the determinant, but is annihilated at once. This shows that creation and annihilation operators also anticommute,

$$[c_n, c_m^\dagger]_+ = 0, \quad n \neq m. \tag{19.34}$$

As long as the indices are different, c and $c^\dagger$ all anticommute, so the pairs $c_n c_m, c_n c_m^\dagger, c_n^\dagger c_m$ and $c_n^\dagger c_m^\dagger$ can be carried through any product of creation or annihilation operators in which the indices n, m do not occur.

Next, we note that $c_p^\dagger |vac\rangle \equiv |p\rangle$ is a one-body wave function; $c_p c_p^\dagger |vac\rangle = |vac\rangle$ and $c_p^\dagger c_p c_p^\dagger |vac\rangle = c_p^\dagger |vac\rangle$. Now one can check that

$$n_p \equiv c_p^\dagger c_p \tag{19.35}$$

is the occupation number operator, since it has eigenvalue 1 on any determinant where p is occupied and 0 if p is empty. On the other hand, $c_p c_p^\dagger$ has eigenvalue 0 on any determinant where p is occupied and 1 if p is empty. Thus in any case $c_p c_p^\dagger + c_p^\dagger c_p = 1$. Since this holds on all the complete set it is an operator identity and we may complete the rules with

$$[c_p, c_q^\dagger]_+ = \delta_{pq}. \tag{19.36}$$

Note that $n_p^\dagger = n_p$ and $n_p^2 = n_p$.

It is important to be able to change basis, i.e., to switch from basis set $\{a_n\}$ to a new set $\{b_n\}$; since

$$|b_n\rangle = \sum_k |a_k\rangle \langle a_k | b_n\rangle, \tag{19.37}$$

the rule is

$$b_n^\dagger = \sum_k a_k^\dagger < a_k | b_n >, \quad b_n = \sum_k a_k < b_n | a_k > . \tag{19.38}$$

It is often useful to go from any set $\{u_n\}$ to the coordinate representation introducing the creation and annihilation field operators

$$\begin{cases} \Psi^\dagger(x) = \sum_n c_n^\dagger u_n^\dagger(x) \\ \Psi(x) = \sum_n c_n u_n(x), \end{cases} \tag{19.39}$$

(here, $u_n^\dagger$ denotes the conjugate spinor). Note that $c_p^\dagger | vac \rangle$ is a one-electron state and corresponds to the first-quantized spinor $u_p(x)$; $\Psi^\dagger(y) | vac \rangle$ is a one-electron state and corresponds to the first-quantized spinor with spatial wave function $\sum_n u_n^\dagger(y) u_n(x) = \delta(x - y)$; thus, it is a perfectly localized electron. The rules are readily seen to be

$$[\Psi(x), \Psi(y)]_+ = 0, \quad [\Psi^\dagger(x), \Psi^\dagger(y)]_+ = 0, \tag{19.40}$$

and

$$[\Psi^\dagger(y), \Psi(x)]_+ = \sum_{p,q} [c^\dagger{}_p, c_q]_+ u_p{}^\dagger(x) u_q(y) = \sum_{p,q} u_p{}^\dagger(x) u_p(y) = \delta(x - y),$$

$$\tag{19.41}$$

where the δ also imposes the same spin for both spinors.

A one-body operator $V(x)$ in second-quantized form becomes

$$\hat{V} = \int dx \Psi^\dagger(x) V(x) \Psi(x) = \sum_{p,q} V_{p,q} c_p^\dagger c_q. \tag{19.42}$$

This gives the correct matrix elements between determinantal states, as one can verify.

The above expressions imply spin sum along with the space integrals, although this was not shown explicitly; let me write the spin components, for one-body operators:

$$\hat{V} = \sum_{\alpha,\beta} \int dx \Psi_\alpha^\dagger V_{\alpha,\beta}(x) \Psi_\beta \tag{19.43}$$

For the spin operators, setting $\hbar = 1$, and using the Pauli matrices, $S_z = \frac{1}{2}\sigma_z$, $S^+ = \begin{pmatrix} 0 & 1 \\ 0 & 0 \end{pmatrix}$, and the rule (19.42), one finds

$$S_z = \frac{1}{2} \int dx \left(\Psi_\uparrow^\dagger(x) \Psi_\uparrow(x) - \Psi_\downarrow^\dagger(x) \Psi_\downarrow(x) \right), \quad S^+ = \int dx \Psi_\uparrow^\dagger(x) \Psi_\downarrow(x). \tag{19.44}$$

When using a discrete basis and notation, we shall write

$$S^+ = \sum_k c_{k\uparrow}^\dagger c_{k\downarrow},$$ (19.45)

which is obtained from (19.44) by taking a Fourier transform in discrete notation. A two-body operator $U(x, y)$ becomes

$$\hat{U} = \int dx \int dy \Psi^\dagger(x)\Psi^\dagger(y)U(x, y)\Psi(y)\Psi(x) = \sum_{ijkl} U_{ijkl} c_i^\dagger c_j^\dagger c_l c_k$$ (19.46)

(please note the order of indices carefully). The Hamiltonian for N interacting electrons in an external potential $\varphi(x)$ is the *true* many-body Hamiltonian in the non-relativistic limit that we shall often regard as the full many-body problem for which approximations must be sought. It may be written as

$$H(r_1, r_2, \ldots, r_N) = H_0(r_1, r_2, \ldots, r_N) + U(r_1, r_2, \ldots, r_N),$$ (19.47)

where H_0 is the free part

$$H_0 = T + V_{ext} = \sum_i \left\{ -\frac{1}{2}\nabla_i^2 + V(r_i) \right\} = \sum_i h_0(i),$$ (19.48)

with T the kinetic energy and V_{ext} the external potential energy, while

$$U = \frac{1}{2} \sum_{i \neq j} u_C(r_i - r_i)$$ (19.49)

is the Coulomb interaction. This Hamiltonian may be written in second-quantized form

$$H = H_0 + U,$$
$$H_0 = \sum_\sigma \int dr \Psi_\sigma^\dagger(r) h_0 \Psi_\sigma(r),$$
$$U = \frac{1}{2} \sum_{\alpha,\beta,\gamma,\delta} \int \int dx dy \psi_\alpha^\dagger(x) \psi_\beta^\dagger(y) u_C(x - y)_{\alpha\gamma,\beta\delta} \Psi_\delta(y)\Psi_\gamma(x).$$ (19.50)

Often, the spin indices are understood as implicit in the integrations. It should be kept in mind that relativistic corrections are needed in most problems with light elements and the relativistic formulation is needed when heavy elements are involved. Fortunately, the ideas that we shall develop lend themselves to a direct generalization to Dirac's framework.

19.5.1   Fano and Anderson Models

As an example of the above formalism, consider the following first-quantized Hamiltonian H_{fq}, representing a discrete level $|0\rangle$ with energy ϵ_0, and a continuum of levels $|k\rangle$ with energy ϵ_k;

$$H_{fq} = \sum_k \epsilon_k |k\rangle\langle k| + \epsilon_0 |0\rangle\langle 0| + \sum_k [V_k |k\rangle\langle 0| + V_k^* |0\rangle\langle k|]. \tag{19.51}$$

The last term allows for a mixing between the continuum and the discrete level. This model represents a single particle (Fermi or Bose) that can hop from a localized state to a band of states and back. The second-quantized version of the model is the Fano model

$$H_F = \sum_{k,\sigma} \epsilon_k \hat{n}_{k\sigma} + \epsilon_0 \hat{n}_{0\sigma} + \sum_k [V_k c_{k\sigma}^\dagger c_{0\sigma} + h.c.]. \tag{19.52}$$

(Here, h.c. stands for the Hermitean conjugate of the preceding term.)

This represents the same physics as (19.51) if it is taken to act in single-particle subspace (but now the anti-commuting creation operators are for spinors and σ denotes spin; the solution to this model is very instructive and is deferred to Chap. 24.

The same formalism lends itself to important extensions. It allows us to work with a many-body systems where the electron states are filled up to a Fermi level. Finally, we may add an interaction term:

$$H_F = \sum_{k,\sigma} \epsilon_k \hat{n}_{k\sigma} + \epsilon_0 \hat{n}_{0\sigma} + \sum_k [V_k c_{k\sigma}^\dagger c_{0\sigma} + h.c.] + U n_{0+} n_{0-}. \tag{19.53}$$

This is the model introduced by Anderson to describe magnetic impurities in metals with many interesting many-body complications.

19.6   Quasiparticles: Bosons, Fermions and Anyons

The quantum many-body problem in atoms and condensed matter systems is extremely involved, since the electron-electron interaction is always important, but some excitations allow a simple effective description reminiscent of non-interacting models. For instance, the excitation that corresponds to the threshold of light absorption in a semiconductor, in a rigorous treatment involves all the electrons and the nuclei of the system; yet, it admits a simple, rather accurate and enlightening description as a one-electron interband transition in an independent particle model. However, in this description, the fermion is not simply the electron, but has a renormalized mass, a modified gyromagnetic ratio and a finite lifetime. In other words, it is a quasi-electron. The absence of an electron below the Fermi level gives rise to complicated excited states of the many-body system that are well described by Fermion quasiparticles, the so-called holes. Such a description, originally due to Lev Landau, is borne out by the modern many-body Green's function formalism and is extremely useful for describing complex situations quantitatively and figuring out what happens. Even in strongly correlated systems, the low energy excitations behave as Fermions of Boson weakly interacting particles, that are known as quasi-particles since they have a finite width in energy that can be interpreted as a finite life time. A polariton is an electron dressed by photons in a solid-state environment. Many excitations are quantized boson quasi-particles, like the phonons, charge density waves or spin waves. Some of the higher energy excitations (plasmons, or quantized plasma oscillations) are also oscillator-like and can be treated as Bosons. Their frequency is $\omega_p = \sqrt{\frac{4\pi ne^2}{m}}$, where n is the density, e the electron charge and m the electron mass. Classically, they can be understood as being due to the oscillation of the electrons in a slab against a positive background. A shift η produces a surface density $\sigma = \eta ne$, and the electric field $E = 4\pi\sigma$ causes the oscillations. The plasmons are the quanta of this oscillator. A more quantitative theory is obtained through field theory methods, often built in computer codes. Nevertheless, the qualitative analysis of the problems remains important.

In 2d (two-dimensional systems), quasiparticles can be not only Bosons or Fermions, but, more generally, Anyons. These are quasi- particles that are possible only in 2d. An exchange P_{12} of Anyons 1 and 2 multiplies the wave function of the system by some phase factor different from ± 1. Indeed, a double exchange P_{12}^2 is equivalent to a round trip of particle 1 around particle 2, which leaves the system in the same state. The same state means the same wave function, up to an arbitrary phase factor. We may write $P_{12} = e^{i\phi}$ with real ϕ. In 3d, one can deform any close path continuously to a point, and this implies that $\phi = 0 \Rightarrow P_{12}^2 = 1 \Rightarrow P_{12} = \pm 1$. The only possibilities are Bosons and Fermions, as we know. In 2d, however, one cannot reduce the path to a point without crossing particle 2. Therefore, any ϕ is possible.

One can obtain a semi-classical mental picture of an anyon as a spinless particle with charge q orbiting around a thin solenoid at distance r. If the current in the solenoid vanishes (i = 0), then L_z is an integer. Now turn on the current i. The particle feels an electric field such that

$$\int rot\,E.nd^2r = -\frac{\partial}{\partial t}\int B.nd^2r = -\frac{\partial\phi}{\partial t},$$

where ϕ is the flux within the orbit. The e.m.f. is $E = -\frac{\frac{\partial\phi}{\partial t}}{2\pi r}$ and is directed in a plane, orthogonal to the radius. The electric field changes the angular momentum. The intensity of the torque is $rE = \frac{\frac{\partial\phi}{\partial t}}{2\pi} = \dot{L}_z$. Thus, the angular momentum is modified by $\Delta L_z = -\frac{q}{2\pi}\Delta\phi$.

In a more accurate description, the charge is being switched on with the flux, and the two are proportional. This leads to $\Delta L_z = -\frac{q}{4\pi}\Delta\phi$. The idea behind such formulations is that a time-dependent flux has a chirality and can impart angular momentum; the presence of the flux can be accepted since the anyon is actually a composite quasi-particle object.

Anyons do exist. They are excitations that form at very low temperatures in the electron liquid when a thin semiconductor layer is sandwiched between AlGaAs layers in strong magnetic field; they carry fractional units of magnetic flux in the fractional quantum Hall effect. Two anyons cannot occupy the same quantum state, but several of them can form time-dependent braids, in which the anyons enter with several possible topologies. The different braids contain information. It has been suggested by Alexei Kitaev in 1997 that the topological properties should be robust against the perturbations that produce decoherence (see Sect. 9.5.1), and therefore it should be possible to build logical gates and a quantum computer operating with anyons. Research on the topological quantum computer is ongoing.

Chapter 20
Perturbation Theory

Qualitative and approximate methods are very important, since they allow us to understand trends. The simple techniques in this chapter help one's physical intuition very much.

20.1 Time-Independent Rayleigh–Schrödinger Perturbation Theory

Suppose we can find the bound states of some Hamiltonian H_0 by solving exactly the time-independent Schrödinger equation $H_0\psi_n^{(0)} = E_n^{(0)}\psi_n^{(0)}$. We calculate the discrete energy levels $\{E_1^{(0)}, E_2^{(0)}, E_3^{(0)}, \ldots\}$ and find agreement with experiment. Invariably then, we wish to see what happens when we perturb the system by what we physically feel is a weak perturbation $\hat{V}$. In most cases, we discover that we are unable to solve

$$(H_0 + \hat{V})\psi_n^{(0)} = E_n^{(0)}\psi_n^{(0)}, \tag{20.1}$$

or perhaps we can solve, but the solution is too involved to make sense of, but if the perturbation is *weak*, E_n differing little from $E_n^{(0)}$, the changes are small compared to the gap between different unperturbed eigenvalues. Then, perturbation theory offers a simple approximate solution that is satisfactory for many purposes and allows us to grasp trends and make predictions. The idea is that of expanding in powers of the perturbation. To this end, we need to rewrite the perturbation in the form $\lambda\hat{V}$, where λ is a dimensionless parameter that eventually becomes 1 but serves to keep track of the powers in the various terms of the expansion; eigenvalues and eigenvectors become series in λ, and the powers of λ become the *perturbation order*. In practice, nobody goes beyond second-order or third-order. With increasing order, the approach quickly becomes unwieldy, and one must look for a different route. Besides, in some important problems, the approach fails, and then it emerges that the solution is not an analytical function of λ, so the perturbation series is a bad start. If the spectrum

© Springer International Publishing AG, part of Springer Nature 2018
M. Cini, *Elements of Classical and Quantum Physics*,
UNITEXT for Physics, https://doi.org/10.1007/978-3-319-71330-4_20

is continuous, the perturbation is never small, and one must resort to the field theory approaches which are a smarter form of perturbation theory taken to infinite order. On the other hand, the present approach can be useful in one-body and many-body problems, with or without spin, with Fermions or Bosons.

We expand the perturbed wave function of the level n over the unperturbed basis $\{\psi^{(0)}\}$:

$$\psi_n = \sum_m c_{n,m} \psi_m^{(0)};$$

obviously, for $\lambda \to 0$, $c_{n,m} \to \delta_{mn}$, and $E_n \to E_n^{(0)}$; this is 0th order. In general, $c_{n,m}$ may be thought of as what would be if $\hat{V}$ produced virtual transitions from unperturbed level n to level m. Then, from (20.1),

$$\sum_m c_{n,m} \left(E_m^0 + \lambda \hat{V} \right) \psi_m^{(0)} = E \sum_m c_{n,m} \psi_m^{(0)}.$$

Multiplying by ψ_k^{0*} and integrating, we obtain the exact formal restatement of the problem:

$$(E_n - E_k^{(0)}) c_{n,k} = \lambda \sum_m V_{km} c_{n,m}. \qquad (20.2)$$

Equation (20.1) is rewritten as a discrete infinite system that we are going to solve by successive approximations.

20.1.1   Discrete Non-degenerate Spectrum

In this case (different $\{E_n^{(0)}\}$ for different n), the weak perturbation changes the levels so little (compared to the level separation) that one can tell, for each perturbed energy, the unperturbed one from which it is derived. Then, we look for the solution of (20.2) in the form of a series expansion that starts from $\psi_n^{(0)}$ for $\lambda \hat{V} \to 0$, assuming a series expansion in powers of λ,

$$\psi_n = \psi_n^{(0)} + \lambda \psi_n^{(1)} + \lambda^{(2)} \psi_n^{(2)} + \dots.$$

Therefore, we set

$$c_{n,k} = c_{n,k}^{(0)} + \lambda c_{n,k}^{(1)} + \lambda^{(2)} c_{n,k}^{(2)} + \dots$$
$$E = E_n^{(0)} + \lambda E_n^{(1)} + \lambda^{(2)} E_n^{(2)} + \dots,$$

with $c_{n,k}^{(0)} = \delta_{k,n}$. Then, (20.2) becomes

$$(E_n^{(0)} - E_k^{(0)} + \lambda E_n^{(1)} + \lambda^2 E_n^{(2)} + \ldots)(\delta_{k,n} + \lambda c_{n,k}^{(1)} + \lambda^2 c_{n,k}^{(2)} + \ldots) = \tag{20.3}$$
$$\lambda \sum_m V_{km}(\delta_{m,n} + \lambda c_{n,m}^{(1)} + \lambda^2 c_{n,m}^{(2)} + \ldots).$$

We multiply, then we separate the powers of λ:

$$[E_n^{(0)} - E_k^{(0)}]\delta_{k,n} + \lambda[E_n^{(1)}\delta_{k,n} + c_k^{(1)}(E_n^{(0)} - E_k^{(0)})]$$
$$+ \lambda^2[E_n^{(2)}\delta_{k,n} + c_k^{(2)}(E_n^{(0)} - E_k^{(0)}) + c_k^{(1)}E_n^{(0)}] + \ldots \tag{20.4}$$
$$= \lambda \sum_m V_{km}\delta_{m,n} + \lambda^2 \sum_m V_{km}c_m^{(1)} + \ldots.$$

At order 0, we get the trivial result $(E_n^{(0)} - E_k^{(0)})\delta_{k,n} = 0$. At first order, however,

$$\underbrace{(E_n^{(0)} - E_k^{(0)})c_k^{(1)}} + \underbrace{E_n^{(1)}\delta_{k,n}} = \sum_m V_{k,m}\delta_{m,n} = V_{k,n}.$$

For $k = n$, the first curly bracket vanishes, and one finds

$$E_n^{(1)} = V_{n,n}. \tag{20.5}$$

For $k \neq n$, the second curly bracket vanishes, and one finds the amplitudes of the *virtual excitations* $n \leftrightarrows k$

$$c_{n,k}^{(1)} = \frac{V_{k,n}}{E_n^{(0)} - E_k^{(0)}}, \quad k \neq n,$$

and the first-order correction to $\psi_n^{(0)}$ is found to be

$$\psi_n^{(1)} = \sum_{k \neq n} \frac{V_{k,n}}{E_n^{(0)} - E_k^{(0)}} \psi_k^{(0)}. \tag{20.6}$$

The correction cannot blow up since the spectrum is discrete. Since we need $< \psi^{(1)}|\psi^{(0)} >= 0$ in order to ensure that $\langle \psi|\psi \rangle = 1 + O(\lambda^2)$, that is, $\psi^{(1)}$ is correctly normalized (the error is at second order). It follows that $c_{n,n}^{(1)} = 0$.

From (20.6), it is clear that the first approximation is good if $\frac{V_{k,n}}{E_n^{(0)} - E_k^{(0)}} \ll 1$.

In second order, (20.4) yields

$$\underbrace{(E_n^{(0)} - E_k^{(0)})c_{n,k}^{(2)}} + \underbrace{E_n^{(1)}c_{n,k}^{(1)} + E_n^{(2)}\delta_{k,n}} = \sum_m V_{k,m}c_{n,m}^{(1)}.$$

For $k = n$, the second curly bracket survives; since $c_{n,n}^{(1)} = 0$; the second-order correction to the energy reads as:

$$E_n^{(2)} = \sum_m V_{n,m}c_{nm}^{(1)} = \sum_{m \neq n} \frac{|V_{n,m}|^2}{E_n^{(0)} - E_m^{(0)}}. \tag{20.7}$$

This is always negative for the ground state.

Problem 34 Show that the energy of the H atom in the ground state plunged into a weak uniform electric field $\mathbf{E}$ is $E = E_0 + \frac{1}{2}\alpha\mathbf{E}^2$ and show the order of magnitude of the polarizability α.

Solution 34 The effect on the ground state of the perturbation $\hat{V} = ez\mathbf{E}$ is quadratic, since the operator is odd and $\langle\psi_0|\hat{V}|\psi_0\rangle = 0$. The correction to the energy arises in second order. The polarizability has the dimensions of a volume and is about a_0^3. The experimental value is, $\alpha = 0.66 \ 10^{-24} \ \mathrm{cm}^3$.

Problem 35 The oscillator with Hamiltonian

$$H_0 = \frac{p^2}{2m} + \frac{1}{2}m\omega^2 x^2$$

is perturbed by

$$H' = bx + c.$$

Find the correction to the ground state energy by perturbation theory up to the second order in the case $b = -m\omega^2 a$ with $c = \frac{1}{2}m\omega^2 a^2$. Since this case corresponds to a shift $x \to x - a$, one can compare the perturbation theory approximation with the exact result.

Solution 35 Recall that $\hat{x} = \frac{x_0(a+a^\dagger)}{\sqrt{2}}$, $\hat{p} = -i\hbar\frac{(a-a^\dagger)}{x_0\sqrt{2}}$, where $x_0 = \sqrt{\frac{\hbar}{m\omega}}$, $[a, a^\dagger] =$
1. In first order, the average yields c, and in second order, $E^{(2)} = -\frac{b^2(x_{01})^2}{\hbar\omega} = -\frac{b^2}{\hbar\omega}\frac{x_0^2}{2} = -\frac{1}{2}\frac{b^2}{m\omega^2}$. The correction up to second order gives

$$c - \frac{1}{2}\frac{b^2}{m\omega^2} = 0.$$

This is *exact*, since the effect of H' is $x \to x - a$. All the higher terms of the perturbation series sum up to 0.

Problem 36 For the *anharmonic* oscillator with Hamiltonian

$$\hat{H} = \left(\hat{n} + \frac{1}{2}\right)\hbar\omega + \alpha\left(\frac{x}{x_0}\right)^4,$$

calculate the effect of the perturbation on the ground state in first and second order. When can we say that the second order is small compared to the first?

Solution 36 Recall that $\hat{x} = (x_0(a+a^\dagger))/\sqrt{2}$, $\hat{p} = -i\hbar((a-a^\dagger))/(x_0\sqrt{2})$, where $x_0 = \sqrt{\frac{\hbar}{m\omega}}$, $[a, a^\dagger] = 1$. The first-order correction is $E^1 = \alpha[(\frac{x}{x_0})^4]_{0,0}$. The nonzero matrix elements (16.28) of x are $x_{n+1,n} = x_{n,n+1} = C\sqrt{n+1}$, with $C = \frac{x_0}{\sqrt{2}}$. Acting on x, from $|0\rangle$, one arrives at $|1\rangle$, and so

$$(x^4)_{0,0} = x_{0,1}(x^2)_{1,1}x_{1,0}.$$

Therefore,

$$(x^2)_{1,1} = x_{1,0}x_{0,1} + x_{1,2}x_{2,1}.$$

So,

$$(x^4)_{0,0} = x_{0,1}[x_{1,0}x_{0,1} + x_{1,2}x_{2,1}]x_{1,0} = 3C^4,$$

and the first-order correction is $E^1 = \frac{3}{4}\alpha$.

In working out (20.7), the states with $m > 4$ play no role. Since by parity, $(x^4)_{0,1} = 0 = (x^4)_{0,3}$,

$$\begin{aligned}(x^4)_{0,2} &= x_{0,1}(x^3)_{1,2} = x_{0,1}[x_{1,2}(x^2)_{2,2} + x_{1,0}(x^2)_{0,2}] \\ &= x_{0,1}x_{1,2}(x_{2,1}x_{1,2} + x_{2,3}x_{3,2}) + x_{0,1}x_{1,0}x_{0,1}x_{1,2} \\ &= (\sqrt{2}(2+3) + \sqrt{2})C^4 = 6\sqrt{2}C^4\end{aligned}$$

and

$$(x^4)_{0,4} = x_{0,1}x_{1,2}x_{2,3}x_{3,4} = \sqrt{24}C^4.$$

The second order contribution is:

$$-E_0^{(2)} = \frac{\alpha^2}{x_0^8 \hbar\omega}\left\{\frac{|x_{02}^4|^2}{2} + \frac{|x_{04}^4|^2}{4}\right\} = \frac{\alpha^2 C^8}{x_0^8 \hbar\omega}\left\{\frac{|6\sqrt{2}|^2}{2} + \frac{|\sqrt{24}|^2}{4}\right\}.$$

The final result is

$$E_0 \approx \frac{1}{2}\hbar\omega + \frac{3}{4}\alpha - \frac{21\alpha^2}{8\hbar\omega},$$

and the second order is small if $\alpha \ll \hbar\omega$.

Problem 37 For an anharmonic oscillator with Hamiltonian $\hat{H} = (\hat{n} + \frac{1}{2})\hbar\omega + \alpha(\frac{x}{x_0})^3$, calculate the correction to the energy of the state n in first and second order. Perturbation theory, however, cannot be sound. Why?

Solution 37

$$E_n = \left(n + \frac{1}{2}\right)\hbar\omega - \frac{15\alpha^2}{\hbar\omega}\left(n^2 + n + \frac{11}{30}\right).$$

However, the perturbed potential has no minimum and true discrete levels cannot exist.

Problem 38 A particle in the one-dimensional well $0 < x < L$ is perturbed by the small potential $V(x) = V_0 \cos(\frac{\pi x}{L})\theta(\frac{L}{2} - x)$. Calculate the first-order correction to the levels $n = 1$ and $n = 2$.

Solution 38 Using $\psi_n(x) = \sqrt{\frac{2}{L}}\sin(\frac{n\pi x}{L})$, we obtain

$E'_n = \frac{2V_0}{L}\int_0^{\frac{L}{2}} dx \sin(\frac{n\pi x}{L})^2 \cos(\frac{\pi x}{L}) = \frac{2V_0}{\pi}\int_0^{\frac{\pi}{2}} dt \sin(nt)^2 \cos(t)$. For $n = 1$, $E'_1 = \frac{2V_0}{3\pi}$, and for $n = 2$, $E'_2 = \frac{8V_0}{\pi}\int_0^{\frac{\pi}{2}} d(t) \sin(t)^2 \cos^3(t)$. Integrating in $d\sin(t)$, one finds $= 8\frac{V_0}{\pi}\int_0^1 dx(x^2 - x^4) = \frac{16V_0}{15\pi}$.

20.1.2 Degenerate Levels

Let the eigenvalue $E_n^{(0)}$ be g times degenerate; let

$$\{\psi_\nu^{(0)}\}, \quad \nu = 1, \ldots, g$$

denote an orthogonal basis for the subspace of degenerate eigenfunctions. At order 0, Eq. (20.2) reduces to $(E_n^{(0)} - E_k^{(0)})c_k = 0$. Hence, $c_k^{(0)} = 0$ for $E_n^{(0)} \neq E_k^{(0)}$. A perturbation $\hat{V}$ small compared to the energy separation from the other levels produces a weak mixing with the corresponding eigenfunctions; yet, in the $E_n^{(0)}$ subspace, it may have strong effects. The consequences are particularly significant if the degeneracy is (totally or partially) removed.

First Order

Therefore, to calculate the first-order corrections E_n^1, we ignore the mixing with the eigenfunctions outside the subspace. In the subspace,

$$H_{\mu,\nu} = E_n^{(0)}\delta_{\mu,\nu} + V_{\mu,\nu}, \quad \mu, \nu = 1 \ldots g;$$

we have to diagonalize $\hat{V}$. Letting $E = E_n^0 + E_n^1$, we must solve the secular equation

$$E_n^1 c_\mu = \sum_\nu V_{\mu,\nu}c_\nu. \tag{20.8}$$

The eigenvectors are the eigenfunctions of the 0th approximation, which are linear combinations of the unperturbed $\psi_\nu^{(0)}$.

Second Order

The other levels are relevant to the second and higher approximations. Usually, this is important when the degeneracy is not resolved in the first approximation.

In such cases, to find the correction $E^{(2)}$, one starts from

$$(E - E_n^{(0)})c_\mu = E^{(2)}c_\mu = \sum_m V_{\mu m}c_m$$

and determines the amplitude of the state m outside the degenerate level n by $(E_n - E_m)c_m = \sum_\nu V_{m\nu}c_\nu$, which is accurate to the first order. Substituting,

$$E^{(2)}c_\mu = \sum_m \frac{V_{\mu m}}{E_\mu - E_m} \sum_\nu V_{m\nu}c_\nu.$$

Therefore, the second-order correction is given by condition:

$$det\left|\sum_m \frac{V_{m\mu}V_{\nu m}}{E_\mu - E_m} - E^2\delta_{\mu\nu}\right| = 0. \tag{20.9}$$

The change with respect to the previous case is $V_{\mu\nu} \to \sum_m \frac{V_{m\mu}V_{\nu m}}{E_\mu - E_m}$.

Problem 39 Recall (from Sect. 13.5) that for the plane rigid rotor, with momentum of inertia I and Hamiltonian $H = \frac{\hat{L}_z^2}{2I}$, the eigenfunctions are $\psi_k(\phi) = \frac{e^{ik\phi}}{\sqrt{2\pi}}$ with eigenvalues $k = 0$, and $k = \pm m\hbar$, $m = 1, 2, \ldots$. The energy eigenvalues $E_k = \hbar^2\frac{k^2}{2I}$ are degenerate, except $k = 0$. Find the effect of the directional perturbation

$$\hat{V} = 2\pi\lambda\,\delta(\phi - \phi_0),$$

where λ, ϕ_0 are constants.

Solution 39 The ground level $k = 0$ gets shifted at first order in λ. For the excited states,

$$\langle\phi_m|V|\phi_m\rangle = \lambda\int_0^{2\pi} d\phi\, e^{-im\phi}\delta(\phi - \phi_0)e^{im\phi} = \lambda$$

$$\langle\phi_{-m}|V|\phi_m\rangle = \lambda\int_0^{2\pi} d\phi\, e^{im\phi}\delta(\phi - \phi_0)e^{im\phi} = \lambda e^{2im\phi_0},$$

and the perturbation mixes the functions i ψ_{-m}, ψ_m, $m > 0$. The matrix reads as:

$$V = \begin{bmatrix} \lambda & \lambda e^{2im\phi_0} \\ \lambda e^{-2im\phi_0} & \lambda \end{bmatrix}.$$

The eigenvalues give the first-order corrections.

The eigenvalue $E^1 = 0$ corresponds to the eigenvector $\frac{1}{\sqrt{2}}\begin{pmatrix} -e^{im\phi_0} \\ e^{-im\phi_0} \end{pmatrix}$ of V, which is

$$\psi_- = -\frac{e^{im\phi_0}}{\sqrt{2}}\phi_{-m} + \frac{e^{-im\phi_0}}{\sqrt{2}}\phi_m = i\frac{\sin[m(\phi - \phi_0)]}{\sqrt{\pi}}$$

and is not affected by the perturbation. The eigenvalue $E^1 = 2\lambda$ corresponds to the eigenvector $\frac{1}{\sqrt{2}}\begin{pmatrix} e^{im\phi_0} \\ e^{-im\phi_0} \end{pmatrix}$, that is, to $\psi_+ = \frac{\cos[m(\phi - \phi_0)]}{\sqrt{\pi}}$.

Problem 40 Explain the *Stark–Lo Surdo effect* for the H atom with $n = 2$ (the H atom plunged in an electric field $\mathcal{E}$).

Solution 40 The perturbation

$$V = ez\mathcal{E}$$

acts in first order *(linear Stark–Lo Surdo effect)*. The wave functions $\psi_{nlm} = R_{nl}Y_{lm}(\theta, \phi)$ involved are:

$$
\begin{cases}
\psi_{200} & R_{20}(r) = 2(\frac{1}{2a_0})^{\frac{3}{2}}(1 - \frac{r}{2a_0})e^{-\frac{r}{2a_0}} & Y_{00} = \frac{1}{\sqrt{4\pi}}, \\
\psi_{210} & R_{21}(r) = \frac{1}{\sqrt{3}}(\frac{1}{2a_0})^{\frac{3}{2}}\frac{r}{a_0}e^{-\frac{r}{2a_0}} & Y_{10} = \sqrt{\frac{3}{4\pi}}\cos(\theta), \\
\psi_{211} & R_{21}(r) = \frac{1}{\sqrt{3}}(\frac{1}{2a_0})^{\frac{3}{2}}\frac{r}{a_0}e^{-\frac{r}{2a_0}} & Y_{11} = -\sqrt{\frac{3}{8\pi}}\sin(\theta)e^{i\phi}, \\
\psi_{21-1} & R_{21}(r) = \frac{1}{\sqrt{3}}(\frac{1}{2a_0})^{\frac{3}{2}}\frac{r}{a_0}e^{-\frac{r}{2a_0}} & Y_{1-1} = -\sqrt{\frac{3}{8\pi}}\sin(\theta)e^{-i\phi}.
\end{cases}
$$

We need the matrix of the perturbation on this basis. Since V is odd, the diagonal elements vanish, and since V does not depend on ϕ, the matrix elements between functions with different m also vanish. ψ_{200} gets mixed with ψ_{210}, and effectively the matrix that we must diagonalize is 2×2. Using $\cos(\theta) = \frac{z}{r}$, one finds

$$\langle \psi_{200}|V|\psi_{210}\rangle = 3ea_0\mathcal{E}$$

and

$$
\begin{cases}
E_-^{(1)} = 3ea_0\mathcal{E} & \frac{\psi_{200} - \psi_{210}}{\sqrt{2}} \\
E_+^{(1)} = -3ea_0\mathcal{E} & \frac{\psi_{200} + \psi_{210}}{\sqrt{2}}
\end{cases}.
$$

The degenerate level splits into 3.

20.2   Time-Dependent Perturbations

Transitions between quantum states occur under the influence of some time-dependent external perturbation $\hat{V}(t)$. This is what one normally does to study a system. In order to test the properties of a molecule or a superconductor, one uses a probe like a weak field that produces excitations, but still allows us to classify the states in terms of the stationary states of the unperturbed system, while allowing transitions between them. The present section is therefore fundamental for all spectroscopies. We wish to solve

$$i\hbar\frac{\partial\psi}{\partial t} = \left[H_0 + \hat{V}(t)\right]\psi \tag{20.10}$$

in terms of the full orthonormal set of solutions of the *unperturbed* problem

$$ i\hbar \frac{\partial \phi_n}{\partial t} = H_0 \phi_n(t); $$

evolving according to

$$ \phi_n(t) = \phi_n(0) \exp\left(\frac{-i\epsilon_n t}{\hbar}\right). $$

Let us suppose that the system described by H_0 is prepared in the eigenstate ϕ_n, but under the action of the *weak* perturbation $\hat{V}(t)$, it makes a transition to other unperturbed states. First, let us expand[1] on the unperturbed basis with $\psi = \sum_k a_k(t)\phi_k(t)$; substituting into (20.10), we find

$$ i\hbar \sum_k \dot{a}_k(t)\phi_k(t) = \sum_k a_k(t)\hat{V}\phi_k(t). $$

Scalar multiplication by $\phi_m(t)$ gives us

$$ i\hbar\dot{a}_m = \sum_k V_{mk}(t)a_k(t), \tag{20.11} $$

where

$$ V_{mk}(t) = \langle \phi_m(t)|\hat{V}(t)|\phi_k(t)\rangle = \langle \phi_m(0)|\hat{V}(t)|\phi_k(0)\rangle e^{i\omega_{mk}t}, $$

$$ \hbar\omega_{mk} = \epsilon_m - \epsilon_k. \tag{20.12} $$

Note the double time-dependence of $V_{mk}(t)$ via the operator and the phase factor. From the *initial conditions*

$$ \psi(0) = \phi_n(0) \implies a_k(0) = \delta_{kn}, \tag{20.13} $$

we can write an exact formal solution, in the presence of $\hat{V}$ and obtain the transition amplitude to any final state m. For clarity,

> we shall denote the amplitude to go to m by $a_{n\to m}$.

In view of (20.12), if $\omega_{kn} > 0$, the system is *promoted* to a higher level m, while for $\omega_{mn} < 0$, we speak of a *decay* to a lower level. The system energy is not conserved, since H is time-dependent.

So far we have made no approximations, but usually the exact solution is prohibitively difficult or expensive. Now we introduce the assumption that the perturbation is so weak, that the probability of exciting the system are small. In this case,

---

[1]P.A.M. Dirac, 1926.

we feel licensed not only to use perturbation theory, but even to stop after one step, or *first approximation*. Simply, we put into the r.h.s. of (20.11) the initial condition (20.13) $\psi(t) \sim \phi_n(t)$, or, in other words, $a_k(0) \sim \delta_{kn}$. So,

$$i\hbar \frac{d}{dt} a_{n \to m}(t) = V_{mn}(t),$$

which is solved by

$$a_{n \to m}(t) = -\frac{i}{\hbar} \int_{t_{min}}^{t_{max}} V_{mn}(\tau) d\tau. \tag{20.14}$$

The integration limits t_{min}, t_{max} depend on the problem. For instance, if $\hat{V}$ is switched on at time 0 and off at a later time T, we write $\int_0^T$, and so on.

Problem 41 A particle is confined between $x = 0$ and $x = a$ in a deep one-dimensional well. At time $t = 0$, it is in the ground state.

a. Calculate the first-order transition probability P_m to the m-th excited state under the action of a potential $V(t) = \lambda \delta(x - \frac{a}{2})\theta(t)\theta(T - t)$.
b. State the validity criterion of the approximation.
c. Find T such that the transition probability to the second excited level $m = 3$ is maximum.

Solution 41 Using the eigenfunctions (11.3) $\psi_n(x) = \sqrt{\frac{2}{a}} \sin\left[n\pi\frac{x}{a}\right]$, one finds

$$V_{m1}(t) = \frac{2\lambda}{a} \int dx \sin\left(\frac{m\pi x}{a}\right) \sin\left(\frac{\pi x}{a}\right) \delta\left(x - \frac{a}{2}\right) e^{i\omega_{m1}\tau},$$

with

$$\omega_{m1} = \frac{\hbar\pi^2(m^2 - 1)}{2ma^2}.$$

Thus, $P_m = |\frac{2\lambda}{\hbar a} \sin(m\pi/2) \int_0^T e^{i\omega_{m1}\tau} d\tau|^2$. Moreover, $P_m \ll 1$.
Since $|\int_0^T e^{i\omega_{m1}\tau} d\tau|^2 = 2\frac{1-\cos(\omega_{m1}T)}{\omega_{m1}^2}$, the condition is $\cos(\omega_{31}T) = -1$, which implies that $\omega_{31}T = \pi, 3\pi, 5\pi, \ldots$

Problem 42 A harmonic oscillator is in the ground state at time $t = 0$. Calculate the probability of transition to level n at time T under the action of the weak perturbation $V(t) = (a + a^\dagger)V_0 T\delta(T - t)$.

Solution 42 It is only possible to jump to $n = 1$. The probability is $|\frac{V_0T}{\hbar}|^2$.

20.2.1 Periodic Perturbations and Fermi Golden Rule

The interaction of charges with a monochromatic electromagnetic wave is most often well described by in lowest-order perturbation theory, unless one wants to study nonlinear optics by means of intense laser radiation. In most cases, one introduces the minimal coupling rule (2.94), neglecting the $\vec{A}^2$ term on the grounds that it is quadratic. Then setting $A(t) = A_0 e^{i\omega t} + A_0^* e^{-i\omega t}$, one considers a periodic perturbation

$$\hat{V}(t) = \hat{W} e^{i\omega t} + \hat{W}^\dagger e^{-i\omega t}, \quad \omega > 0. \tag{20.15}$$

where $\hat{W} = \vec{A}_0 \cdot \vec{p}$. We shall see that the first terms produces emission and the second absorption of a photon. Since the integrand is periodic, we use (20.14) with integration limits $-\infty, \infty$. Working out

$$a_{n \to m}(t) = -\frac{i W_{mn}}{\hbar} \int_{-\infty}^{\infty} \left[e^{i(\omega_{mn} + \omega)\tau} + e^{i(\omega_{mn} - \omega)\tau} \right] d\tau,$$

using $\int_{-\infty}^{\infty} e^{imx} = 2\pi \delta(x)$, we get

$$a_{n \to m} \equiv a_m(t \to \infty) \overset{?}{=} -\frac{i W_{mn}}{\hbar} 2\pi \left[\delta\left(\omega_{mn} + \omega\right) + \delta\left(\omega_{mn} - \omega\right) \right].$$

If $\omega_{mn} < 0$ (emission), only the first term contributes, if $\omega_{mn} > 0$ (absorption) only the second. In this way, one obtains delta-like lines with the resonance condition $\pm\omega_{mn} = \omega$ for absorption and emission. Is this OK? No! This is a transition amplitude, and in order to find the probability $P_{n \to m} = a_{n \to m}^* a_{n \to m}$, we must square; alas, $\delta(x)^2$ is a mathematical nonsense. We made a mistake, but where?

Actually, it is not a mathematical error, but an oversimplified scheme. A perfectly monochromatic excitation should last for a time $T \to \infty$ and cannot exist. Often in Physics problems, one avoids trouble by introducing a cutoff, that is, by replacing zero with a small quantity and infinity with a large one. Even here, we get out of trouble if we introduce the duration T of the excitation. For the absorption of a wave that lasts a time T, we find

$$a_{n \to m}(t) = -\frac{i W_{mn}}{\hbar} \int_0^T e^{i(\omega_{mn} - \omega)\tau} d\tau.$$

and so

$$P_{n \to m} = \frac{i W_{mn}^*}{\hbar} \int_0^T e^{-i(\omega_{mn} - \omega)\tau_1} d\tau_1 \times \left(-\frac{i W_{mn}}{\hbar} \right) \int_0^T e^{i(\omega_{mn} - \omega)\tau_2} d\tau_2.$$

Since $\int_0^T e^{ix\tau} d\tau = \frac{e^{ixT} - 1}{ix}$, now the lines have some width $\sim \frac{1}{T}$. The problem of the squared delta is solved. Then, we can decide to neglect the finite width and simplify the result by approximating one of the integrals with $\int_0^T \sim \int_{-\infty}^{\infty} = 2\pi\delta(\omega_{mn} - \omega)$; then, the other integral equals T. In this way,

$$P_{n \to m} = \left| \frac{W_{mn}}{\hbar} \right|^2 T \times 2\pi\delta(\omega_{mn} - \omega).$$

This probability grows linearly with the duration T of the experiment. With increasing T, one first violates the condition $P \ll 1$ which is needed for first-order perturbation theory, and eventually one finds $P > 1$ which makes no sense. One solution would be: evaluate a realistic duration T and plug it into the formula. Actually, one can avoid this step, by defining the transition rate

$$R_{n \to m} = \frac{\partial P_{n \to m}}{\partial T}$$

which is the actually measured quantity. In this way, one arrives at the *Fermi golden rule*:

$$R_{n \to m} = \frac{1}{\hbar} |W_{mn}|^2 \, 2\pi\delta(\epsilon_m - \epsilon_n - \hbar\omega). \tag{20.16}$$

Problem 43 Let $H_0 = \epsilon\sigma_z$; $\hat{V} = \lambda(e^{i\omega t} + e^{-i\omega t})\sigma_x$, with $\psi(0) = \binom{0}{1}$; calculate the rate R.

Solution 43 $R = \frac{2\pi\lambda^2}{\hbar} \delta(2\epsilon - \hbar\omega)$.

Selection Rules for the H Atom

The H atom can absorb a photon and promote the electron to a higher level; when excited, it emits a photon while the electron jumps down to a lower level. The transition rate can be obtained in the dipole approximation assuming a uniform electric field; this is good in the visible or near ultraviolet, since the wave length is on the order of hundreds of nanometers and is large compared to the Bohr radius. To find the transition rate, one computes the matrix elements of $\vec{r} = (x, y, z)$ between the initial and final wave functions. The transition occurs if at least one between

$$A_x = \langle \psi_{n'l'm'} | x | \psi_{nlm} \rangle, \ A_y = \langle \psi_{n'l'm'} | y | \psi_{nlm} \rangle, \ A_z = \langle \psi_{n'l'm'} | z | \psi_{nlm} \rangle,$$

does not vanish. Since the dipole is odd, this requires that the initial and final wave functions have opposite parity. Since the parity of the harmonic $Y_{lm}(\theta, \phi)$ is $(-1)^l$, this requires that l e l' have opposite parity (*Laporte's rule*).

Besides the parity, the total angular momentum (including the photon) is conserved. This leads to the rules

$$\boxed{m' = m \text{ or } m' = m \pm 1, l' = l \pm 1.} \tag{20.17}$$

These rules forbid the transitions for which the matrix element vanishes. Indeed, up to a constant, $z \sim Y_{10}(\theta, \phi)$, $(x \pm iy) \sim Y_{1\pm 1}(\theta, \phi)$. Thus the angular factor in the matrix element reads as

$$\int d\Omega\, Y^*_{l'm'}(\theta,\phi) Y_{1q}(\theta,\phi) Y_{lm}(\theta,\phi) = \sqrt{\frac{3}{4\pi}\frac{2l+1}{2l'+1}}\,\langle l100|l'0\rangle\langle l1mq|l'm'\rangle,$$

and the forbidden transitions depend on vanishing Clebsh–Gordan coefficients. The result shows that the Photon spin is 1.

Chapter 21
Variational Principle for Schrödinger–Pauli Theory

Variational principles are ubiquitous in Theoretical Physics and are very useful mathematical tools. This one depends on the existence of a ground state.

21.1 The Ground State and the Absolute Minimum of Energy

The Schrödinger–Pauli theory is characterized by the fact that every system must have a ground state, whose energy is a lower bound to the energies of all states. This is not true in Classical Mechanics, in which, for instance, a H atom could have any energy.

Let ϕ denote any wave function of an arbitrary quantum system, with one or several particles, duly normalized with

$$N = < \phi|\phi >= 1. \tag{21.1}$$

The expectation value of the energy

$$E = < \phi|H|\phi > \tag{21.2}$$

cannot be lower than the energy ϵ_0 of the ground state. To see that, let us expand ϕ in eigenvectors ψ_n of H:

$$|\phi\rangle = \sum_n |\psi_n\rangle\langle\psi_n|\phi\rangle.$$

Since $\epsilon_n \geq \epsilon_0$,

$$E = \sum_n |\langle\psi_n|\phi\rangle|^2 \epsilon_n \geq \epsilon_0. \tag{21.3}$$

© Springer International Publishing AG, part of Springer Nature 2018
M. Cini, *Elements of Classical and Quantum Physics*,
UNITEXT for Physics, https://doi.org/10.1007/978-3-319-71330-4_21

So, the ground state can be sought by looking for the energy minimum. We can get close to the real ground state if we start from a correct qualitative idea of the form of the ground wave function. To explore this possibility, let us consider a familiar case, namely, the ground state of the harmonic oscillator ($H = \frac{p^2}{2m} + \frac{1}{2}m\omega^2 x^2$). We can pick the exact solution among the normalized trial functions, depending on the unknown parameter b,

$$\phi(x) = Ae^{-bx^2}, \quad A = \left(\frac{2b}{\pi}\right)^{1/4}.$$

We use the familiar integrals $\int_\infty^\infty dx \exp\left[-\alpha x^2\right] = \sqrt{\frac{\pi}{\alpha}}$, $\int_\infty^\infty dx x^2 \exp\left[-\alpha x^2\right] = \frac{1}{2}\sqrt{\frac{\pi}{\alpha^3}}$. Using $\langle\phi|x^2|\phi\rangle = \frac{1}{4b}$, $\langle\phi|\frac{d^2}{dx^2}|\phi\rangle = \langle\phi|(-2b + 4b^2x^2)|\phi\rangle = -b$, one finds that

$$E = \frac{\hbar^2 b}{2m} + \frac{m\omega^2}{8b}.$$

The first term is kinetic energy and the second is potential energy. $E(b)$ has one minimum. $\frac{dE}{db} = \frac{\hbar^2}{2m} - \frac{m\omega^2}{8b^2} = 0$ yields $b = \frac{m\omega}{2\hbar}$, $E = \hbar\omega/2$. In this way, one arrives at the exact solution, since the solution was one of the trial functions.

The energy (21.2) is a quadratic *functional*[1] of ϕ. Let us see the effect on E of a variation of ϕ; the variation must be infinitesimal but otherwise arbitrary. We shall denote the variation as $\phi \to \phi + \alpha\eta$, where η is an arbitrary complex function of the same variables as ϕ, while $\alpha \to 0$ is a complex parameter with arbitrary phase. The variation of energy $E = <\phi|H|\phi>$, which follows from $\phi \to \phi + \alpha\eta$, in first order in α, is therefore

$$\delta E = \alpha^* < \eta|H|\phi > + \alpha < \phi|H|\eta > + O(\alpha^2),$$

where $O(\alpha^2)$ is negligible. The condition for a stationary E is

$$E \text{ is stationay} \iff \{\delta E = 0, \text{ arbitrary } \eta\}. \tag{21.4}$$

The condition (21.4) is too strong to be interesting, since we cannot accept the variations of ϕ that violate the normalization condition. The variation changes the norm by

$$\delta N = \alpha^* < \eta|\phi > + \alpha < \phi|\eta > + O(\alpha^2).$$

Therefore, the extremum must be conditioned. Using the Lagrange method we write

$$\delta(E - \lambda N) = 0.$$

---

[1] A functional of ϕ is an integral that has ϕ in the integrand; it can be considered as a function of infinitely many variables, which are the values taken by ϕ in the field of integration. This one is quadratic, since ϕ appears in bra and ket.

Note that this condition is obtained from the unconstrained minimum condition $\delta E = \delta \langle \phi | H | \phi \rangle = 0$ by the substitution $H \to H - \lambda$. In this way, $|\phi >$ depends on the multiplier λ, which is fixed eventually by setting $N = < \phi(\lambda) | \phi(\lambda) >= 1$.

We obtain the same result more simply if we vary only the bra, that is. redefine δE and δN as

$$\delta E = \alpha^* < \eta | H | \phi >, \quad \delta N = \alpha^* < \eta | \phi > . \tag{21.5}$$

By the Lagrange method, we obtain the constrained minimum condition

$$\alpha^* < \eta | H - \lambda | \phi >= 0.$$

Since η is arbitrary, it follows that

$$(H - \lambda) | \phi >= 0.$$

Thus, the condition coincides with the stationary state equation and λ coincides with the eigenvalue E. In summary,

$$\{H\phi = E\phi, \ < \phi | \phi >= 1\} \Leftrightarrow \{\delta(E - \lambda N) = 0, \lambda = E\} \Leftrightarrow \{\delta(E) = 0, \ N = 1\}.$$

The extremum condition is another way to state the stationary state Schrödinger equation. It holds for all the eigenstates, and the Lagrange multiplier is the energy eigenvalue.

Example

Consider the Hamiltonian matrix

$$H = \begin{pmatrix} -3 & 1 & 1 & 1 & 1 \\ 1 & 0 & 0 & 0 & 0 \\ 1 & 0 & 0 & 0 & 0 \\ 1 & 0 & 0 & 0 & 0 \\ 1 & 0 & 0 & 0 & 0 \end{pmatrix}.$$

Use the variational method to find the eigenvectors of the following form:

$$\psi = \begin{pmatrix} \alpha \\ \beta \\ \beta \\ \beta \\ \beta \end{pmatrix}.$$

The normalization requires $N = \langle \psi | \psi \rangle = \alpha^2 + 4\beta^2 = 1$; the energy is $E = \langle \psi | H | \psi \rangle = 8\alpha\beta - 3\alpha^2$: the principle requires that we extremize $f(\alpha, \beta) = E - \lambda N$. One finds:

$$\begin{cases} \frac{\partial f}{\partial \alpha} = 0 \implies 4\beta = (3+\lambda)\alpha \\ \frac{\partial f}{\partial \beta} = 0 \implies \alpha = \lambda\beta. \end{cases}$$

The condition $\lambda(3+\lambda) = 4$, gives the roots $\lambda = -4$, $\lambda = 1$. For $\lambda = -4$ from $\alpha = -4\beta$, we get

$$\psi_{-4} = \frac{1}{\sqrt{20}} \begin{pmatrix} -4 \\ 1 \\ 1 \\ 1 \\ 1 \end{pmatrix},$$

which is the exact ground state with eigenvalue $\epsilon = -4$.

For $\lambda = 1$, $\alpha = \beta$ gives us $\psi_1 = \frac{1}{\sqrt{5}} \begin{pmatrix} 1 \\ 1 \\ 1 \\ 1 \\ 1 \end{pmatrix}$, which is the exact excited state with

eigenvalue $\epsilon = 1$.

21.2 Variational Approximations

The great theoretician Richard Feynman wrote that in order to make progress, a theoretical physicist must understand a given problem in several different ways. The variational principle is equivalent to the stationary state Schrödinger equation, and aside from the mathematical interest, it is a very useful tool, for generating approximations of otherwise intractable problems. The real power of this method has been demonstrated by its application to hard many-body problems.

One can invent trial functions $\phi(x, \{\lambda_1, \lambda_2, \dots \lambda_n\})$ that depend on a number of parameters $\{\lambda_1, \lambda_2, \dots \lambda_n\}$. Then, the method requires a minimization of the energy as a function of the parameters. If ϕ is not already normalized, the normalization constraint can be imposed via a Lagrange multiplier. If the exact ground ϕ happens to be in the class of trial functions, the exact result is gained. Otherwise, it is good to know that the approximate E is higher than the exact value. The approximation gets better if we improve or enlarge the class of functions and increase the number of parameters. Typically, one accurate estimates of E, even with relatively poor $\phi(x)$.

Even the excited states correspond to extrema of the functional, but the method has a limited applicability to them.

> *The trouble is that the true eigenstates must be
> orthogonal, but this property cannot be granted for
> the f that belong to a limited class of functions.*

The orthogonality is granted in the LCAO (= linear combination of atomic orbitals) method for finding the molecular orbitals in a simple independent-particle scheme, but this is an exception. We cannot dispense from orthogonality; if an excited state is not orthogonal to the ground state, we cannot give it any meaning and if we orthogonalize, the variational principle is not satisfied. In general this method is not suited to excited states. However, the lowest state of each symmetry type can be sought through the variational method. For instance, the non-relativistic atomic states are labelled by L, m (angular momentum and its z component); states with different angular momenta are automatically orthogonal. So, if, for instance, the approximate ground state has $L = 0$, and the approximate first excited state has $L = 1$, we can use the variational principle in this case. Any symmetry works, since by symmetry, we mean operator X, which is unitary (that is $XX^\dagger = 1$, see Sect. 13.2), such that $[H, X]_- = 0$. Eigenstates of an unitary operator X belonging to different eigenvalues are orthogonal. Indeed, if $X\phi_1 = e^{i\alpha}\phi_1$ and $X\phi_2 = e^{i\beta}\phi_2$,

$$(\phi_1, \phi_2) = (\phi_1, X^\dagger X\phi_2) = e^{i(\beta-\alpha)}(\phi_1, \phi_2),$$

and, if $\alpha \neq \beta$, this requires that $(\phi_1, \phi_2) = 0$.

Problem 44 For the quartic anharmonic oscillator of Problem 36 find the variational condition for a trial function $\phi(x) = Ae^{-bx^2}$, $A = (\frac{2b}{\pi})^{1/4}$, using the identity $\int_{-\infty}^{\infty} dx\, x^4 \exp(-\alpha x^2) = \frac{3}{4}\sqrt{\frac{\pi}{\alpha^5}}$.

Solution 44 $E = \frac{\hbar^2 b}{2m} + \frac{m\omega^2}{8b} + \frac{3\alpha}{16x_0^4 b^2} = \text{minimum}$.

There is no general method for choosing the trial functions; a proper choice is important, and that is the really hard initial step of the variational theory. Some solutions are very famous, like the Hartree–Fock method; the theory of superconductivity by Bardeen, Cooper and Schrieffer was initially formulated in terms of a variational solution.

21.3 Hartree–Fok Method

The non-relativistic N-electron Hamiltonian for an atom, molecule or solid $H(r_1, r_2, \ldots, r_N) = H_0 + V$, consists of a free part and an interaction term. The free part is of the form

$$H_0 = T + H_w = \sum_i h(i) \equiv \sum_i \left\{ -\frac{1}{2}\nabla_i^2 + w(r_i) \right\}, \tag{21.6}$$

where $w(r_i)$ is the electrostatic potential of the nuclei acting on electron i. This part, taken alone, would be relatively easy to solve, in terms of a determinantal state

$$\Psi(1, 2, \ldots N) = \frac{1}{\sqrt{N!}} \begin{pmatrix} u_1(1) & u_2(1) & \ldots & u_N(1) \\ u_1(2) & u_2(2) & \ldots & u_N(2) \\ \ldots & \ldots & \ldots & \ldots \\ u_1(N) & u_2(N) & \ldots & u_N(N) \end{pmatrix} \tag{21.7}$$

written in terms of spin-orbitals $u_i(i)$ to be determined. The great complication arises from the interaction term

$$V = \frac{1}{2} \sum_{i \neq j} \frac{e^2}{|r_i - r_j|}. \tag{21.8}$$

The Hartree–Fock (HF) Method, or mean field approximation, seeks the best variational determinantal wave function. Using the rules of Chap. 25, the expectation value of H is found to be

$$E_N = \sum_i^N I_i + \frac{1}{2} \sum_{i \neq j}^N \left[C_{ij} - E_{ij} \right], \tag{21.9}$$

where $I_i = \langle u_i | h(i) | u_i \rangle$ and C_{ij}, E_{ij} denote the Coulomb and Exchange integrals. The first term

$$C_{ij} = \int d^3r_1 d^3r_2 \frac{|u_i(r_1)|^2 |u_j(r_2)|^2}{r_{12}} \tag{21.10}$$

has a clear electrostatic meaning, and is called the *Coulomb term*; the second contribution vanishes for anti-parallel spins, and for parallel spins reads as

$$E_{ij} = \int d^3r_1 d^3r_2 \frac{u_i^*(r_1) u_j^*(r_2) u_i(r_2) u_j(r_1)}{r_{12}}. \tag{21.11}$$

Looking for the extremum of energy constrained by normalization, one finds the Hartree–Fock (HF) equations. We introduce the direct potential

$$V^d = \sum_i^N V_i^d(\mathbf{r}), \quad V_i^d(\mathbf{r}) = \int d\mathbf{r}' \frac{|u_i(\mathbf{r}')|^2}{|\mathbf{r} - \mathbf{r}'|}$$

summed over all electrons, and the exchange potential

$$V^{ex} = \sum_{i}^{\uparrow\uparrow} V_i^{ex}(\mathbf{r}), \quad V_i^{ex}(\mathbf{r}) f(\mathbf{r}) = u_i(\mathbf{r}) \int d\mathbf{r}' \, \frac{u_i(\mathbf{r}')^* f(\mathbf{r}')}{|\mathbf{r} - \mathbf{r}'|}, \qquad (21.12)$$

where the summation $\sum_{i}^{\uparrow\uparrow}$ runs over the spin-orbitals with the same spin as i. For an atom with atomic number Z, the Fok operator

$$\hat{f} = \frac{p^2}{2m} - \frac{Z}{|\mathbf{r}|} + V^d(\mathbf{r}) - V^{ex}(\mathbf{r}) \qquad (21.13)$$

allows us to write the HF equations in the one-body form

$$\hat{f} \, u_i(\mathbf{r}) = \varepsilon_i \, u_i(\mathbf{r}). \qquad (21.14)$$

This form looks simple, but the problem is highly nonlinear, and the solution is usually sought numerically by iteration. The HF equations have a complete orthonormal set of solutions. The lowest N spin-orbitals allow us to build the determinantal wave functions. The rest are called *virtual orbitals*; they are not directly related to any experiment involving excited states, but they are often useful for generating multi-determinantal developments, like the Configuration Interaction expansion. The presence of the exchange term is essential in order to explain, for instance, the covalent bond and the cohesive energy of metals. A conduction electron feels the attraction of a piece of metal, which is neutral, because of the exchange contribution, which physically means that in the metal an electron manages to keep the electrons of the same spin far enough away to produce a sort of *Fermi hole* where the positive charge prevails.

This is just the beginning of the many-body theory. The typical error in the energy of valence electrons in the HF method is on the order of 1 eV, which may be relatively small, compared to the atomic ground state energies, but is comparable to the ionization potentials and to the energy of a typical chemical bond. In order to proceed one must go beyond Hartree–Fok. Green's function methods and many-body perturbation theories, as well as Density Functional are used for that. In addition, in the case of heavy elements, the effects of relativity on the inner electrons and also indirectly on the valence electrons are so strong that one must start with the relativistic formulation.

Chapter 22
Discrete Models

Many problems are easier to grasp if we can make them discrete.

22.1 Matrices and Useful Models

In some problems the states that mix in a significant way are few; then, the method of matrices lends itself to insightful (if qualitative) descriptions. The simplest model is 2×2:

$$H = \begin{pmatrix} E_a & V, \\ V^* & E_b \end{pmatrix}.$$

This has many applications: diatomic molecules, etc. For example, if E_a, E_b are atomic levels and V a mixing term, which may represent a covalent bond, you get a lower *bonding* level and an *anti-bonding* one. This is a model for a molecule such as H_2^+. This theoretical model is rough, since it accounts for the energy gain of the electron which is delocalized but ignores interaction and correlation effects; however, qualitatively it works, while no explanation of the covalent bond is classically possible. It has been widely used for a rough description of polyatomic molecules, especially in organic chemistry, and is called LCAO (Linear Combination of Atomic Orbitals). For its great simplicity and its ability to explain some experimental trends, has quite a favorable cost-benefit ratio.

As a second example, we consider neutrino oscillations. The problem originated in the '60's when it was found that the solar electron neutrinos are definitely fewer than the expectations based on the energy output. Neutrino oscillations have since been observed using reactors and accelerator experiments. There are 3 neutrino flavors, and we denote their spinors by $|\nu_\alpha\rangle$, with $\alpha = 1, 2$ and 3, respectively, for electron,

© Springer International Publishing AG, part of Springer Nature 2018
M. Cini, *Elements of Classical and Quantum Physics*,
UNITEXT for Physics, https://doi.org/10.1007/978-3-319-71330-4_22

muon and tauon neutrinos. The Hamiltonian is not diagonal on this basis, but on the basis $|\nu_i\rangle$ of mass eigenstates, with i = 1, 2, 3. On this basis, $H = \text{diag}(m_1, m_2, m_3)$, where the masses m_i are obviously relativistic (the rest masses are a fraction of 1 eV, while the observed energies are several Mev.) The energy eigenstates (also called mass eigenstates) for the neutrinos ν_i, i = 1, 2, 3 are related to the flavor eigenstates by a 3 × 3 unitary matrix called the Pontecorvo, Maki, Nagakawa, Sakada matrix $U_{\alpha i}$. When only two flavors are important, this unitary matrix depends only on one parameter β and may be written as

$$U = \begin{pmatrix} \cos(\theta) & \sin(\theta) \\ -\sin(\theta) & \cos(\theta) \end{pmatrix}.$$

The probability that a particle created at time $t = 0$ as a flavor α neutrino is detected at time t by a detector for flavor β is readily found to be $P = |\langle \nu_\beta(t)|\nu_\alpha\rangle|^2$. This is a simple, phenomenological, but working, model of neutrino oscillations.

22.2 Gauge Changes in Lattice Models

In order to deal with charged particles on discrete lattices, one must be able to switch fields and make gauge changes (Sect. 15.1.5) like in continuous descriptions. A mini-lattice of two sites, with site energies 0 and ϵ, is enough to show how this works. One can introduce an electric field $\mathbf{E}$ across a bond in the lattice by introducing a scalar potential difference between the sites, or alternatively, by introducing a vector potential. We can introduce the vector potential as follows:

$$H = \begin{pmatrix} 0 & \tau \\ \tau & \epsilon \end{pmatrix} \rightarrow H(t) = \begin{pmatrix} 0 & \tau e^{i\omega t} \\ \tau e^{-i\omega t} & \epsilon \end{pmatrix}. \tag{22.1}$$

Here, $\omega = \frac{2\pi}{\phi_0}\int_{\text{bond}} \mathbf{A}.d\mathbf{r}$, $\phi_0 = \frac{hc}{e}$ is the flux quantum, the vector potential is $\mathbf{A} = c\mathbf{E}t$ and the integral extends over the bond. Since the time-dependence is due to the vector potential, I append a label A to the amplitudes. The electron amplitudes a_A and b_A on the two sites are found by solving the time-dependent S.E.; one finds

$$\begin{cases} i\dot{a}_A = \tau b_A(t)e^{i\omega t} \\ i\dot{b}_A = \epsilon b_A(t) + \tau a_A(t)e^{-i\omega t}. \end{cases}$$

We can also represent the same field with a scalar potential V producing an energy shift $e\Delta V = \hbar\omega$ of the second site relative to the first. Now, I append a label V to the amplitudes in this gauge. It turns out that

$$a_V = a_A(t) \exp\left[\frac{-i\omega t}{2}\right],$$

$$b_V = b_A \exp\left[\frac{i\omega t}{2}\right].$$

Differentiating, one finds that these amplitudes obey the S.E. with a time-independent Hamiltonian. Now the rule (22.1) is replaced by:

$$H = \begin{pmatrix} 0 & \tau \\ \tau & \epsilon \end{pmatrix} \rightarrow H_V = \begin{pmatrix} \frac{\omega}{2} & \tau \\ \tau & \epsilon - \frac{\omega}{2} \end{pmatrix}.$$

The change produces a space-dependent and time-dependent phase factor of the amplitudes, as in the continuous case.

22.3 Flux Quantization

A neat example shows the flux quantisation. The model Hamiltonian

$$H = \begin{pmatrix} 0 & \tau_{12} & 0 \\ \tau_{21} & 0 & \tau_{23} \\ 0 & \tau_{32} & 0 \end{pmatrix} \tag{22.2}$$

with $\tau_{12} = \tau_{23} = 1$ may represent the molecular orbitals of a symmetric triangular molecule. If we now set $\tau_{12} = \tau_{21}^* = e^{i\gamma}$ with real γ, the complex bond bears a vector potential, and consequently, the model is pierced by a magnetic flux. The ground state energy of the model is a periodic function of the flux, and the period is a flux quantum (Fig. 22.1).

Fig. 22.1 The ground state energy of the triangular model as a function of γ

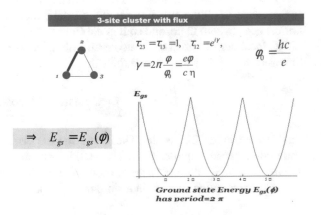

3-site cluster with flux

$$\tau_{23} = \tau_{13} = 1, \quad \tau_{12} = e^{i\gamma},$$

$$\gamma = 2\pi\frac{\varphi}{\varphi_0} = \frac{e\varphi}{c\eta}$$

$$\varphi_0 = \frac{hc}{e}$$

$$\Rightarrow E_{gs} = E_{gs}(\varphi)$$

E_{gs}

Ground state Energy $E_{gs}(\varphi)$
has period=2π

22.4   Infinite Discrete Models and the Local Density of States

Infinite discrete models can have continuous spectra and model quantum wires, surfaces or even solids. The infinite chain is a prototype model of an infinite 1d solid. The sites of the chain are labeled by integers and the Hamiltonian describes nearest-neighbor hopping.

$$H|n\rangle = \tau(|n+1\rangle + |n-1\rangle). \tag{22.3}$$

Expanding the wave functions on the basis of the sites, $|\psi\rangle = \sum_{n=-\infty}^{\infty} \psi_n |n\rangle$, one finds for the eigenstate of energy ϵ_q the S.E. $H|\psi\rangle = \sum_{n=-\infty}^{\infty} \psi_n H|n\rangle = \epsilon_q |\psi\rangle$, which leads to the recursion relation $\epsilon_q \psi_n = \tau(\psi_{n+1} + \psi_{n-1})$. This is solved by $\epsilon_q = 2\tau \cos(q)$ and $\psi_n = e^{iqn}$. The local density of states (LDOS) defined by

$$\rho(\omega) = \sum_k |\langle 0|k\rangle|^2 \delta(\omega - \epsilon_k) \tag{22.4}$$

gives the probability of finding the particle at site 0 with energy $\hbar\omega$ (but it is the same on all sites in this model.) $\rho(\omega)$ contains much information and allows us to visualize the continuum. Using the fact that $\sum_k = \frac{1}{2\pi} \int_{-\frac{\pi}{a}}^{\frac{\pi}{a}}$ one easily does the integral with the δ and finds that the density of states vanishes outside the band $-2t_h \leq \omega \leq 2t_h$, and within the band,

$$\rho(\omega) = \frac{1}{\pi\sqrt{4t_h^2 - \omega^2}}. \tag{22.5}$$

This is shown in Fig. 22.2 left. The divergence at the band edges is topological, i.e. it is common to all the one-dimensional models. Indeed, the group velocity of the particle goes to zero there, and so it is likely that the particle is found nearby.

The local density of states is also easily obtained as an integral in two and three dimensions. In 2 dimensions, one finds that

$$\rho_2(\omega) = \frac{1}{2\pi} \int_{-\pi}^{\pi} dk \rho_1(\omega - 2t_h \cos(k)). \tag{22.6}$$

This is shown in Fig. 22.2 center. The band edge singularity has become a finite jump, while a divergence of the derivative at $\omega = 0$ is an example of the so-called Van Hove singularities. In Solid State Physics, such singularities play a role in the analysis of the optical absorption spectra. In a similar way, one finds for the 3d case,

$$\rho_3(\omega) = \frac{1}{2\pi} \int_{-\pi}^{\pi} \rho_2(\omega - 2t_h \cos(k)). \tag{22.7}$$

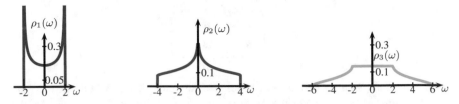

Fig. 22.2 Density of states of the 1d tight-binding chain (left), square lattice (centre) and cubic lattice (right). Note the Van Hove singularities at the band edges and, in the 3d case, at $\omega = \pm 2$

Fig. 22.3 Slices of Bipartite graphs in d, 2d and 3d

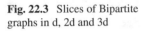

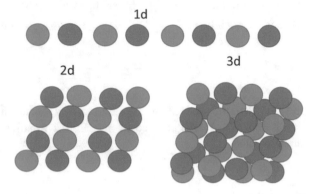

This is shown in Fig. 22.2 right. The band edge singularities at $\omega = \pm\omega_{edge}$ with $\omega_{edge} = 6t_h$ are of the square root type, and this is typical of three-dimensionality.

The densities of states are all symmetric around $\omega = 0$. This is readily understood, since all three lattices are bipartite, that is, they have the property that all sites can be painted red or blue, with every red site having only blue nearest neighbors and every blue site having only red nearest neighbors. This implies that a change $t_h \rightarrow -t_h$ is just a gauge change, equivalent to changing the sign of all the red sites; no physical observable can be affected. On the other hand the Hamiltonian changes sign. Therefore, the spectrum is symmetric for $\omega \rightarrow -\omega$, and $t_h \rightarrow -t_h$ brings an eigenfunction to an eigenfunction corresponding to an opposite eigenvalue (Fig. 22.3).

22.5   Exactly Solved Non-periodic Infinite Model

The above models are easily solved thanks to the symmetry. A few more models that can be solved, even for large or infinite systems, are naturally of interest. We shall find the following simple model very enlightening. Consider the Hamiltonian

$$H = \begin{pmatrix} \epsilon_1 & g & g & \cdots \\ g & \epsilon_2 & g & \cdots \\ g & g & \epsilon_3 & \cdots \\ \cdots & \cdots & \cdots & \cdots \end{pmatrix} \qquad (22.8)$$

For the moment, we may take it as a model with site energies ϵ_n, where the electron can hop from one site to any other with the same amplitude g. This model can be solved for any size of the system. Let v_n denote the nth component of any eigenvector. The S.E. gives us $\epsilon_1 v_1 + g \sum_{j \neq 1} v_j = E v_1$, where E is the eigenvalue. Setting $S = \sum_j v_j$, we obtain $v_1 = \frac{gS}{E+g-\epsilon_1}$. Since the equation for the n-th component is of the same form, namely, $v_n = \frac{gS}{E+g-\epsilon_n}$, we readily arrive at the eigenvalue equation

$$\frac{1}{g} = \sum_n \frac{1}{E + g - \epsilon_n} \qquad (22.9)$$

and the problem is reduced to the solution of a transcendental equation. This solution is versatile. It will be used In Sect. 25.6.1 to solve the Richardson model of superconductivity.

22.6 The Current Operator on a Discrete Basis

Consider the following prototype second-quantized model for a quantum wire:

$$H = t_h \sum_i (c_{i+1}^\dagger c_i + h.c.); \qquad (22.10)$$

the corresponding expression in first quantization would read as

$$H = t_h \sum_i (|i + 1\rangle \langle i| + h.c.)$$

and would describe the hopping of one electron. Instead, (22.10) is a many-body Hamiltonian that represents electrons doing the same but obeying the anti-commutation relations.

The continuum representation of the current density (Eq. (10.20)) cannot apply, but we can build a new one still based on the continuity equation; the difference is that now the density of spin σ at site i is represented by

$$\rho(i, \sigma) = e\hat{n}_{i\sigma} = c_{i\sigma}^\dagger c_{i\sigma}.$$

The spin index can be understood when it does not play a role. In writing the continuity equation for the current J this discrete model, we may set

$$\frac{dJ}{dx} \rightarrow J_m - J_{m-1}$$

and so

$$\frac{\partial \rho}{\partial t} = J_m - J_{m-1}.$$

The time derivative is obtained from the Heisenberg equation of motion. In this way we arrive at the current operator

$$J_m = \frac{et_h}{i\hbar}(c^\dagger_{m-1}c_m - c^\dagger_m c_{m-1}). \tag{22.11}$$

An expression like $J_m = \frac{et_h}{i\hbar}(c^\dagger_m c_{m+1} - c^\dagger_{m+1}c_m)$ is equally correct since it is physically equivalent. In Chap. 25 we shall see applications of this result.

Chapter 23
Pancharatnam Phase and Berry Phase

Parametric Hamiltonians convey the indirect influence of the rest of the Universe on the system under study, and this influence is somewhat similar to a special magnetic field in abstract parameter space.

23.1 Pancharatnam Phase

The Indian physicist S. Pancharatnam, working in quantum Optics in 1956, introduced[1] the novel concept of a geometrical phase. Let $H(\xi)$ denote a Hamiltonian that depends on some parameters ξ, with ground state $|\psi(\xi)\rangle$. One can define a phase difference $\Delta\varphi_{12}$ between two ground states $|\psi(\xi_1)\rangle$ and $|\psi(\xi_2)\rangle$:

$$\langle\psi(\xi_1)|\psi(\xi_2)\rangle = |\langle\psi(\xi_1)|\psi(\xi_2)\rangle|e^{-i\Delta\varphi_{12}}.$$

(Of course, we are assuming that the two ground states are not orthogonal.) This result cannot have a physical meaning, since the phase of any quantum wave function is arbitrary and, for a charged particle, depends on the choice of a gauge. Next, consider 3 points ξ in parameter space and calculate the total phase γ in $\xi_1 \to \xi_2 \to \xi_3 \to \xi_1$. Now,

$$\gamma = \Delta\varphi_{12} + \Delta\varphi_{23} + \Delta\varphi_{31}.$$

The phase of each ψ can be changed at will by a gauge transformation, but strikingly, all these arbitrary changes cancel in the calculation of γ. Actually, γ is gauge independent! This result holds in any closed circuit defined by more than 2 points. Now there is no excuse for declaring γ a physically irrelevant quantity. It can be an observable. *Some observables are not eigenvalues of Hermitean operators.*

---

[1] S. Pancharatnam, Proc. Indian Acad. Sci. A **44**, 247 (1956).

© Springer International Publishing AG, part of Springer Nature 2018
M. Cini, *Elements of Classical and Quantum Physics*,
UNITEXT for Physics, https://doi.org/10.1007/978-3-319-71330-4_23

They originate in a subtle way from the parameters in the Hamiltonian that stand for the influence of the external world.

For example, $|\psi_a\rangle$, $|\psi_b\rangle$, $|\psi_c\rangle$ could be ground states for an electron in the sites a, b, c of a discrete model, representing a molecule; H has matrix elements connecting them. Let τ_{ab} be the matrix element connecting sites a and b, which allows electron hopping. This is just the LCAO model of a molecule (see Sect. 22.1).

Following a prescription by R. Peierls, we can *switch on* a vector potential $\vec{A}$ with

$$\tau_{ab} \rightarrow \tau_{ab} e^{\frac{2\pi i}{\phi_0} \int_a^b d\vec{r} \cdot \vec{A}}, \tag{23.1}$$

where $\phi_0 = \frac{hc}{e} = 4 \times 10^{-7} \mathrm{Gauss\, cm^2}$ is the flux quantum or *fluxon*. In the case of a biatomic molecule like H_2, this change has no consequences, but with 3 or more atoms, the physical meaning is that a magnetic flux ϕ pierces the molecule. Changing ϕ modifies all the energy levels, except that changing ϕ by a fluxon is a gauge change (see Chap. 22).

23.2 Berry Phase

Consider the case when the Hamiltonian depends on two or more parameters, and consider a number N of states obtained by letting the parameters form a closed ring. What does the Pancharatnam phase become in the limit of a continuous change of the parameters over a continuum of states? To answer this question, consider a large number N of states in two or more dimensions, forming a finely spaced necklace (Fig. 23.1). If the wave functions are computed on the same footing, their phases will

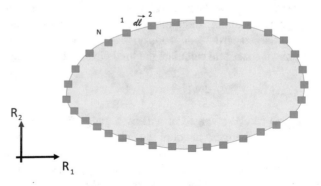

Parameter space

Fig. 23.1 Necklace of states in two-dimensional parameter space

also behave smoothly in going from 1 to 2; ψ denotes the wave function at 1, and we expect to find in 2 the function $(\psi + \delta\psi)e^{-i\delta\phi_{12}}$, where ϕ_{12} is the phase change. Thus,

$$\langle\psi_1|\psi_2\rangle = \langle\psi|\psi + \delta\psi\rangle e^{i\delta\phi_{12}},$$

and since the first-order change $\delta\psi$ must be orthogonal to ψ, $\langle\psi_1|\psi_2\rangle \sim e^{-i\phi_{12}} \sim 1 - i\delta\phi_{12}$.

On the other hand, to first order, $\psi_2 = \psi_1 + \nabla\psi.d\mathbf{r}$. Therefore, we conclude that $\delta\phi_{12} = i\langle\psi|\nabla\psi\rangle$. In the continuum limit, we find that

$$\phi = i \oint \langle\psi|\nabla\psi\rangle d\mathbf{R} \qquad (23.2)$$

is the phase collected between 1 and N; this may be labeled with the quantum number of the states under consideration. Evidently, the so-called Berry connection

$$\mathbf{A}_n = \langle\psi_n|\nabla\psi_n\rangle \qquad (23.3)$$

is a sort of vector potential; $rot\,\mathbf{A}_n = \mathbf{B}_n$ is a would-be magnetic field, which takes the name of a curvature; some curvature must create a flux through the yellow region of Fig. 23.1 if the Berry phase is different from 0.

The Berry phase,[2] which was introduced in 1983, is very fashionable for the important developments in various fields of physics. For instance, the modern theory of the polarization of solids is based on the use of the Berry phase.[3]

An example of its physical importance was discovered long ago[4] and is known as the molecular Aharonov–Bohm effect. Consider an approximate molecular wave function $\Phi = \psi_{el}(\xi, x)\psi_{nuc}(\xi)$, where ψ_{el} and ψ_{nuc} are electron and nuclear wave functions and x, ξ electron and nuclear coordinates. Such wave functions do not really exist in principle, since electrons and nuclei are entangled (see Chap. 27), but they are useful in approximate treatments. Now suppose that we want to study the roto-vibrational spectrum of the molecule. To this end, we must determine the nuclear motion, averaged over the electronic degrees of freedom. Let $p_\alpha = -i\hbar\frac{\partial}{\partial x_\alpha}$ denote the operator of component α of the nuclear momentum. The *effective* nuclear momentum π acting on $\psi_{nucl}(\xi)$ must be obtained by averaging over the electronic wave functions. So, it has components

$$\pi_\alpha\psi_{nucl} = \langle\psi_{el}|p_\alpha|\Phi\rangle = -i\hbar\frac{\partial}{\partial\xi_\alpha}\psi_{nucl}(\xi) + \langle\psi_{el}(\xi, x)|\frac{\partial\psi_{el}(\xi, x)}{\partial\xi_\alpha}\rangle\psi_{nucl}(\xi).$$

[2] M.V. Berry, Proc. R. Soc. Lond. A392, 45 (1984).

[3] See e.g. R. Resta, J. Phys.: Condens. Matter **12** R107 (2000).

[4] Longuet-Higgins H C, O pik U, Pryce M H L and Sack R A 1958 Proc. R. Soc. A 244 1.

Now,

$$A_\alpha = \langle \psi_{el} | p_\alpha \psi_{el} \rangle = -i\hbar \frac{\partial}{\partial \xi_\alpha} \psi_{nucl}(xi) + \langle \psi_{el}(\xi, x) | \frac{\partial \psi_{el}(\xi, x)}{\partial \xi_\alpha} \psi_{el}(\xi, x) \rangle$$

looks very much like a vector potential and is actually a Berry connection. If the electron wave functions can be taken as real, the Berry connection vanishes, since $A_\alpha = \frac{\partial}{\partial \xi_\alpha} \int dx |\psi_{el}(\xi, x)|^2$, and the electron wave functions are normalised to 1. Then, from the nuclear wave function one can derive a rotational and vibrational spectrum. The vibrational spectrum is well approximated by a collection of harmonic terms. The rotational spectrum is that of a rotator, and we saw in Sect. 13.5 that the eigenvalues are labeled by integer quantum numbers.

This is what happens in most cases, but there is trouble in some geometries, like the so-called $e \times \epsilon$ symmetries, when both electron and nuclear wave functions transform under the symmetry operations like the pair (x, y). By symmetry, they are doubly degenerate. It turns out that if the electron wave function are taken to be real, then the nuclear wave functions transform as the components of a spin 1/2 and must change sign under a 2π rotation. This is absolutely not acceptable, since the nuclear wave functions must be single valued! The only possibility then is to assume complex electron wave functions, despite the absence of a magnetic field. Now the single valued wave functions are obtained. but the Berry phase produces half integer rotational quantum numbers, which are in fact observed experimentally.

Chapter 24
Fano Resonances

*Quantum Mechanics replaces smooth trends of Classical
physics by sharp leaps, but the discrete energy levels are always
resonant. This means that we have missed something general
and important up to now.*

All the solvable models that we have seen in previous chapters lead to continuum
spectra for unbound particles and discrete spectra for bound states; for example,
the H atom gave us an infinity of discrete states and the electron-proton continuum.
While this may be a useful first approach to the Physics of bound states, in reality,
all excited states have one or several decay mechanisms. For example, the 2p level
of Hydrogen is discrete, but is degenerate with a system comprising the atom in the
1s state + a photon, whose energy belongs to a continuum. Including the coupling
between the atom and the radiation field, we obtain a finite lifetime and a width to
all excited states, in accord with the uncertainty principle. But the excited levels do
not simply broaden. They acquire structure when coupled to continua. One can add
many more examples of resonances to those enlisted in Sect. 11.6:

1. The Feshbach resonances of molecular Physics are peaks in the electron-molecule
   elastic cross-section. For instance, in electron SF_6 scattering there is a resonance
   when the electron De Broglie wave length is close to the Sulphur-Fluorine distance. The big F ions make a sort of cage around S and when the electron wave
   function fits in the cage it takes some time before it escapes.
2. Phenomena of autoionization in atoms, like the Auger effect (see Chap. 26)
3. In particle physics, most particles have a finite life time and are resonances, like
   the Δ^{++} resonance discovered by Fermi in $\pi-$proton scattering.

© Springer International Publishing AG, part of Springer Nature 2018
M. Cini, *Elements of Classical and Quantum Physics*,
UNITEXT for Physics, https://doi.org/10.1007/978-3-319-71330-4_24

The discrete spectra often arise when a partial Hamiltonian is solved, but are not exact eigenstates when a more complete Hamiltonian is considered. In Chap. 17, when dealing with the one-electron atom, we ignored the coupling to the transverse electromagnetic waves; because of this coupling, all the excited states are resonances.

In order to make progress, let us start with one discrete state that we denote by $|0\rangle$ and one non-degenerate continuum $\mathcal{C}$ of states that we denote by $|k\rangle$. The continuum states would be energy eigenstates with energy ϵ_k, in the absence of the mixing with $|0\rangle$ and a density of states, we shall regard $\rho^c(\omega)$ as a known quantity. Hence, $\int_{E_1}^{E_2} d\omega \rho^c(\omega)$ is the number of states in the interval $E_1 < \omega < E_2$; it may diverge in the continuum limit, when $N \to \infty$. In order to convert the sum into an integral, we need a suitable measure, like $\sum_k \to \frac{\Omega}{(2\pi)^3} \int d^3k$, where Ω is the volume of the system. In the end, all physical quantities have a finite value when $\Omega \to \infty$.

The discrete state would have a sharp energy ϵ_0 in the absence of the mixing; ϵ_0 is supposed to be known from the start.

Some kind of coupling (denoted by $\hat{V}$) changes the discrete delta-like level into a *virtual level*; this has a width, and can have structure. The λ set is complete, hence $\int_{-\infty}^{\infty} d\omega \rho_0(\omega) = 1$.

This is the projected density of states

$$\rho_0(\omega) = \sum_\lambda |\langle 0|\lambda\rangle|^2 \delta(\omega - \epsilon_\lambda), \tag{24.1}$$

where $H|\lambda\rangle = \epsilon_\lambda|\lambda\rangle$; the eigenstates $|\lambda\rangle$ are unknown (for the moment) and the eigenvalues ϵ_λ can be continuous or discrete. We already encountered this function in Chap. 22. The unperturbed quantity is $\rho_0^{(0)}(\omega) = \delta(\omega - \epsilon_0)$.

The projected density of states is built in into the resolvent, or Green operator

$$G(\omega) = (\omega - H + i\delta)^{-1}, \tag{24.2}$$

where δ stands for a positive infinitesimal, or $\delta = +0$ for short. $G(\omega)$ is the Fourier transform of the causal (i.e., proportional to $\theta(t)$) operator

$$G(t) = e^{-iHt}\theta(t).$$

It is causal and analytical in the upper half plane. Now, inserting a complete set, one finds

$$G_{00}(\omega) \equiv \langle 0|G(\omega)|0\rangle = \sum_\lambda \frac{|\langle 0|\lambda\rangle|^2}{\omega - \epsilon_\lambda + i\delta}, \tag{24.3}$$

and the relation to the local DOS is:

$$\rho_0(\omega) = -\frac{1}{\pi} Im\, G_{00}(\omega). \tag{24.4}$$

Here, it is understood that $\sum_\lambda \equiv \int_\mathcal{C} d\epsilon_\lambda + \sum_{\lambda \in \mathcal{D}}$, where $\mathcal{D}$ is the set of the discrete eigenvalues outside the set $\mathcal{C}$ of energies of the unperturbed continuum. While ρ_c is an input quantity of the problem, ρ_0 must be found, and at this point, we must specify a model Hamiltonian.

24.1 Fano Model

It follows from the above consideration that the model we want must have the structure of the Fano model,[1] that we have encountered in Sect. 19.5.1, Eq. (19.51). The model is often written in second-quantized form, where the operators are meant to represent electrons and are meant to anticommute; it reads:

$$H = H_0 + H_h, \ H_0 = \sum_{k \in \mathcal{C}, \sigma} \epsilon_{k,\sigma} n_{k,\sigma} + \sum_\sigma \epsilon_{0,\sigma} n_{0,\sigma} \tag{24.5}$$

while

$$H_h = \sum_{k,\sigma} \{V_k a_{k\sigma}^\dagger a_{0\sigma} + h.c.\} \tag{24.6}$$

This model does not specify the mixing mechanism. Since the problem arises in particle Physics, in Atomic as well as in Solid State Physics, the most diverse mechanisms may be operating: the Fano model simply says that something eventually mixes the continuum states with the localized state. In condensed matter physics $|0\sigma\rangle$ might represent an atomic spin-orbital and $|k\sigma\rangle$ a free-particle state.

This is a one-particle model, so it is exactly solvable and is very useful for a qualitative analysis of real situations. The spin actually plays no role in the simplest version of the model; the many-body extension creates no difficulty since the determinantal eigenstates of H_h simply are formed with the eigen-orbitals of the one-body model the multiplied by spin functions. There are no interaction terms (involving 2 creation and 2 annihilation operators) in the Fano model. The assumption of a non-degenerate continuum can be removed later by a direct extension of the present treatment.[2]

We proceed to the solution of the Fano model in its one-body form (19.51). The Schrödinger equation $H|\lambda\rangle = \epsilon_\lambda |\lambda\rangle$ readily leads to the infinite system

$$(\epsilon_\lambda - \epsilon_0)\lambda|0\rangle - \sum_k \lambda|k\rangle V_{0k} = 0$$

$$(\epsilon_\lambda - \epsilon_k)\langle\lambda|k\rangle - \lambda|0\rangle V_{k0} = 0, \tag{24.7}$$

however, $\langle\lambda|k\rangle$ must be obtained as a distribution; by setting

---

[1] Ugo Fano, Phys. Rev. 124, pp. 1866–1878 (1961).
[2] L.C. Davis and L.A. Feldkamp, Phys. Rev.**B 15** 2961 (1977).

$$\langle\lambda|k\rangle = \langle\lambda|0\rangle V_{k0}\left[\frac{P}{\epsilon_\lambda - \epsilon_k} + Z(\epsilon_\lambda)\delta(\epsilon_\lambda - \epsilon_k)\right],\qquad(24.8)$$

where Z is an unknown, we may substitute in the first of (24.7) which yields, setting $\epsilon_\lambda = \omega$,

$$Z(\omega) = \frac{\omega - \epsilon_0 - Re(\Sigma(\omega))}{Im(\Sigma(\omega))},\qquad(24.9)$$

where

$$\Sigma(\omega) = \sum_k \frac{|V_{0k}|^2}{\omega - \epsilon_k + i\delta}\qquad(24.10)$$

is the self-energy. Σ is an analytic function of $z = \omega + i\delta$, outside the real axis. If $C = \{a \le \omega \le b\}$, then there is a cut just below the axis, with a and b as branch points. We must still find $\langle\lambda|0\rangle$ and appreciate the meaning of the self-energy. This is best done in terms of the density of states, as follows.

24.2 Working in a Subspace: The Self-energy Operator

In many problems, it is convenient to put the Hamiltonian in a block form,

$$H = \begin{pmatrix} H_{AA} & H_{AB} \\ H_{BA} & H_{BB} \end{pmatrix},\qquad(24.11)$$

where A and B are different subspaces of the Hilbert space, that is, the Hilbert space is $A \bigcup B$; typically they denote, respectively, the low energy and high energy parts. Then, a low energy phenomenon occurs essentially in the A subspace but is subject to indirect influences from B. The resolvent matrix G is defined by

$$(\omega - H + i\delta)G(\omega) = 1,\qquad(24.12)$$

where $i\delta$ is a small imaginary part added to the frequency in order to avoid divergences on the real axis; however, the imaginary part can be understood. More explicitly,

$$\begin{pmatrix} \omega - H_{AA} & -H_{AB} \\ -H_{BA} & \omega - H_{BB} \end{pmatrix}\begin{pmatrix} G_{AA} & G_{AB} \\ G_{BA} & G_{BB} \end{pmatrix} = \begin{pmatrix} 1 & 0 \\ 0 & 1 \end{pmatrix}.\qquad(24.13)$$

This is readily solved. The solution in A subspace reads as:

$$G_{AA} = (\omega - H_{AA} - H_{AB}(\omega - H_{BB})^{-1}H_{BA})^{-1}.\qquad(24.14)$$

Here, $-H_{AB}(\omega - H_{BB})^{-1}H_{BA}$ is a matrix self-energy that conveys the effects of the B subspace on the resolvent in the A subspace. That is, one can solve the problem within subspace A with an effective Hamiltonian $H = H_{AA} + \Sigma_{AA}$ where we may write symbolically $\Sigma_{AA} = H_{AB}\frac{1}{\omega - H_{BB}}H_{BA}$. Now we apply this result to the Fano model, setting

$$H_{AA} = \epsilon_0 \hat{n}_0 \equiv \epsilon_0 |0\rangle\langle 0|, \quad H_{BB} = \sum_k \epsilon_k |k\rangle\langle k|, \quad H_{AB} = \sum_k V_{0k} |0\rangle\langle 0|.$$

So,

$$H_{AB}\frac{1}{\omega - H_{BB}}H_{BA} = \sum_k |0\rangle\langle k|\frac{1}{\omega - \sum_{k'}|k'\rangle\epsilon_{k'}\langle k'|}\sum_{k''}|k''\rangle V_{k''0},$$

and finally, we obtain the exact Green's function

$$G_{00} = \frac{1}{\omega - \epsilon_0 - \Sigma(\omega)}, \tag{24.15}$$

where $\Sigma(\omega)$ is the *self-energy* (24.10). The other elements of the resolvent matrix can also be derived in this way.

This is a complex function of ω, in which $\sum_k$ stands for an integral;

$$\Sigma_1(\omega) \equiv \mathrm{Re}\Sigma(\omega) = P\sum_k \frac{|V_{k0}|^2}{\omega - \epsilon_k}$$

$$\Sigma_2(\omega) \equiv \mathrm{Im}\Sigma(\omega) = -\pi\sum_k |V_{k0}|^2\delta(\omega - \epsilon_k). \tag{24.16}$$

Outside $\mathcal{C}$, $\Sigma_2 = 0$. Σ is a **Herglotz** function, that is, $-\pi^{-1}\mathrm{Im}\Sigma(\omega) \geq 0$; it follows that $G_{00}(\omega)$ is also Herglotz, which is important to ensure that $\rho_0 \geq 0$.

In order to obtain new insight on the wave functions, we change variables in Eq. (24.3)

$$G_{00}(\omega) = \int d\epsilon_\lambda \rho_0(\epsilon_\lambda)\frac{|\langle 0|\lambda\rangle|^2}{\omega - \epsilon_\lambda + i\delta}, \tag{24.17}$$

where we used the one-to-one correspondence between states and energies, valid thanks to the assumption of a non-degenerate continuum. In this way, with λ_ω denoting the eigenstate at $\epsilon = \omega$,

$$\mathrm{Im}G_0(\omega) = -\pi\int d\epsilon_\lambda \rho(\epsilon_\lambda)|\langle 0|\lambda\rangle|^2\delta(\omega - \epsilon_\lambda) = -\pi\rho(\omega)|\langle 0|\lambda_\omega\rangle|^2, \tag{24.18}$$

and so,

$$|\langle \lambda_\omega|0\rangle|^2 = \frac{-1}{\pi\rho(\omega)}\mathrm{Im}G_{00}(\omega) = \frac{\rho^0(\omega)}{\rho^c(\omega)}.$$

Since the phase can be chosen at will, we are free to set

$$\langle \lambda_\omega | 0 \rangle = \sqrt{\frac{\rho^0(\omega)}{\rho^c(\omega)}}. \tag{24.19}$$

Together with Eq. (24.8), this result yields all the amplitudes. Besides,

$$\Sigma_2(\omega) = -\pi\rho^0 V_\omega^2, \tag{24.20}$$

where, following Fano, I have set $V_\omega = V_{0k}$ with $\epsilon_k = \omega$.

This technique is useful when it is desirable to work explicitly in a subspace A of the Hilbert space (in this case represented by the discrete state) while taking into account of the remainder B through a self-energy.

24.3   Fano Line Shapes

This model gives a successful qualitative description of resonances like the state $2s2p^1P$ of He. Like all the states of He with both electrons excited, this one is unbound; it can be considered as a temporary bound state, with an energy of about 60 eV, well above the ionization threshold at about 25 eV above the ground state. It broadens by interacting with the auto-ionization continuum. Such situations are common in spectroscopies (e.g. electro energy loss, optical absorption). A common feature of many such spectra is a characteristic skew resonant line shape sitting on a non-resonant continuum. The Green's function G_{00} may be rewritten as

$$G_{00} = \frac{e^{i\Delta}}{\sqrt{[\omega - \epsilon_0 - \Sigma_1]^2 + \Sigma_2^2}},$$

where $\Sigma_1 = Re\Sigma(\omega)$, $\Sigma_2 = Im\Sigma(\omega)$ and

$$\tan(\Delta) = \frac{\Sigma_2}{\omega - \epsilon_0 - \Sigma_1}.$$

Thus, Δ is the phase of G_{00}. Since for large $|\omega|$ $G_{00} \sim \frac{1}{\omega}$, Δ changes by π across the resonance. We may rewrite Eqs. (24.19) and (24.8) in terms of Δ:

$$\langle 0 | \lambda_\omega \rangle = \frac{\sin(\Delta)}{\pi\rho V_\omega}$$

$$\langle k | \lambda_\omega \rangle = \frac{1}{\rho^c}\left[\frac{V(\epsilon_k)\sin(\Delta)}{\pi V(\epsilon_\lambda)(\epsilon_\lambda - \epsilon_k)} - \cos(\Delta)\delta(\epsilon_\lambda - \epsilon_k)\right]. \tag{24.21}$$

Let $\hat{T}$ denote the operator of the transition from an initial state (for instance, the ground state of He) to the resonant H eigenstate $|\lambda_\omega\rangle$. One finds that:

$$\langle i|\hat{T}|\lambda_\omega\rangle = -\cos(\Delta(\omega))\langle i|\hat{T}|k_\omega\rangle + \frac{\sin(\Delta(\omega))}{\pi\rho^c V_\omega}\langle i|\hat{T}|\tilde{0}\rangle, \qquad (24.22)$$

where

$$|\tilde{0}\rangle = |0\rangle + P\sum_k \frac{V(\epsilon_k)}{\omega - \epsilon_k}|k\rangle$$

is the discrete state with a *halo* of continuum. As ω goes from below to above the resonance, $\tan(\Delta)$ diverges, $\sin(\Delta)$ is even and $\cos(\Delta)$ is odd; therefore, the line shapes can be asymmetric. For a simple qualitative analysis, Fano proposed considering

$$q = \frac{\langle i\hat{T}|\tilde{0}\rangle}{\pi\rho^c V_\omega\langle i|\hat{T}|k_\omega\rangle} \qquad (24.23)$$

as a constant. The *reduced line shape* is defined as

$$f = |\frac{\langle i|\hat{T}|\lambda_\omega\rangle}{\langle i|\hat{T}|k_\omega\rangle}|^2. \qquad (24.24)$$

It turns out that

$$f = \sin^2(\Delta)|-\cot(\Delta) + q|^2 = \frac{|-\cot(\Delta) + q|^2}{1 + \cot^2(\Delta)}. \qquad (24.25)$$

Away from the resonance, the denominator represents the background due to transitions to the unperturbed continuum. Rather than plotting versus ω, Fano proposed plotting versus the *reduced energy*

$$E = -\cot(\Delta) = \frac{\omega - \epsilon_0 - \Sigma_1}{\pi\rho^c V_\omega^2}; \qquad (24.26)$$

in this way, one can compare the shapes of resonances centered at different energies with different widths. We arrive at the celebrated Fano line shape. For q = 0, is shows an *anti-resonance*, for q = 1, it becomes step-like, and for q = 3 it is a resonance with

Fig. 24.1 The Fano line shapes. Black, q = 0; Green, q = 1; Red, q = 3

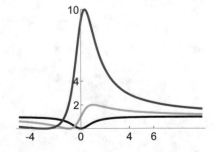

the typical asymmetry. Note that for $E = -q$, the transition is forbidden. For large q, the transition to the discrete state dominates and the shape tends to be symmetric. The Fano treatment is an example of the art of extracting the essential information from a phenomenological model that is by nature qualitative or semi-quantitative (Fig. 24.1).

Chapter 25
Quantum Statistical Mechanics

The classical approach by Gibbs (Sect. 5.20) in the quantum context fully reveals its power.

25.1 Density Matrix

In general, the process that we use to prepare a system of atoms, molecules or whatever we want for a measurement produces many copies, but not all in the same quantum state. A *pure state* in which all the molecules, say, are in the same state, is a limiting case. In general, the system will be in a mixed state. One reason for that is that thermal excitations are unavoidable. Let the possible states be vectors of a Hilbert space with a basis $\{\langle\psi_n|\}$. The expectation value of an operator $\hat{A}$ is

$$\langle\hat{A}\rangle = \sum_n P_n\langle\psi_n|\hat{A}|\psi_n\rangle, \tag{25.1}$$

where P_n is the (classical) probability of finding the system in $\langle\psi_n|$.

Then, one cannot assign to the system a wave function, but rather a density operator

$$\hat{\rho} = \sum_n P_n|\psi_n\rangle\langle\psi_n|, \tag{25.2}$$

and the mean (25.1) is given by

$$\langle\hat{A}\rangle = Tr(\hat{\rho}\hat{A}). \tag{25.3}$$

Evidently, $Tr(\hat{\rho}) = 1$ and $\hat{\rho}^\dagger = \hat{\rho}$. For example, let $|k\rangle$ be the De Broglie wave with momentum $p = \hbar k$; then, we can make a quantum superposition of $\langle x|k\rangle$ and $\langle x| - k\rangle$; the wave function might be $\psi(x) = \frac{1}{\sqrt{\pi}}\cos(kx)$. This state can be obtained by shooting particles with a gun against a potential wall at $x = 0$. The wave

© Springer International Publishing AG, part of Springer Nature 2018
M. Cini, *Elements of Classical and Quantum Physics*,
UNITEXT for Physics, https://doi.org/10.1007/978-3-319-71330-4_25

function vanishes by quantum interference when kx is an integer times π. This may be contrasted with the mixed state given by $\rho = \frac{1}{2}(|k\rangle\langle k|+|-k\rangle\langle-k|)$, which may be obtained by shooting particles with two guns (k from the left and -k from the right). A detector picks one or the other with the same probability, and no interference takes place.

25.2   Quantum Canonical Ensemble

In Quantum Mechanics, the reasoning leading from the microcanonical to the canonical ensemble is quite the same as in the classical theory. In the canonical ensemble, at temperature T the probability that the system energy is one of the eigenvalues ϵ_n is $\frac{1}{Z}e^{-\frac{\epsilon_n}{KT}}$, where $\frac{1}{Z}$ is a normalization factor, in perfect analogy with the classical theory. While in classical Statistical Mechanics, we defined the canonical partition function (5.42) as

$$Z = \int e^{-\beta H_S(p,q)} d\Gamma,$$

in terms of a phase space integral, the quantum *partition function* is

$$Z = Tr[e^{-\beta \hat{H}}], \tag{25.4}$$

that is, $Z = \sum_{n=0}^{\infty} e^{-\frac{\epsilon_n}{KT}}$, where $\hat{H}$ is the Hamiltonian. So, $\frac{1}{Z}e^{-\frac{\epsilon_n}{KT}}$ is the probability of having a system at level n.

According to (5.43), the internal energy U is given by:

$$\overline{H} = -\frac{1}{Z}\frac{\partial Z}{\partial \beta} = -\frac{\partial \ln(Z)}{\partial \beta}.$$

Adding a constant to H_S modifies Z but leaves the physics unchanged. Z allows us to calculate averages. The mean energy is

$$\overline{E} = \frac{\sum_{n=0}^{\infty} \epsilon_n e^{-\frac{\epsilon_n}{KT}}}{\sum_{n=0}^{\infty} e^{-\frac{\epsilon_n}{KT}}} = \frac{Tr[He^{-\beta H}]}{Z} = -\frac{\partial}{\partial \beta} Log Z. \tag{25.5}$$

Let us work out this formula for the harmonic oscillator. Z_{osc} is the sum of a geometrical series:

$$Z_{osc} = \sum_{n=0}^{\infty} e^{-\frac{n\omega\hbar}{KT}} = \left[1 - e^{-\beta\hbar\omega}\right]^{-1}.$$

25.3   Ideal Fermi Gas and Ideal Bose Gas

Suppose we have 2 particles and each of them has the same probability of being in one of two single-particle states A and B. What is the probability P of having both in the same state? It depends on the statistics.

Classical Physics considers only *distinguishable particles*, since (at least, in principle) one can always mark them somehow and follow their trajectories with continuity. There are 4 cases:

$$\begin{cases} \psi_A(1)\psi_A(2) \\ \psi_A(1)\psi_B(2) \\ \psi_B(1)\psi_A(2) \\ \psi_B(1)\psi_B(2) \end{cases} \implies P = \frac{1}{2}.$$

In Quantum Physics, identical particles are *indistinguishable* and cannot be identified when observed again. This makes a lot of difference.

For *Fermions*, there is one possibility:

$$\{\psi_A(1)\psi_B(2) - \psi_B(1)\psi_A(2) \implies P_{together} = 0.$$

For *Bosons*, there are 3 cases:

$$\begin{cases} \psi_A(1)\psi_A(2) \\ \psi_A(1)\psi_B(2) + \psi_B(1)\psi_A(2) \\ \psi_B(1)\psi_B(2) \end{cases} \implies P_{together} = \frac{2}{3}.$$

In the case of photons, the enhanced probability of finding two of them in the same quantum state rather than in different states is called *photon bunching*. In terms of the electric field $E(r, t)$, the correlation function of Eq. (4.11) has a constant modulus in the case of a perfectly coherent wave $E(r, t) = \exp(i(kr - \omega t))$ but is generally peaked at $\tau = 0$ and this is a classical way to describe the underlying photon bunching.

Now suppose we have a gas of $N \gg 1$ identical independent particles (perfect gas). Using the microcanonical ensemble, we suppose it is isolated with total energy E. The system is in a statistical (not quantum) superposition of quantum states, and those leading to the same macroscopic behavior have the same probability. The principle is the one formulated by Gibbs:

> *The basic assumption of Statistical Mechanics is :*
>
> *all the microstates that correspond to the same*
>
> *macroscopic parameters have the same probability*
>
> *in equilibrium.*

One can measure the energies ϵ of individual molecules in the gas. They turn out to be distributed according to a law $n(\epsilon)$.

To define a microstate, we work in the μ space, which is the phase space of the single particle, with coordinates x, y, z, p_x, p_y, p_z. We make this space discrete, that is, we divide it into small cells. More precisely, the size of a cell must be macroscopically negligible, yet it must be large enough to accommodate the coordinates of many molecules at any given instant of time, and the ith cell must contain $g_i \gg 1$ degenerate quantum states for a single-particle. We also make the single particle energies discrete by dividing the ϵ axis into small intervals. Next, we consider all the possible ways to distribute the N gas molecules among the cells with n_i molecules in the ith cell, with the constraint that the number of molecules and the total energies are given, that is,

$$\sum n_i = N, \tag{25.6}$$

$$\sum \epsilon_i n_i = E. \tag{25.7}$$

Since the particles are indistinguishable, the exchange of particles between different states makes no sense. We first seek to find how many microstates realize a given distribution $n_1, n_2, n_3 \ldots$ and then we obtain the most probable distribution. For classical Boltzmann particles, one obtains

$$\bar{n}_r = g_r e^{-\beta(\epsilon_r - \mu)},$$

where μ is the chemical potential. We get different solutions for Bose and Fermi cases.

25.3.1 Bose–Einstein Statistics

The statistics assigns the equilibrium value $\bar{n}_i$ of the number of Bosons in cell i. A microstate is specified when all the numbers of bosons in each quantum state are assigned. The equilibrium condition corresponds to the maximum number W of microstates that can be attained by varying the population n_r of cell r compatibly with the above conditions is therefore

$$\frac{\partial}{\partial n_r} [\ln W - \alpha N - \beta E] = 0.$$

where α and β are Lagrange multipliers. Now,

$$W = \prod_r W_r,$$

where W_r is the number of ways one can distribute n_r particles in cell r among the degenerate g_r states. This is the same problem as finding the number of ways to put n_r marbles in g_r drawers. For 2 marbles in 3 drawers, one finds 6 ways:

$$\| \cdot \cdot \qquad | \cdot | \cdot \qquad | \cdot \cdot |$$

$$\cdot \| \cdot \qquad \cdot | \cdot | \qquad \cdot \cdot \|$$

The first diagram puts 2 marbles in drawer 3, the second diagram puts one in the second and one in the third, and so on. The positions of the vertical lines | correspond to the pairs $(1, 2)$, $(1, 3)$, $(1, 4)$, $(2, 3)$, $(2, 4)$, $(3, 4)$. In general, one obtains a configuration for each way of aligning the dots with the $g_r - 1$ vertical lines, so

$$W_r = \binom{n_r + g_r - 1}{g_r - 1} \approx \binom{n_r + g_r}{g_r},$$

since $g_i \gg 1$. Then,

$$\frac{\partial}{\partial n_i} \ln W = \frac{\partial}{\partial n_i} \sum [(n_r + g_r) \ln(n_r + g_r) - n_r \ln n_r - g_r \ln g_r] = \ln \frac{n_i + g_i}{n_i},$$

and the condition

$$\ln \frac{n_i + g_i}{n_i} - \alpha - \beta \epsilon_i = 0$$

yields the Bose–Einstein distribution

$$\bar{n}_i = \frac{g_i}{e^{\alpha + \beta \epsilon_i} - 1} \equiv \frac{g_i}{e^{\beta(\epsilon_i - \mu)} - 1}.$$

The parameter α is fixed by $\sum n_i = N$, and μ is the chemical potential. By analogy with the Boltzmann distribution, which is the limit for $\beta \epsilon_i \gg 1$, $\beta = \frac{1}{K_B T}$.

Ideal Gas of Atoms

For an ideal gas of atoms having integer spin s, since $\bar{n}_i$ must be positive, μ must be below the lowest of the levels ϵ_i. Let this level be $\epsilon_m = 0$, that is, we shall take ϵ_m as the origin of energies. In line with Lagrange's method, it must be determined by fixing the number of particles in a given volume V.

Above a critical temperature T_C, the particles occupy all the energy states in such a way that every single state gives a negligible contribution, in analogy with the Boltzmann distribution; however, T_C is the onset of the Bose–Einstein condensation. This marks a dramatic departure from the classical behavior. Below T_C, a number n_0 of particles, which is a finite fraction of N, is in the state with $\epsilon = 0$. This is the condensate. As a consequence, when we work out the sum over states as an integral, we must be ready to sort the population of $\epsilon = 0$ out of the integral.

Let us consider $T > T_C$ first. In order to obtain the number of atoms in the cell in phase space, whose volume is h^3, one must multiply the Bose distribution by $\frac{(2s+1)d^3p\,dV}{h^3}$, and the condition (25.6) is readily found by integrating over the angles:

$$N = \frac{4\pi V(2s+1)\sqrt{2m^3}}{h^3} \int_0^\infty \frac{\sqrt{\epsilon}\,d\epsilon}{e^{\beta(\epsilon-\mu)}+1}. \tag{25.8}$$

μ is still unknown, but we can find it soon. To this end, we introduce the De Broglie's thermal wave length

$$\lambda(T) = \frac{h}{\sqrt{2\pi m K_B T}}, \tag{25.9}$$

then define the fugacity $z = e^{\beta\mu}$, and define the integral

$$g_{\frac{3}{2}}(z) = \frac{2}{\sqrt{\pi}} \int_0^\infty \frac{\sqrt{x}\,dx}{z^{-1}e^x - 1}. \tag{25.10}$$

Now we can rewrite the condition (25.6) in the form

$$\frac{N}{V(2s+1)} = \frac{g_{\frac{3}{2}}(z)}{\lambda^3(T)} \tag{25.11}$$

This is the equation that must be solved for z and so yields μ. When T decreases towards T_C, we reach the critical point where $\mu = 0, z = 1$ and $g_{\frac{3}{2}}(1) = \zeta(\frac{3}{2}) \approx 2.612$, where ζ is the Riemann zeta function. Hence, the critical temperature is given by:

$$T_C = \left(\frac{N}{(2s+1)V\zeta(\frac{3}{2})}\right)^{\frac{2}{3}} \frac{h^2}{2\pi m K_B}. \tag{25.12}$$

For $T < T_C$, Eq. (25.8) becomes

$$N - n_0 = \frac{4\pi V(2s+1)\sqrt{2m^3}}{h^3} \int_0^\infty \frac{\sqrt{\epsilon}\,d\epsilon}{e^{\beta(\epsilon-\mu)}+1}, \tag{25.13}$$

where $n_0 = \frac{1}{e^{-\beta\mu}-1} = \frac{z}{z-1}$.

Liquid He 4

Liquid He 4 exists in two distinct phases. Above a critical temperature ($T_C = 2.19\,^\circ$K at low pressures, but $2\,^\circ$K at 10 atmospheres), He I behaves as an ordinary liquid. Below T_C, HeII is still a liquid, but a very unusual one. Pyotr Kapitsa, John F. Alen and Don Misener discovered, 1n 1937, that He 4 becomes a superfluid, with zero viscosity and zero entropy. This is a manifestation of the Boson condensation. It does not become a solid at lower temperatures, because the zero point oscillations keep the would-be solid consistently molten unless under a pressure of about 25 atmospheres. Moreover, He II is a superfluid, with quantized vortices. It has a very high thermal conductivity and heath propagates as *second sound* waves. He II has a complex spectrum of elementary excitations. The properties of the ideal Bose gas are somewhat spoiled by the interactions, but a partial condensation occurs and a fraction of the atoms go to a single state, the one of lowest energy *(Bose condensation)*.

The Bose–Einstein condensation of supercool gases was obtained experimentally in the '90s with the introduction of laser cooling techniques. William Phillips, Claude Cohen-Tannoudji and Steven Chu received the 1997 Nobel prize for this and for opening a new chapter of Physics at ultra-low temperatures.

25.4   Black Body and Photons

In the case of Photons, the condition $\sum n_i = N$ is missing (the number of photons is variable), and the Lagrange multiplier is $\alpha = 0$. So, the mean number of photons in each normal mode (energy $\epsilon_\omega = \hbar\omega$) is

$$\overline{n}_\omega = \frac{1}{e^{\beta\epsilon_\omega} - 1},$$

and each normal mode contains the energy

$$\overline{E}_\omega = \frac{\epsilon_\omega}{e^{\beta\epsilon_\omega} - 1}. \tag{25.14}$$

The k points form a lattice in reciprocal space, and each corresponds to two normal modes (two polarizations). In the continuum limit, the number of modes in a volume of k space is obtained dividing by the volume associated with each mode:

$$\sum_k \rightarrow \frac{L^3}{(2\pi)^3} \int_k d^3k. \tag{25.15}$$

Let $V = L^3$ denote the volume of the photon gas; the number of modes in dk is $2\frac{V}{8\pi^3}4\pi k^2 dk$. Since $\omega = ck$, the number of modes in $d\omega$ is $\frac{V\omega^2 d\omega}{\pi^2 c^3}$. Then, the energy at ω is

$$d\overline{E}_\omega = \hbar\omega \frac{1}{e^{\beta\epsilon_\omega} - 1} \frac{V\omega^2 d\omega}{\pi^2 c^3} \equiv V u(\omega) d\omega.$$

Hence, we get Planck's law for the energy density:

$$u(\omega) = \hbar\omega \frac{1}{e^{\beta\epsilon_\omega} - 1} \frac{\omega^2}{\pi^2 c^3}.$$

or

$$u(\nu) = \frac{8\pi^2 h\nu^3}{c^3} \frac{1}{e^{\frac{h\nu}{K_B T}} - 1},$$

which was the starting point of Quantum Theory in the year 1900. This simple formula, which holds true from radio waves to γ rays, has the remarkable property that when the black body is in motion relative to the observer, the spectrum is Doppler shifted, but still appears as a black body spectrum, albeit, at a different temperature (hotter if the body is approaching, cooler if it is receding). The reason is simple: if it is black, it remains black. A series expansion gives $u(\nu) = \frac{8\pi\nu^2 K_B T}{c^3} - \frac{4\pi h\nu^3}{c^3} + \frac{2\pi h^2 \nu^4}{3c^3 K_B T} + O(\nu^5)$ and the classical formula la (5.47) is the first term.

The maximum of the distribution ω_{max} is proportional to the temperature (Wien's law). Integrating over frequency, we obtain the Stefan-Boltzmann: law:

$$U = \int u d\omega = \frac{\hbar}{\pi^2 c^3} \int_0^\infty \frac{\omega^3 d\omega}{\exp[\hbar \frac{\omega}{K_B T}] - 1}.$$

Setting $x = \beta \hbar \omega$,

$$U = \frac{\hbar}{\pi^2 c^3} \left(\frac{K_B T}{\hbar} \right)^4 \int_0^\infty \frac{x^3 dx}{e^x - 1}.$$

Now,

$$\int_0^\infty \frac{x^3 dx}{e^x - 1} = \frac{\pi^4}{15}$$

and $U \sim T^4$, but the constant of proportionality, which could not be determined by thermodynamics, is fixed by the Quantum theory.

25.4.1   Einstein A and B Coefficients and the Principle of the LASER

Already in 1917, A. Einstein used Planck's law to establish relations between absorption and emission of radiation by a molecule at temperature T. The rate of stimulated absorption is $W_a = B_a u(\omega)$, where B_a is the absorption coefficient. The rate of stimulated emission is $W_{se} = B_{se} u(\omega)$, where B_{se} is the stimulated emission coefficient. Letting A denote the spontaneous emission rate, the total emission rate is $W_{tot} = A + B_{se} u(\omega)$. Introducing the numbers of molecules N_u and N_l in the upper and lower levels, respectively, the Boltzmann distribution requires that $\frac{N_u}{N_l} = \exp(-\frac{h\nu}{K_B T})$. In equilibrium, $N_l B_{sa} u(\omega) = N_u (A + B_{se} u(\omega))$; this gives us $u(\omega) = \frac{\frac{A}{B_{sa}}}{\frac{N_l}{N_u} - \frac{B_{se}}{B_a}}$. Comparing with Planck's law, one finds that $B_a = B_{se}$ and $A = \frac{8\pi h\nu^3}{c^3} B$. Normally, a light wave is absorbed by matter, but if an excited state is more populated than the ground state, the stimulated emission prevails on absorption and the light is amplified. The population inversion can be generated by optical or electric excitation. The Light Amplification Stimulated Emission of Radiation is based on this principle, but only started to operate in 1960. The stimulated emission is produced in a cavity and the frequency is one of the resonant frequencies; in this way all the photons produced by the laser correspond to a single mode and are in phase, with the result that the laser beam has a high degree of coherence and directionality. Initially it was regarded as a useless machine, just an interesting toy for scientists; now, the applications in surgery and in everyday life are countless.

25.4.2 Specific Heats of Solids

The classical equipartition theorem assigns to every oscillator in 3d the energy $E_{osc} = 3K_B T$. Therefore, the contribution of each oscillator to the specific heat $C_V = \frac{\partial E_{osc}}{\partial T}$ should be $3K_B$. This is *grosso modo* true around room temperature (the Dulong and Petit law), but specific heaths drop at low temperatures.

By 1907, Einstein had already proposed a quantum model of the specific heats of solids, in which the quantized atomic vibrations (phonons) give to the energy a contribution

$$E_{phon} = \frac{n\hbar\omega}{e^{\beta\hbar\omega} - 1},$$

where n is the number of oscillators and ω their pulsation; the phonons are Bosons. The specific heath is given by:

$$C_V \equiv \frac{\partial E_{phon}}{\partial t} = \frac{n(\hbar\omega\beta)^2 K_B e^{\beta\hbar\omega}}{(e^{\beta\hbar\omega} - 1)^2}.$$

For $\beta \to 0$ (high temperatures) this agrees with the classical Dulong and Petit law. Below the Einstein temperature $T_E = \frac{\hbar\omega}{K_B}$, the drop $C_V \to 0$ for $T \to 0$ is in qualitative agreement with experiment.

25.5 Fermi-Dirac Statistics

We divide the phase space into macroscopically small cells, but such as to contain a large number of quantum states; so, cell i will contain $g_i \gg 1$ states. Since the exchange of identical Fermions has no meaning, a micro-state is assigned once the numbers $n_1, n_2, n_3, \ldots$ of particles in the cells of energy $\epsilon_1, \epsilon_2, \epsilon_3 \cdots$ is given.

The number W of ways we can realize the distribution must be maximised under the constraints $\sum_i n_i = N$, $\sum_i \epsilon_i n_i = E$.

Since the cells are statistically independent, e $W = \prod_i W_i$. Cell i contains n_i particles in g_i states, and the occupation numbers can only be 0 or 1. Hence,

$$W_i = \binom{g_i}{n_i},$$

$$W = \prod_i W_i = \prod_i \frac{g_i!}{n_i!(g_i - n_i)!}$$

and Stirling's formula $\ln n! \approx n \ln n$ gives us:

$$\ln W \approx \sum_i \{ g_i \ln g_i - n_i \ln n_i - (g_i - n_i) \ln(g_i - n_i) \}.$$

We need the maximum for fixed E and N. By the method of Lagrange's multipliers, we impose the vanishing of

$$\frac{\partial}{\partial n_r} \left[\sum_i \{ g_i \ln g_i - n_i \ln n_i - (g_i - n_i) \ln(g_i - n_i) \} - \alpha \sum n_i - \beta \sum \epsilon_i n_i \right].$$

Since 1 is negligible compared to n_i,

$$- \ln n_r - (-) \ln(g_r - n_r) - \alpha - \beta \epsilon_r = 0.$$

Hence, $\ln \left[\frac{g_r - \bar{n}_r}{n_r} \right] = \alpha + \beta \epsilon_r$ and this implies the Fermi-Dirac distribution

$$\bar{n}_r = \frac{g_r}{e^{\alpha + \beta \epsilon_r} + 1} \equiv \frac{g_r}{e^{\beta(\epsilon_r - \mu)} + 1}, \tag{25.16}$$

where μ is the chemical potential. For large $\epsilon_r - \mu$, this reduces to the Boltzmann distribution with $\beta = \frac{1}{K_B T}$. Only a sign in the denominator distinguishes the Fermi from the Bose distribution.

At absolute zero, the Fermi function $f(x) = \frac{1}{e^{\beta x} + 1}$ becomes a step function, and $\mu = E_F$, since the step is at the Fermi level, or highest occupied level. The step becomes gradual at finite temperatures, but for a typical metal $E_F =$ some eV, and many thousand degrees are needed to round the step significantly (room temperature corresponds to 25 meV).

25.5.1 Fermi Gas

The Fermi distribution describes the occupation of electronic states in solids. The *Fermi gas* is the perfect gas of Fermions with a given uniform density ρ. If interactions are considered, it becomes a *Fermi liquid*, characterized by plasma oscillations (e.g. in metals) and in He 3 by the propagation of hydrodynamic waves called zero sound, having frequency of the order of $q v_F$, where q is the wave vector and v_F is the Fermi momentum divided by the electron mass.

1 dimension (d = 1)

Suppose there are $N \gg 1$ electrons in a *box* of length L with periodic boundary conditions[1] ($x = L$ is identified with $x = 0$; if you like, you can think of a ring). The density is $\rho = \frac{N}{L}$ and the wave functions have the space-dependent factor

$$\psi_k(x) = \frac{1}{\sqrt{L}} e^{ikx};$$

the boundary conditions $e^{ikL} = 1$ allow for the wave vectors

$$k_n = \frac{2\pi n}{L}. \tag{25.17}$$

The interval between two allowed wave vectors is

$$\Delta k = \frac{2\pi}{L}.$$

The N electrons in the ground state of the Fermi gas fill the levels according to the *aufbau* principle, up to the *Fermi level*; it corresponds to the wave vector $k = k_F$, to the momentum $p_F = \hbar k_F$ and to the energy $\epsilon_k = \frac{p_F^2}{2m}$. Between $k = 0$ and $k = k_F$ there are N spin-orbitals, that is, $N/2$ orbitals:

$$N = 2 \sum_{k < k_F} 1.$$

To compute k_F, one goes over to a continuous description and

$$N/2 = \sum_{k \leq k_F} \rightarrow \int_0^{k_F} \frac{dk}{\Delta k} = \frac{k_F}{\Delta k}. \tag{25.18}$$

therefore, k_F is given by:

$$k_F = \Delta k \frac{N}{2} = \frac{\pi N}{L} = \pi \rho. \tag{25.19}$$

d = 3

In 3 dimensions, this becomes the old *Sommerfeld theory* of metals: let us consider a gas with density $\rho = \frac{N}{L^3} \equiv \frac{N}{V}$. Instead of (25.18), in $3d$, one writes:

---

[1] For large L, the properties become independent of the size of the box.

$$\sum_{k<k_F} = \frac{L^3}{(2\pi)^3} \int_{k<K_F} d^3k. \tag{25.20}$$

Again, we put 2 electrons in every orbital in order of increasing energy untill all electrons have been used (the *aufbau* method); in spherical coordinates,

$$N = 2\sum_{k<k_F} = 2\frac{V}{(2\pi)^3} \int_{k<k_F} d^3k = \frac{2V4\pi}{8\pi^3} \int_0^{k_F} k^2 dk,$$

and so

$$N = \frac{Vk_F^3}{3\pi^2}, \tag{25.21}$$

that is,

$$k_F = (3\pi^2\rho)^{\frac{1}{3}} \sim 3.093 \ \rho^{\frac{1}{3}}.$$

Fermi Gas Properties of Simple Metals

Elements with s or p valence electrons like Na, K and Al, are called simple because qualitatively several properties can be described as by the Sommerfeld model, or by Jellium model, that I illustrate briefly in this paragraph. The Fermi gas with density ρ has a characteristic length, and the mean radius per particle r_0 is defined by

$$\frac{1}{\rho} = \frac{4\pi}{3}r_0^3. \tag{25.22}$$

Therefore, $r_0 \sim \frac{1}{k_F}$. Usually it is measured in Bohr radii introducing the Wigner-Seitz radius $r_s = \frac{r_0}{a_B}$. The mean distance between an electron and its nearest neighbor is $\sim r_0$. Cs has $r_s \sim 6$, while for Al, $r_s \sim 2$; the other metallic elements are in this range. The Fermi energy is

$$E_F = \frac{\hbar^2}{2m}(3\pi^2\rho)^{\frac{2}{3}}.$$

For $\rho = 0.1$ a.u. the Fermi energy is[2] $E_F \sim 1.03a.u. \sim 28\,\text{eV}$; the Fermi energies of metallic elements are typically 5–10 eV. The total energy of the gas is

$$E_{tot} = 2\sum_{k<K_F} \hbar^2\frac{k^2}{2m} = \hbar^2\frac{V}{2\pi^2 m} \int_0^{K_F} k^4 dk = \hbar^2\frac{V}{10\pi^2 m}k_F^5, \tag{25.23}$$

and the kinetic energy density

$$\frac{E_{tot}}{V} \sim \rho^{\frac{5}{3}} \sim r_0^{-5} \tag{25.24}$$

---

[2]In Atomic Units (a.u.) we set $\hbar = 1$, lengths are in Bohr radii (0.529 ρA) energies in Hartrees (27.2 eV).

is a steep function of the density. Using (25.21), the mean kinetic energy per electron is found to be

$$\bar{\epsilon} = \frac{E_{tot}}{N} = \frac{3}{5}E_F.$$

For $\rho = 0.1 a.u.$, the energy density is about 0.06 a.u.

The Sommerfeld theory explains why the electronic contribution to the specific heath of metals grows linearly with the temperature T.

However, the energy density of the Fermi gas is positive (there is only kinetic energy) and the electrons must be confined in the metal by some box, otherwise they would escape and the metal should be unstable. While it is clear that electrostatic forces must be responsible somehow, the cohesion of metals is harder to understand than that of ionic crystals, which is readily explained classically. The Sommerfeld theory is not enough.

To describe metals, the *Jellium* model is widely used; in it, the electrons move over a uniform positive charge background. The negative charge of the electrons is exactly compensated by the background, and there is no direct electrostatic contribution to the ground state energy. However, there is also an exchange contribution. One can show that this can explain the cohesion of the metal in a limited range of ρ, which is close to the observed one. One finds that

$$\bar{\epsilon} \sim \frac{2.21}{r_s^2} - \frac{0.916}{r_s} \tag{25.25}$$

where the energy is in (1 Rydberg $= 13.59$ eV). The first term is kinetic energy, while the second is exchange. Extra corrections, which become unimportant at small r_s, arise from correlation (that is, from the fact that electrons can reduce the mean repulsion by adopting a many-body wave function that is a mixture of many Slater determinants).

The above oversimplified picture should not conceal the fact that in solids, widely different phenomena can be caused by tiny energy differences. Magnetism is a very large and growing chapter of solid state physics. A weak effective attraction between the Fermions near the Fermi level can be the result of effective screening of the repulsion and small lattice and/or magnetic effects. The Fermions tend to form pairs called Cooper pairs at low temperature. This leads to superconductivity in which a macroscopic quantum wave function is formed (Sect. 25.6.1). The Cooper pairs are behave like, bosons, except that they do not produce a Bose condensate but a more subtle many-body ground state with perfect diamagnetism and zero resistance.

Equation (25.25) shows that, unlike the classical gas, the Fermi gas tends to be perfect at high density, because the kinetic energy dominates. Everything becomes a metal under high enough pressure, and the interior of the planet Jupiter is believed to contain a deep ocean of metallic Hydrogen.

The Meitner–Auger Effect

The Fermi distribution is the ground state of the electron liquid (i.e., interacting electron gas). Any state with a hole in a single- electron level below the Fermi

Table 25.1 The atomic subshells are denoted by principal quantum number, the symbol s, p, d, $\cdots$ of the orbital angular momentum and a suffix showing the total one-electron angular momentum for one electron, as shown in the upper row. The corresponding spectroscopic notations for the same subshells are reported in the lower row

$1s_{\frac{1}{2}}$	$2s_{\frac{1}{2}}$	$2p_{\frac{1}{2}}$	$2p_{\frac{3}{2}}$	$3s_{\frac{1}{2}}$	$3p_{\frac{1}{2}}$	$3p_{\frac{3}{2}}$	$3d_{\frac{3}{2}}$	$3d_{\frac{5}{2}}$	$4s_{\frac{1}{2}}$	$4p_{\frac{1}{2}}$	$4p_{\frac{3}{2}}$	$4d_{\frac{3}{2}}$	$4d_{\frac{5}{2}}$	$4f_{\frac{5}{2}}$	$4f_{\frac{7}{2}}$
K	L_1	L_2	L_3	M_1	M_2	M_3	M_4	M_5	N_1	N_2	N_3	N_4	N_5	N_6	N_7

level is an intrinsically unstable excited state and must decay somehow. Indeed, it can decay by a mechanism driven by the Coulomb interaction itself through a self-ionizing process. Let $|\Phi_i\rangle = c_h|\Phi_{gs}\rangle$ denotes the state obtained from the N-electron ground state $|\Phi_{gs}\rangle$ by removing an electron in a level below the Fermi energy; this excited state can decay by shooting an electron out of the system and leaving the system with a pair of holes nearer in energy to the Fermi energy. The final state will be of the form $|\Phi_f\rangle = c^\dagger(k)c_{h1}c_{h2}|\Phi_{gs}\rangle$, where $c^\dagger(k)$ creates a free electron, that is not interacting with the rest of the system and $h1, h2$ are hole states. The transition is allowed by energy conservation if $|\Phi_f\rangle$ is degenerate with $|\Phi_i\rangle$ and the Coulomb interaction has a matrix element connecting $|\Phi_i\rangle$ with $|\Phi_f\rangle$. The German Gregor Wentzel discovered this mechanism and obtained the transition rate

$$P_{if} = \frac{2\pi}{\hbar}|\langle \Phi|H_C|\Phi_f\rangle|^2 \qquad (25.26)$$

through the Fermi Golden rule; the effect had been discovered experimentally in atoms in 1923 by Lise Meitner and clarified in 1925 by Pierre Auger. There may be many alternative hole states available to a system for decay and the final-state holes can decay themselves by the same mechanism until the Fermi distribution is established. A plot of the Auger electron current versus kinetic energy is called the Auger spectrum; indeed, Auger spectroscopy is rich in information on the chemical state of the atom and on the local physics of the system. The Auger effect is in competition with radiative decay, which is generally faster in heavy atoms (X-ray emission). In modern theories, the production and the various decay channels of a core-hole in atoms is described as a coherent quantum process. In Auger spectra, one can observe the atomic multiplets of the two final-state holes, as shown in Fig. 25.1. The red spectrum originates from the decay of a M_4 hole and the blue from a M_5 hole. The multiplet symbols that label each peak refer to the two-hole final state of the Auger transition. For example, $^3P_2$ means that the two holes are in a triplet state, with orbital angular momentum L = 1 and total angular momentum $J = 2$ (Fig. 25.2).

Degeneracy Pressure

The Fermi distribution is a sharp step at absolute zero; with increasing K_BT, the step becomes a gradual drop. The chemical potential μ corresponds to a $\frac{1}{2}$ occupancy. At $K_BT > \mu$, many electrons are in the exponential tail of the distribution, and the Fermi gas resembles the classical one. In metals, under normal conditions, the step is sharp and the gas is *degenerate*. Its equation of state is quite unlike a classical gas.

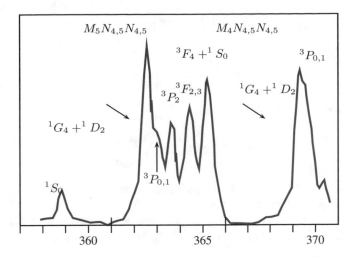

Fig. 25.1 Sketch of the $M_5N_{4,5}N_{4,5}$ (red) and $M_4N_{4,5}N_{4,5}$ (blue) spectrum of Cd vapor, in arbitrary units, from measurements by H. Aksela and S. Aksela. Many of the multiplet terms are well-resolved and the intensities also convey useful information on the hole wave functions. The labels denote the final-state two-holes states, in the notation explained in Table 25.1. The assignments were done by intermediate coupling calculations of line positions and intensities. The intermediate coupling calculations include effects of the Coulomb interaction and of the relativistic spin-orbit coupling

Fig. 25.2 Thick curve: Fermi function for $K_BT = 0.1\mu$; thin curve: Fermi function for $K_BT = 0.3\mu$

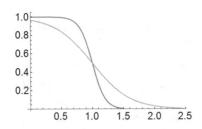

For fixed N, the energy (25.23) depends on volume:

$$E_{tot} = \hbar^2 \frac{V}{10\pi^2 m}(3\pi^2 N)^{\frac{5}{3}} V^{\frac{-5}{3}} = \eta\, V^{-\frac{2}{3}}, \tag{25.27}$$

where

$$\eta = \frac{\hbar^2}{10\pi^2 m}(3\pi^2 N)^{\frac{5}{3}}. \tag{25.28}$$

So, even at absolute zero, the Fermi gas has pressure, which is of quantum nature. One can evaluate this pressure through a thought experiment. When the volume expands adiabatically[3] by dV, the gas does a work $\delta L = PdV$; its energy changes by

---

[3]By the first principle, $dE_{tot} = \delta Q + \delta L$, and here, $\delta Q = 0$.

$$dE_{tot} = -PdV. \tag{25.29}$$

From (25.27),

$$\frac{dE_{tot}}{dV} = -\frac{2}{3}\frac{E}{V}.$$

So, the *degeneracy pressure* is

$$P = \frac{2}{3}\frac{E_{tot}}{V} = \frac{2}{3}\hbar^2 \frac{(3\pi^2\rho)^{\frac{5}{3}}}{10\pi^2 m}. \tag{25.30}$$

The equation of state of the classical perfect gas reads as: $P = NK_B T\rho$; the pressure vanishes at absolute zero and grows linearly with the density. Equation (25.30) predicts a very different trend, with a steeper increase of pressure, even at the lowest temperatures. This explains why metals are highly incompressible.

Degenerate Stars

The same law explains the stability of the degenerate stars (white dwarfs and neutron stars). White dwarfs are stellar remnants with a mass on the order of a solar mass but a radius comparable to the Earth radius. The forces arising from the degeneracy pressure are among the strongest in Nature, and the electron gas can keep the star stable against gravitational collapse up to the Chandrasekar limit (1.4 solar masses). The neutron stars are even more extreme, with a similar mass in a sphere of a few km across. In both cases, the work of Chandrasekar requires the relativistic extension of the theory.

He 3

The isotope 3 of He is produced from the decay of artificially produced Tritium; it could be a convenient fuel for fusion reactors. Here I stress that it is a Fermion and does not become superfluid like He 4. At 2.49 mK temperatures, however, the weak attractive forces produce Cooper pairs and at still lower temperatures a superfluid phase occurs.

25.6 Superconductivity

In 1911, Kamerlingh Onnes discovered superconductivity, and if I say that the discovery was unexpected, this is an understatement. It was a mystery, something unbelievable, just it could be reproduced at will by measuring the resistance of a piece of Hg versus temperature. Below a critical temperature T_C, which in conventional superconductors is no more than a few degrees Kelvin, the electrical resistance is nothing. Please believe me, I do not mean that it is small; it is just 0. Moreover, the superconductor is a perfect diamagnet, expelling the entire magnetic field from the bulk. If a piece of superconductor is put in a field, this is allowed within a thin penetration depth, and repelled. The repulsion levitates objects. Magnetically levitated trains can travel at 500 Km per hour. This phenomenon is one of the first discovered examples of the *magic of Quantum Mechanics,* but several decades passed before it was understood. In superconductivity, most of the electrons behave as they do in ordinary paramagnetic metals, but a strict minority of them forms pairs, thus producing a drastic change of the macroscopic properties; the most important changes are zero resistance and perfect diamagnetism. The most complete theory is the one credited to Bardeen, Cooper and Schrieffer (BCS). The ingredients are bare electrons, the Coulomb repulsion, phonons, and the electron-phonon interaction. BCS show that the electrons close to the Fermi level with opposite spin and momentum form bound pairs and produce the observed physical effects. This is the mechanism of the so-called conventional superconductors, i.e., metals and alloys that start to supercon-duct below a critical temperature T_C which is of the order of several degrees Kelvin at most. The high-T_C superconductors discovered in 1986 by Georg Berdnoz and Alex Müller (Nobel prize 1987) reach 138°. The mechanism in this case could be different, but there is a strong disagreement about it.

25.6.1 Richardson Model for Superconductivity

Any microscopic theory is necessarily involved, but a much simpler and phenological description credited to Richardson[4] has the merit of being readily applicable to finite systems like nanoscopic clusters and even atomic nuclei, where a similar pairing between nucleons exists. The Hamiltonian is taken to be

$$H = \sum_{k,\sigma} \epsilon_k \hat{n}_{k,\sigma} + g \sum_{k,h} c^\dagger_{k+} c^\dagger_{-k-} c_{-h-} c_{h+}, \tag{25.31}$$

where the first term is kinetic energy, while the interaction term scatters a pair of Fermions of opposite spin and wavevector $\vec{h}$ to a similar pair with a different wavevector. This model Hamiltonian is tantamount to an assumption that some pairing mechanism is in action. The unpaired electrons remain spectators, and we forget about them; the dynamics involves pairs, and we rewrite the remaining pair model as

$$H_{pair} = \sum_{kh} (2\epsilon_h \delta_{hk} - g) b^\dagger_h b_k.$$

The pair operators $b^\dagger_k = c^\dagger_{k,\sigma} c^\dagger_{-k,-\sigma}$ have commutator relations in the pair space $b^{\dagger 2}_j = 0$, $[b_h, b^\dagger_k] = \delta_{hk}(1 - 2b^\dagger_k b_k)$, $[b^\dagger_k b_k, b^\dagger_h] = \delta_{hk} b^\dagger_k$. These are called hard core boson relations. Physically, one can consider that these relations are Boson relations in almost all cases, and it is a good idea to treat the b operators as if they were hundred per cent Bosons, with $[b_k, b^\dagger_h] = \delta_{hk}$ instead of being too choosy. The

---

[4]See R.W. Richardson, Phys. Lett. 3, 277 (1963); also Jan Von Delft and Fabian Braun, cond-mat/9911058.

model Hamiltonian then becomes the solved model of Sect. 22.5. The Hamiltonian
is diagonalized in the form

$$\tilde{H} = \sum_k E_k B_k^\dagger B_k \tag{25.32}$$

and the interacting ground state is just the ground state of a system of renormalized
bosons $B_k^\dagger$; one finds that

$$B_k^\dagger = gC_k \sum_h \frac{b_h^\dagger}{2\epsilon_k - E_h}$$

and

$$\frac{1}{(gC_k)^2} = \sum_h \frac{1}{2\epsilon_h - E_k}.$$

So, the ground state is made up of pairs and the ground state energy for n pairs is the
sum of n eigenvalues found in Sect. 22.5. This is negative, so the Fermi distribution
is superseded by the paired state. Incidentally, the Richardson model with hard core
Bosons is also solved exactly and does not modify the qualitative picture.

25.6.2   Why Perfect Diamagnetism and Zero Resistance?

It is rather common that students go through the intricacies of the BCS theory but fail
to gain an intuitive understanding of the basic properties. Here is a simple explanation.
Consider the current

$$J_0(k) = \frac{e\hbar}{2mi} (\phi_k^\dagger \nabla \phi_k - \phi_k \nabla \phi_k^\dagger)$$

calculated over a plane-wave spinor $\phi_k = \frac{\exp(ikx)}{\sqrt{\Omega}} \chi$, where Ω is the volume and χ
a spin function. It is $J_0(x) = \frac{ek}{m\Omega}$; this is a vector, and its average over the ground
state of a superconductor that has no privileged direction vanishes.

Let us switch an external field with $H' = -\frac{e}{mc} \vec{A} \cdot \vec{p}$, in the gauge $\text{div} \vec{A} = 0$.
The average of the momentum operator $\vec{p}$ over a pair state $(k, -k)$ is obviously
zero; we know (Sect. 19.3) that $\vec{p}$, being a one-body operator, has vanishing matrix
elements between pairs with different k. So, there are no matrix elements between
the ground state and the low (paired) excited states; one has to break pairs in order
to excite the system, but this costs an energy exceeding the gap. So, there is no
first-order correction to the wave function, and the ground state is not changed by
the application of a weak external field. The wave function of the superconductor
is rigid; and from this very rigidity, the characteristic properties of superconductors
descend. It is known as London Rigidity, after a theory developed by brothers Fritz
and Heinz London. The electron current density in the presence of the field is

$$J(x) = J_0(x) - \frac{e^2}{mc} \Psi^\dagger A \Psi;$$

the extra term is diamagnetic. In a normal metal, the carriers are electrons, and a magnetic field couples to their spin; so, the first order correction to the many-body wave function is responsible for the fact that in the presence of an external field, a paramagnetic current flows and prevails over the diamagnetic contribution. In superconductors,[5] however, the carriers are spinless pairs, $\langle J_0(x) \rangle = 0$ even in the presence of the field, and

$$J(x) = -\frac{ne^2}{mc} A(x); \tag{25.33}$$

Equation 25.33 is known as London equation and describes the phenomenology of the superconductor, namely, zero resistance and perfect diamagnetism. A curl gives us

$$rot\, J = -\frac{ne^2}{mc} B; \tag{25.34}$$

a second curl (using rot rot = grad div $- \nabla^2$) leads to

$$\nabla^2 B = \frac{4\pi ne^2}{mc^2} B. \tag{25.35}$$

This yields the perfect diamagnetism, because solving for a half space of a super-conductor with a magnetic field applied on the vacuum side, the field is found to decay exponentially inside the superconductor. The London penetration depth is $\Lambda = \sqrt{\frac{mc^2}{4\pi ne^2}}$. A time derivative of (25.34) then yields

$$rot \frac{\partial J}{\partial t} = -\frac{ne^2}{mc} \frac{\partial B}{\partial t}.$$

Since $J = nev$ and $\frac{\partial B}{\partial t} = c\nabla \times E$, this implies that $m\frac{\partial v}{\partial t} = eE$, the electric field accelerates the charges and there is no damping. The resistance vanishes.

25.7 Quantum Gravity, Unruh Effect and Hawking Radiation

The Quantum Theory has been extended to comply with Special Relativity. The need for a relativistic formulation led Dirac to the prediction of antimatter, later verified experimentally, and Pauli to the Spin-Statistics Theorem. Dirac's theory of

---

[5]I leave aside the rare, more involved case of triplet superconductors, like Sr_2RuO_4.

the electron and Quantum Field Theory give an extremely successful description of
Electromagnetism, as well as Weak and Strong Interactions. Condensed Matter and
Nuclear Physics are based on a firm foundation. The gravity field is not something that
propagates in space-time, but is a corrugation of the space-time geometry and does
not lend itself to a canonical quantization procedure like the one in use for the gauge
fields. Possibly, it should not be quantized! At any rate, all the attempts at a quantum
theory of gravity have engaged great minds but have run into trouble so far. The
existence of a quantised gravity wave, the graviton, is doubtful.[6] Roger Penrose[7]
argued that the linearity of Quantum Theory is a problem, when the gravitational
effects of matter are included, since General Relativity is non-linear. The best that
one can do at the present time is to write Dirac's equation and replace the derivatives
by covariant derivatives, thus enabling the electron to move in a curved space-time.
This approach is called semiclassical gravity theory. However, the observer introduced by Einstein to emit or receive signals at a point of space-time is inherently
classical. It is probably a close relative of the classical apparatus that is needed in
Quantum Mechanics to make wave functions collapse. Nobody can tell. In view of
the central role of Quantum Mechanics in our present understanding of Nature and
its overwhelming success in all other fields, the current state of Theoretical Physics
suffers from schizophrenia. Considering the scientific caliber of the investigators in
the field (starting with Einstein) the problem must be really hard!

   Steven Hawking proved an important theorem, stating that any sort of process
can only increase the area of the event horizon of a black hole. For instance, if
two black holes collide and merge, the area of the resultant black hole must exceed
the sum of the areas of the two colliding holes. There is an evident analogy with
the second law of Thermodynamics. But if the black hole has an ever-increasing
entropy, it must also possess a temperature. A temperature implies the exchange of
heat with a thermal bath, and the emission of thermal radiation. On the other hand,
the emission of radiation seems to contradict the very definition of a black hole.
To see how Quantum Mechanics can circumvent this problem, let us consider an
observer in a space ship in free fall near the event horizon of a Schwarzschild black
hole. He sees nothing special around him, in his inertial system, since there is just
a vacuum. Naturally enough, the observer decides to escape from the black hole to
infinity, and turns on powerful rockets, which give strong acceleration to the space
ship. While the space ship escapes, the observer sees particle-hole pairs popping out
of the vacuum by the Unruh effect. Indeed, such a process was predicted by William
Unruh in 1976. An accelerated observer with acceleration a in a Minkowsky vacuum
should observe a thermal bath at temperature

$$T = \frac{\hbar a}{2\pi c K_B};$$ (25.36)

---

[6]"Is a Graviton Detectable?", Freeman Dyson, International Journal of Modern Physics A 28 (2013)
1330041.

[7]See his book: "Fashion, Faith and Fantasy in the new Physics of the Universe",
ISBN9781400880287.

this effect has not yet been observed experimentally. At large distances from the black hole, these particles are seen as Hawking radiation emitted by the black hole. A detailed description of the process is quite demanding and requires approximations, so I come to the conclusions accepted in the Literature. A Schwarzschild black hole with mass M emits a black body radiation at temperature

$$T = \frac{\hbar c^3}{8\pi GMK_B}. \tag{25.37}$$

This formula is striking, because it is the first to contain both G and $\hbar$, and so it smells like a start of the unified theory which is greatly desired. In this way a perfectly isolated black hole should lose mass by radiating a power $P = 3.5 \frac{10^{32}}{M^2}$ Watts, where M is in Kg, and it should evaporate in a time of about $8.4 \ 10^{-17} M^3$ seconds. The process should start slowly, but speed-up and produce a final bang of X-Rays. So the black hole would not be the final fate of the matter falling in, which should be simply waiting for a resurrection in the form of thermal photons, but after a very, very long time. For a solar mass ($2 \ 10^{30}$ Kg) the evaporation time has been evaluated as 10^{67} years, but its present temperature should actually be much lower than the background temperature of the Universe, and so it should start growing. If the Big Bang at the origin of the Universe produced black holes of the right mass, it would be so kind of them to produce their own small bang right now, thus giving an experimental support to this theory. No observational confirmation of evaporating black holes has been obtained so far, and by the way, we also know that the current theory cannot be fully correct.

25.7.1 Hanbury Brown and Twiss Effect: Bosons and Fermions

One speaks about Hanbury Brown and Twiss effect when the correlation in the simultaneous detection of pairs of identical particles shows that the particles follow a non-classical statistics. The original discovery[8] was done with photons. The experiment was aimed at the measurement of the angular radius of Sirius, which turned out to be 3 billionths of a radiant. In Sect. 4.5.1, a classical account of the effect is briefly presented, and it is shown that the effect is borne out by the Maxwell equations. It is also interesting to discuss the effect from the quantum viewpoint, in terms of the photons that obey Bose statistics. While the simultaneous arrival of pairs of photons, one in each detector, was not measured, the correlation function (4.12) represents an average thereof, and would have been constant if the photons arrived at random. Since different photons are emitted from different atoms, say, 1,000,000 km apart, were it not for the permutation symmetry principle, one would expect that

---

[8]R. Hanbury Brown and R.Q. Twiss, "A test of a New Type of Stellar Interferometer on Sirius", Nature **178**:1046–1048 (1956).

the very existence of photons leads to no correlation at all. However, this is not the case. As pointed out by Fano, the correct reasoning is in terms of pairs of photons. If two atoms a and b emit two photons that are collected by detectors 1 and 2, there are two amplitudes: $A(a \rightarrow 1, b \rightarrow 2)$ and $A(a \rightarrow 2, b \rightarrow 1)$; the probability is $A(a \rightarrow 1, b \rightarrow 2) \pm A(a \rightarrow 2, b \rightarrow 1)$, with the upper sign holding for Bosons and the lower sign for Fermions.

A direct experimental verification of the above ideas was made possible[9] by the use of ultra-cold He* atoms at $0.5 \, \mu°$K. Here, He* is a symbol for excited atoms in a 1s2s (singlet or triplet) configurations; such configurations are metastable, because the 2s→1s decay is strictly forbidden with one photon. The atoms were prepared and dropped at a hight L above a detector, where they fell by gravity. The detector allowed for the construction of the normalised correlation function $g^{(2)}(\Delta r)$, i.e., the probability of joint detection at two points separated by Δr, divided by the product of the single detection probabilities at each point. Statistically independent detection events would give $g^{(2)}(\Delta r) = 1$. A value larger than 1 indicates bunching, that is, the Bose-like tendency to have a larger-than-classical population of the phase-space cell; a value less than 1 is evidence of Fermion-like anti-bunching. The experiment was done with ^{4}He* (bosons) and ^{3}He* (fermions), showing that $g^{(2)}(\Delta r) = 1$ beyond a correlation length on the order of 1mm, with a clear increase at short distances for Bosons and a decrease for Fermions. The correlation length was $l = \frac{\hbar t}{ms}$, where t is the falling time, m the mass and s the vertical size of the source.

The anti-bunching correlation has been experimentally observed by Kiesel et al. in 2002 and by Kimble et al. for Na atoms in 1977. A Fermionic Hanbury Brown and Twiss experiment[10] has been performed at 2.5 °K by partitioning an electron beam by means of a metallic gate in a two-dimensional electron gas in the quantum Hall effect regime. The fluctuations in the currents in the two partial beams were found to be anti-correlated, as expected.

---

[9]"Comparison of the Hanbury Brown and Twiss effect for boson and Fermions", by T. Jeltes et al., http://archiv.org/abs/cond-mat/0612278.

[10]M. Henny et al. Science 284, 296 (1999).

Chapter 26
Quantum Transport and Quantum Pumping

> *This Chapter is different, in that we deal with many-body and off-equilibrium problems.*

The usual laws of the circuits (Ohm's law, Kirchoff's law, and so on) are valid in the macroscopic world. However, when the linear dimensions are less than an electron mean free path, which may be on the order of 10 nm (22 in Al and 55 in Cu), the motion of the electrons is ballistic, that is, the circuit obeys quantum laws. While there is a lot of interest in nano-circuits, these laws have only been partly understood. It must be stressed that in the quantum world, the dimensionality has a much stronger meaning than in classical Physics. In thin films that are extended in the $x - y$ plane, there are excitations along z, but if the thickness is L, the creation of such excitations takes energies on the order of $\frac{\hbar}{2mL^2}$, where m is the electron mass; therefore, the low-energy and low-temperature Physics is almost completely 2d. In the same way, nano-wires and Carbon Nano-tubes can be really well described by 1d models. The theoretical analysis of such systems is made more difficult by an enhanced importance of interactions. In low-capacitance tunnel junctions under small bias, one observes the *Coulomb blockade*: the presence of an electron in a Quantum Dot prevents other electrons from jumping, and one observes an increased resistance. Needless to say, nano-devices are extremely fashionable today in science and technology.

26.1 Characteristics and Transients

One can work with continuous and discrete models of biased systems in parallel[1]; for a discrete model, the operator for the current is (22.11),

---

[1]M. Cini, Phys. Rev.B 22,5887 (1980) and Phys. Rev. B 89, 239902(E) (2014).

© Springer International Publishing AG, part of Springer Nature 2018
M. Cini, *Elements of Classical and Quantum Physics*,
UNITEXT for Physics, https://doi.org/10.1007/978-3-319-71330-4_26

373

$$J_m = \frac{e\tau_{mn}}{i\hbar}(c_m^\dagger c_{m-1} - c_{m-1}^\dagger c_m). \qquad (26.1)$$

Assume an independent-particle model in which the system is initially in equilibrium at some temperature T, the one-body states $|q\rangle$ are occupied with probability given by the Fermi distribution $f_q(T)$ and the average current vanishes. At time $t = 0$, the bias is switched on and the system starts evolving with a final-state Hamiltonian $H_f(t)$. The system was initially a determinant of orthonormal spin-orbitals labelled by q. At time t, it becomes a determinant of evolved states $q(t)$. Are they still orthonormal? The answer is yes. The evolution operator $U(t)$ is unitary, so it preserves normalization and orthogonality. Hence,

$$\langle J_{mn}(t)\rangle = \sum_q f_q \langle q|U^\dagger(t) J_{mn} U(t)|q\rangle. \qquad (26.2)$$

We can also put the Fermi distribution inside the matrix element, as an operator:

$$\langle J_{mn}(t)\rangle = \sum_q \langle q|U^\dagger(t) J_{mn} U(t) \hat{f}|q\rangle, \qquad (26.3)$$

with $\hat{f} = (\exp(\beta(H_0 - \mu) + 1)$ and with H_0 the unbiased Hamiltonian. The states that were occupied before the switching of the bias are those that carry the current at all times. Thus, the exact current through site m in a discrete model with hopping parameter t_h is given by:

$$\langle J_m\rangle = \frac{2et_h}{\hbar} \text{Im} \sum_q f_q g_{m,q}^r(t) g_{m-1,q}^r(t)^*, \qquad (26.4)$$

where the retarded Green's function is given by:

$$i g_{m,j}^r(t) = \theta(t)\langle j|U(t)|i\rangle, \qquad (26.5)$$

and $|i\rangle$ is the one-electron wave function localized at site i. In this way, the problem reduces to a one-body time-dependent calculation. It is instructive to consider the case when the initial Hamiltonian $H = H_i$ represents an infinite wire (in a discrete or continuous model), while at time $t = 0$ the Hamiltonian jumps to $H_f = H_i + \hat{V}$ where $\hat{V}$ is a potential that raises the left hand half of the wire by V. Then, for $t > 0$, $U(t) = \exp(-iH_ft)$.

In this way, one can solve, for instance, for the transient current following the onset of the bias; also, letting $t \to \infty$, one can use the asymptotic expansion techniques to derive the dependence of J on V in the steady state; $J(V)$ is called the current-voltage characteristics of the model. A similar treatment is also available for continuous models. In Fig. 26.1, we see that the current is about linear for small V but it changes slope suddenly and then it starts decreasing. This is evidently an effect of

Fig. 26.1 Current-Voltage
characteristic of the 1d
continuous model

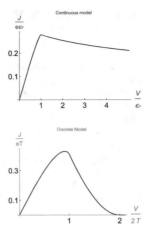

Fig. 26.2 Current-Voltage
characteristic of the 1d
discrete model

dimensionality. Incidentally, the quantum of conductance[2] is $G = \frac{2e^2}{h} \sim 7.7 \times 10^{-5}$ S, and an experiment by B.J. Van Wees has nicely shown this quantization experimentally in 1988. In Fig. 26.2 the characteristic of the discrete model is shown. We see a similar trend initially, but when the bias exceeds the bandwidth, the current vanishes, since the Fermi level of the left half wire is outside the right continuum. Besides, this formalism allows us to follow the transient effects after the bias is switched off; in the above examples the current start from zero, then overshoots the asymptotic $t \to \infty$, oscillates around it a few times and converges to the final value with a characteristic time given by the inverse band width. Currently man groups are engaged in developing codes to deal with electron-electron interactions effects in realistic situations. The development of magnetic and relativistic effects has not received much attention to date, but will be a challenge in the near future.

26.2 Quantum Pumping

Before closing this topic, it is worth mentioning that among the non-classical phenomena discovered so far, there are several mechanisms of pumping. By this term, one means the possibility of achieving a current across a circuit by acting locally on some region of the device, without applying a potential difference between the ends of the circuit. A first example of this phenomenon was reported by Thouless in 1983, but several different mechanism has been invented later for sending a charge current or even a spin current by pumping. Brower[3] has shown that there is a connection between a class of pumping mechanisms, based on varying two parameters in the Hamiltonian, and the Berry phase. However there are also non-adiabatic mechanisms

---

[2] Apart from the factor 2, the existence of this quantum is evident on dimensional grounds.
[3] P.W Brower Phys. Rev. B58, 10135 (1998).

Fig. 26.3 Top: in a laterally connected ring, the choice of a chirality also defines a direction along the wire. Bottom left: a hexagonal ring and the flux piercing it as a function of time. The time is measured in units of $\frac{\hbar}{\tau}$, where τ is the matrix element between neighboring sites. Bottom right: charge pumped from left to right in the wire in the process; each fluxon through the ring produces a step

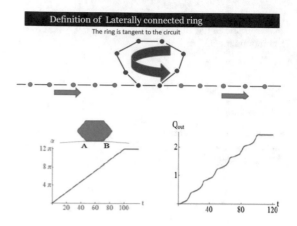

of pumping. A simple example[4] is shown in Fig. 26.3. The model is a simple discrete ring laterally connected to a wire. As shown in the figure, a laterally connected ring carrying a counterclock current will pump charge to the right. For each flux quantum swallowed by the ring, there is an amount of charge shifted from left to right. However this amount decreases if the flux is inserted slowly, and this shows the non-adiabatic character of the process. Many groups are engaged in the research of spin polarised quantum pumping that one day could replace electronics with spintronics.

[4]Michele Cini and Enrico Perfetto, PRB 84 245201 (2011).

Chapter 27
Entanglement, Its Challenges and Its Applications

> *The theory makes predictions that shock our idea of reality, but the experimental tests designed to show the failure of Quantum Mechanics have shown that it works over large distances and promises new applications.*

27.1 What is the Electron Wave Function in the H Atom?

When we dealt with the H atom in Chap. 17, we started from the classical formulation in terms of an effective particle having a reduced mass (see Sect. 2.5.1). The same argument works in Quantum Mechanics, with the momenta replaced with the quantum operators, starting from the Hamiltonian

$$H = -\frac{\hbar^2}{2m_e}\nabla_e^2 - \frac{\hbar^2}{2m_p}\nabla_p^2 + V(\rho);$$

here, ∇_e acts on r_e, ∇_p acts on r_p, and $\vec{\rho} = \vec{r_e} - \vec{r_p}$. The wave function $\psi = \psi(\vec{r_e}, \vec{r_p})$ depends on the coordinates and spins of both particles, however we omit spins, since H does not depend on them in the non-relativistic limit. Next, we introduce the center of mass

$$\vec{R} = \frac{m_e\vec{r_e} + m_p\vec{r_p}}{m_e + m_p},$$

the reduced mass μ with

$$\frac{1}{\mu} = \frac{1}{m_e} + \frac{1}{m_p}$$

© Springer International Publishing AG, part of Springer Nature 2018
M. Cini, *Elements of Classical and Quantum Physics*,
UNITEXT for Physics, https://doi.org/10.1007/978-3-319-71330-4_27

and the total mass $M = m_e + m_p$. For hydrogen,[1] $\mu \sim m_e$. Using the transformed Hamiltonian

$$H = -\frac{\hbar^2}{2M}\nabla_R^2 - \frac{\hbar^2}{2\mu}\nabla_\rho^2 + V(\rho),$$

the problem is *separable* in the sense that

$$\psi(\overrightarrow{r_e}, \overrightarrow{r_p}) = \phi(\overrightarrow{R})\psi(\overrightarrow{\rho}); \tag{27.1}$$

the first factor represents the free motion of the center of mass. Once the quantum mechanical problem is solved, one is interested in the motions of the electron and of the proton. Classically one can solve for $\overrightarrow{r_e}$ and $\overrightarrow{r_p}$ and find the trajectories of both particles. Since in the quantum problem there are no orbits, one would rather ask for their wave functions. However, $\psi(\overrightarrow{r_e}, \overrightarrow{r_p})$ is not *separable* in the form $\phi(\overrightarrow{r_e})\psi(\overrightarrow{r_p})$, that is, cannot be written as a product of an electron wave function times a product wave function. The particles are entangled, and such individual wave functions do not exist! For the same reason, one can speak of electron states in molecules and solids only in the Born-Oppenheimer approximation, and this approximation may be useful or poor, depending on the problem.

   This situation is really amazing. In everyday life, in order to describe the state of a system, one has to specify the state and the position of the components. In the motor of a car, every screw has a location, and the mechanic checks that everything is in place. In Quantum Mechanics, one who knows the state of the system can only speak of probability of the parts being in a given state if they are entangled.

27.2 EPR Paradox

Another example of entanglement is enlightening. Consider a spin $\frac{1}{2}$ and the eigenfunctions α and β of S_z. The two-body basis contains the 4 states

$$\alpha(1)\alpha(2), \alpha(1)\beta(2), \beta(1)\alpha(2), \beta(1)\beta(2).$$

Now, the state $\alpha(1)\beta(2)$ has particle 1 in α and particle 2 in β, but S^2 has no sharp value. On the other hand, the singlet

$$|S, M\rangle = |0, 0\rangle = \frac{1}{\sqrt{2}}(|\alpha(1)\beta(2)\rangle - |\beta(1)\alpha(2)\rangle) \tag{27.2}$$

is entangled: the spin of one particle depends on the state of the other. This is not simply very unusual, but has astonishing consequences. Suppose the system is pre-

---

[1] In Positronium, the proton is replaced by a positron, the antiparticle of the electron, having the same mass, and $\mu = \frac{m_e}{2}$. It decays into photons.

pared in the state $|S, M\rangle = |0, 0\rangle$ and then the spin of particle 1 is measured. The results are random, but if the spin is α, then the spin of 2 is granted to be β, and if it is β, particle 2 must have spin α. This behavior seems to require some sort of hidden interaction between the particles, but this interaction does not exist. Worse still, the act of measurement produces an instantaneous collapse of the wave function $|0, 0\rangle$, and the interaction (if it existed) should be instantaneous, no matter how large is the distance between the particles at the time of measurement. In principle this distance can be large, and entangled photons emitted by a satellite have recently been found to be entangled at a distance of 1,200 km. According to Relativity, any reasonable interaction should imply a distance-dependent delay, but the collapse does not take time.

This feature of Quantum Mechanics perplexed the very founders of the subject. In 1935, it was already clear that Quantum Mechanics was enormously successful, but the fact that it implied such an instantaneous *spooky action at a distance,* as Einstein called it, was a paradox. This is now called the E.P.R. paradox put forth in 1935 by Einstein, Podolsky e Rosen[2] in a very enlightening paper, which was of enormous benefit to Science for the crisis it produced. The criticism of EPR is based on two main points: the principle of realism and on the principle of locality. The principle of realism says that any measurable magnitude has a well-defined value regardless of the fact that it is measured or not. Put it in another way: the moon is there even if nobody looks at it. This is something that does not appear to be questionable. The principle of locality states that if observers A and B are causally disconnected, in the sense of relativity, no measurement made by A cannot affect a measurement performed by B. There is no way to do an experiment that produces an instantaneous effect on the Andromeda Galaxy. Now, I show that Quantum Mechanics contradicts both these apparently bomb-proof assumptions. EPR proposed that Quantum Mechanics was providing a description of reality that contained much truth, but not all the truth. The missing truth could be in *hidden variables.* Including such variables, one should render the theory deterministic and complete. The nature of the hidden variables was debated for many years, while the triumph of Quantum Mechanics was complete in all fields of Physics from elementary particles to Astrophysics. Inevitably, this debate on the hidden variables appeared to be in stand-by, as a philosophical, verbose issue. Suddenly, in 1964, John Stewart Bell, a Northern Irish physicist, discovered a fundamental theorem[3] showing that the predictions of any hidden variable theories and those of quantum mechanics can be compared and the issue can be decided experimentally. Many experiments since then have confirmed the instantaneous collapse of the wave function, thus supporting Quantum Mechanics. I shall now outline the physical point of the Bell theorem.

The Nobel laureate Max Born gave an alternative formulation of the EPR paradox in which Charlie produces a beam of singlet pairs $|\psi\rangle = \frac{|+-\rangle - |-+\rangle}{\sqrt{2}}$ of spin $\frac{1}{2}$ Fermions and sends a member of each pair to Alice and a member to Bob. If Alice finds $\sigma_z^A = +1$, the state collapses to $|+-\rangle$ and Bob measures $\sigma_z^B = -1$, while if Alice

[2] A. Einstein, B. Podolsky and N. Rosen, Phys. Rev. **47**, 777 (1935).

[3] J.S. Bell, Physics **1**, 195 (1964).

finds $\sigma_x^A = -1$, the state collapses to $|-+\rangle$ and Bob measures $\sigma_x^B = 1$. Alice's measurement *instantly* changes the state of Bob's particle, whatever the distance. Alice can choose which spin component to measure after the particles have separated. Moreover Relativity says that for some observers Bob's measurement occurs before Alice's; for this observer her decision influences the past!

Alice, Bob and Charlie manage to have parallel axes. We shall denote by σ_x^A, σ_z^A the results from Alice and by σ_x^B, σ_z^B those from Bob.

After recording a large number of results, the two lists are combined to compute a quantity M. This is chosen in such a way that the principle of realism is enough to make predictions. According to the principle of realism, all the σ values are properties of the particles that exist independently of the measurement, and since each of them is ± 1, it is clear that both $(\sigma_z^A + \sigma_x^A)$ and $(\sigma_z^A - \sigma_x^A)$ may be 0 or ± 2. Moreover, if the sum is 0, the difference is ± 2, and if the sum is ± 2, the difference is 0. For this reason we choose the following quantity:

$$M = (\sigma_z^A + \sigma_x^A)\sigma_z^B + (\sigma_z^A - \sigma_x^A)\sigma_x^B. \tag{27.3}$$

With this choice, each pair yields ± 2.

Then it is clear that the average over many pairs must be such that

$$|\langle M \rangle| \le 2. \tag{27.4}$$

This is Bell's inequality in this case, and is the result of the principle of realism. Let us compare this with the quantum result, using the Pauli matrices. Since

$$(\sigma_z^A + \sigma_x^A)(-|1_A - 1_B\rangle) = -|1_A - 1_B\rangle - |-1_A - 1_B\rangle,$$

$$(\sigma_z^A + \sigma_x^A)(-|1_A 1_B\rangle) = |-1_A - 1_B\rangle - |1_A - 1_B\rangle,$$

and so on, one finds $\langle \psi | M | \psi \rangle = -2$. This is striking, but still marginally in agreement with Bell's inequality.

Next, consider a variant of the Born formulation of the EPR paradox in which the reference used by Bob is not parallel to Alice's. The situation becomes crisp when Alice's reference is parallel to Charle's, but Bob has tilted axes. For the sake of definiteness, assume that, expanding on Alice's basis, the z axis unit vectors are $\tilde{n}_z = \frac{(1,0,1)}{\sqrt{2}}$ and $\tilde{n}_x = \frac{(-1,0,1)}{\sqrt{2}}$. For each pair, both Alice and Bob may decide to measure either the x or the z component of the spin of their particle, and both choose at random and independently, and record the sequence of ± 1 values they obtain (in units of $\frac{\hbar}{2}$, of course). The quantum result is (see next problem)

$$\langle M \rangle = 2\sqrt{2}. \tag{27.5}$$

This result definitely disagrees with the Bell inequality. If the principle of realism holds, Quantum Mechanics is wrong. Thus, experiment can decide, and the experimental results support Quantum Mechanics (otherwise the reader would have read the news in the newspaper).

Problem 45 Prove Eq. 27.5.

Solution 45 Since

$$\sigma_x^B = \sigma.n_x^B = \frac{-\sigma_x^A + \sigma_z^A}{\sqrt{2}} = \frac{1}{\sqrt{2}}\begin{pmatrix} 1 & i \\ -i & -1 \end{pmatrix},$$

and

$$\sigma_z^B = \begin{pmatrix} 1 & -i \\ i & -1 \end{pmatrix},$$

one contribution is

$$\langle\psi|\sigma_z^A\sigma_z^B|\psi\rangle = \left[\frac{\langle+-|-\langle-+|}{\sqrt{2}}\right]\sigma_z^A\sigma_z^B\left[\frac{|+-\rangle-|-+\rangle}{\sqrt{2}}\right].$$

Now let σ_z^A act on the left spin; since it is diagonal, one obtains

$$\langle\psi|\sigma_z^A\sigma_z^B|\psi\rangle = \frac{1}{2}[\langle-|\sigma_z^B|-\rangle - \langle+|\sigma_z^B|+\rangle].$$

Proceeding in this way, one easily gets the result.

27.3 Bell States

Consider a quantum system, such as a spin $\frac{1}{2}$ with a two-dimensional Hilbert space; I shall denote the ortogonal states by $|0\rangle$, $|1\rangle$. A general state of such a qbit is $|\psi\rangle = \alpha|0\rangle + \beta|1\rangle$, whith $|\alpha|^2 + |\beta|^2 = 1$. Now suppose we have two such systems (like, for instance, an electron spin and a proton spin). A basis state for the states of the compound system is provided by the set

$$|0\rangle|0\rangle, |0\rangle|1\rangle, |1\rangle|0\rangle, |1\rangle|1\rangle, \tag{27.6}$$

where the first factor refers to the first subsystem or particle. According to the rules of Quantum Mechanics, one is free to use the entangled basis of Bell states

$$\Phi^+ = \frac{|0\rangle|0\rangle+|1\rangle|1\rangle}{\sqrt{2}} \qquad \Phi^- = \frac{|0\rangle|0\rangle-|1\rangle|1\rangle}{\sqrt{2}}$$
$$\Psi^+ = \frac{|0\rangle|1\rangle+|1\rangle|0\rangle}{\sqrt{2}} \qquad \Psi^- = \frac{|0\rangle|1\rangle-|1\rangle|0\rangle}{\sqrt{2}} \tag{27.7}$$

and to transform back to the unentangled[4] basis by:

$$|0\rangle|0\rangle = \frac{\Phi^+ + \Phi^-}{\sqrt{2}} \qquad |0\rangle|1\rangle = \frac{\Psi^+ + \Psi^-}{\sqrt{2}}$$
$$|1\rangle|0\rangle = \frac{\Psi^+ - \Psi^-}{\sqrt{2}} \qquad |1\rangle|1\rangle = \frac{\Phi^+ - \Phi^-}{\sqrt{2}}. \tag{27.8}$$

The paradox already lurks in the equivalence of the entangled basis. We can know that the state of the system is, say, Φ^+, but then we cannot tell the state of individual particles; this is opposite of the classical situation when it is the knowledge of the state of the individual screws and wheels that enables the mechanic to tell the state of a car motor. If we know that the wave function is Φ^+, then all we can grant is that if and only if the first system is in $|0\rangle$, then the second is also in $|0\rangle$.

27.3.1 The No-cloning Theorem Saves Coexistence with Relativity

Consider an experiment of the EPR type in which there is a source of pairs of spin $\frac{1}{2}$ particles in the singlet states: suppose that one particle in a pair is sent to Alice and the other to Bob. Alice might decide to send a message to Bob, encoded in a sequence of 0 and 1. To transmit a 0 she measures the x component of the spin, and for every 1 the z component. So Bob receives electrons that are eigenstates of the respective components. The collapse occurs instantly. Now, how could Bob read the message? Measuring the spin once along any axis he cannot come to any conclusion. If Bob were able to make numerous clones of each electron he could find out which is the quantization axis, and read the instant message. If it were possible to clone the state of particles, the instantaneous transmission of information at any distance would be possible and Relativity would be violated. So we may say that the No-cloning Theorem (Sect. 15.6) saves the coexistence of Quantum Theory with Relativity. No information travels faster than light. On the other hand, since the experimental results support Quantum Mechanics, the principle of realism does not hold and the principle of locality does not work as expected by E.P.R. Note that Relativity is not involved in the proof of the no-cloning theorem.

27.4 Quantum Computer

In 1982, R.P. Feynman suggested the possibility of a computer based on the superposition principle and on quantum effects. Instead of the classical bits of Sect. 5.6.2 one could consider two-component spins as quantum bits or qbits. In this context, one writes $|0\rangle \equiv |\uparrow\rangle$ and $|1\rangle \equiv |\downarrow\rangle$. Since one can prepare the qbit in any state of

---

[4]I choose the example of an electron spin and a proton spin, because the particles must not be identical, if we wish to be able to choose between entangled and unentangled bases.

the form of a linear combination $\alpha|0\rangle + \beta|1\rangle$, there are many more ways to prepare the qbit than the classical bit, but when the information is read, it must be read only once, and the result is either $|0\rangle$ or $|1\rangle$. For this reason, the information stored in a qbit is the same as in a classical bit.[5] Qbits can be manipulated by quantum logic gates, analogous to the classical ones.

The quantum analogue of the classical logic gates is provided by the evolution of the qbits under the action of some Hamiltonian. The permutation matrices that represent the action of the classical reversible logical gates are replaced by the unitary matrices; for instance, one can use of the quantum evolution operator. Therefore, the quantum logical gates are always reversible and their operation costs no energy. For a single qbit, one can use the Pauli matrices as gates. The quantum Toffoli gate is the same as the classical gate, and has 3 qbits in input and in output. The real advantages of quantum computing comes from the possibility of producing entanglement of qbits. This means that the quantum gate acting on a product of qbits yields a sum of product states.

As an example of a calculation in which a quantum computer should perform better, consider a function $f_\alpha(x)$ for which the calculation is demanding, but it is known that $f_\alpha(0)$ and $f_\alpha(1)$ can be 0 or 1 and that the values depend on the parameter α. Suppose one just needs to know if, for a given α, f_α takes the same value or different values for $x = 0$ and $x = 1$. David Deutsch has shown theoretically that a quantum computer would find the result by one calculation of f. The bonus of quantum computers is the ability to provide global information on the functions to calculate.

Another notable example has important practical applications in cryptography. It is the factorization of large numbers into prime factors. This task requires a number of operations that increases exponentially on a classical computer. P.W. Shor, in 1994, proposed an efficient algorithm for quantum computers that would speed-up the calculation considerably. Virtually, the quantum computer could lead to an enormous progress; however, there are great difficulties.

A basic problem is due to the fact that the quantum system is never perfectly isolated, and sooner or later is subject to the loss of coherence, or *decoherence* (Sect. 19.6). The basic idea is the following. Consider a spin in a state $|\psi\rangle = \alpha|\uparrow\rangle + \beta|\downarrow\rangle$ with $|\alpha| = |\beta| = \frac{1}{2}$. Our quantum computer could use the information contained in the relative phase of α and β and related to the way the spin evolves in an applied magnetic field. However, if the spin is not well insulated, there is an unknown magnetic field. In other words, part of the Hamiltonian is unknown and describes the interaction with some external system X. The *decoherence* consists in the fact that if X interacts in a different way with $|\uparrow\rangle$ and $|\downarrow\rangle$, and behaves as a detector of the spin state, within some characteristic time, the phase relationship is lost. This is analogous to what happens in the double-slit experiment if one tries to reveal which way the particle passes: *ipso facto,* such an attempt destroys the

---

[5]This is the physical meaning of the Holevo theorem which puts an upper limit on the classical information that can be extracted by Bob by performing measurements on a quantum state, after Alice has encoded classical information in it.

Fig. 27.1 Upper picture: initial situation, with Alice that can handle a qbit and a particle in an entangled pair, and Bob that can act on the other particle. Lower picture: after the measurement by Alice, Bob has the qbit and Alice the information needed by Bob to reproduce the initial qbit state

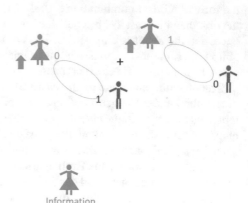

interference and produces a classic behavior. One can say that the classical behavior emerges naturally when the quantum system is so complex that we can not know all the Hamiltonian, and then we lose the effects of interference.

27.5 Teleportation

In 1993 the word teleportation, familiar in science fiction, first appeared in Physical Review Letters[6] in a very often referenced theoretical paper. The thought experiment involves two characters, Alice and Bob (see Fig. 27.1). Alice has a qbit $\psi_A = \alpha|0\rangle_A + \beta|1\rangle_A$; the notation means that only Alice can act on the qbit, which must be transmitted to Bob. Neither Alice nor Bob have any information about α and β, but they share an entangled pair of the Bell Ψ^- type

$$\Psi^- = \frac{|0\rangle_A|1\rangle_B - |1\rangle_A|0\rangle_B}{\sqrt{2}}. \tag{27.9}$$

The 3 particles are in the following state:

$$\Omega_3 = \psi_A \Psi^-. \tag{27.10}$$

Expanding $\Omega_3 = (\alpha|0\rangle_A + \beta|\rangle_A)\frac{|0\rangle_A|1\rangle_B - |1\rangle_A|0\rangle_B}{\sqrt{2}}$, one finds:

$$\sqrt{2}\Omega_3 = \alpha|0\rangle_A|0\rangle_A|1\rangle_B + \beta|1\rangle_A|0\rangle_A|1\rangle_B - \alpha|0\rangle_A|1\rangle_A|0\rangle_B - \beta|1\rangle_A|1\rangle_A|0\rangle_B.$$

---

[6]Charles H. Bennet, Gilles Brassard, Claude Crepeau, Richard Josza, Asher Peres and William K. Wootters, Phys. Rev. Letters 70 1895 (1993).

In terms of Bell states,

$$2\Omega_3 = |\Phi^+\rangle_A (\alpha|1\rangle_B - \beta|0\rangle_B)$$
$$+|\Phi^-\rangle_A (\alpha|1\rangle_B + \beta|0\rangle_B) + |\Psi^+\rangle_A (\beta|1\rangle_B - \alpha|0\rangle_B) +$$
$$|\Psi^-\rangle_A (\beta|1\rangle_B + \alpha|0\rangle_B). \tag{27.11}$$

Now, Alice measures the state of her two particles, thereby performing the instantaneous teleportation according to the following scheme, in which Bob's qbit is rewritten in spinor notation:

Alice's pair	Bob's qbit	Restoring Operator
$|\Phi^+\rangle$	$\alpha|1\rangle - \beta|0\rangle$	σ_y
$|\Phi^-\rangle$	$\alpha|1\rangle + \beta|0\rangle$	σ_x
$|\Psi^+\rangle$	$\beta|1\rangle - \alpha|0\rangle$	σ_z
$|\Psi^-\rangle$	$\beta|1\rangle + \alpha|0\rangle$	1

If Alice finds Ψ^-, then Bob has already got $\alpha|0\rangle_B + \beta|1\rangle_B$, while in the other cases, he must apply a restoring operator, as shown in the table. He can be sure about the state of his qbit after he receives a call from Alice informing him of the result of her measurement. Therefore, faster then light teleportation is not allowed. Alice does not know the original qbit that she destroyed in the measurement. Not even Bob can know α and β and the No-Cloning theorem forbids the fabrication of the copies that would be necessary for this investigation. However, the quantum state has been transferred from Alice to Bob. In 2017, Chinese scientists teleported a photon from Tibet to a satellite in orbit, up to 1,400 km above the Earth's surface (see https://www.space.com/37506-quantum-teleportation-record-shattered.html).

Index

© Springer International Publishing AG, part of Springer Nature 2018
M. Cini, *Elements of Classical and Quantum Physics*,
UNITEXT for Physics, https://doi.org/10.1007/978-3-319-71330-4

Printed in the United States
By Bookmasters